Microfluidic Technology and Applications

MICROTECHNOLOGIES AND MICROSYSTEMS SERIES

Series Editor: **Professor R. Pethig**
University of Wales, Bangor, UK

1. Microfluidic Technology and Applications
Michael Koch, Alan Evans and **Arthur Brunnschweiler**

2. Electrokinetic Technologies for Sub-Micron Particles *
H. Morgan and **N. Green**

* *forthcoming*

Microfluidic Technology and Applications

Michael Koch
Alan Evans
Arthur Brunnschweiler

RESEARCH STUDIES PRESS LTD.
Baldock, Hertfordshire, England

RESEARCH STUDIES PRESS LTD.
16 Coach House Cloisters, 10 Hitchin Street, Baldock, Hertfordshire, England, SG7 6AE

and

325 Chestnut Street, Philadelphia, PA 19106, USA

Marketing:

UK, EUROPE & REST OF THE WORLD
Research Studies Press Ltd.
16 Coach House Cloisters, 10 Hitchin Street, Baldock, Hertfordshire, England, SG7 6AE

Distribution:

NORTH AMERICA
Taylor & Francis Inc.
International Thompson Publishing, Discovery Distribution Center, Receiving Dept., 2360 Progress Drive Hebron, Ky. 41048

ASIA-PACIFIC
Hemisphere Publication Services
Golden Wheel Building # 04-03, 41 Kallang Pudding Road, Singapore 349316

UK, EUROPE & REST OF THE WORLD
John Wiley & Sons Ltd.
Shripney Road, Bognor Regis, West Sussex, England, PO22 9SA

Library of Congress Cataloging-in-Publication Data

Koch, Michael, 1970-
Microfluidic technology and applications / Michael Koch, Alan Evans, Arthur Brunnschweiler.
p. cm. -- (Microtechnologies and microsystems series ; 1)
ISBN 0-86380-244-3 (alk. paper)
1. Fluidic devices. 2. Micromechanics. I. Evans, Alan, 1944 - II. Brunnschweiler, Arthur, 1936- III. Title IV. Series

TJ853 .K63 2000
620.1'06--dc21 99-089448

British Library Cataloguing in Publication Data
A catalogue record for this book is available from the British Library.

ISBN 0 86380 244 3

Printed in Great Britain by SRP Ltd., Exeter

Editorial Foreword

I am delighted to welcome the first book in this new Series on Microtechnologies and Microsystems.

The primary objectives of the Series will be to introduce and teach the basic underlying theories and methods, as well as review the latest advances.

Michael Koch, Alan Evans and Arthur Brunnschweiler, in their enthusiastic response to the challenge they were given, have produced a landmark handbook on microfluidics. They have given us a concise and expert introduction to this rapidly expanding field, as well as new knowledge on key research areas.

Ron Pethig
School of Informatics
University of Wales, Bangor

and

Aura Oncology Systems
Sunnyvale, California

September 2000

Contents

List of Figures

List of Tables

CHAPTER 1

Introduction

A market study of microfluidics and MEMS (Micro Electro Mechanical Systems) devices was conducted in 1994 [2]. The market value in the year 2000 for basic components and systems in microfluidics was estimated to be $2.6 billion and $ 6.5 billion, respectively (see also figure 1.1).

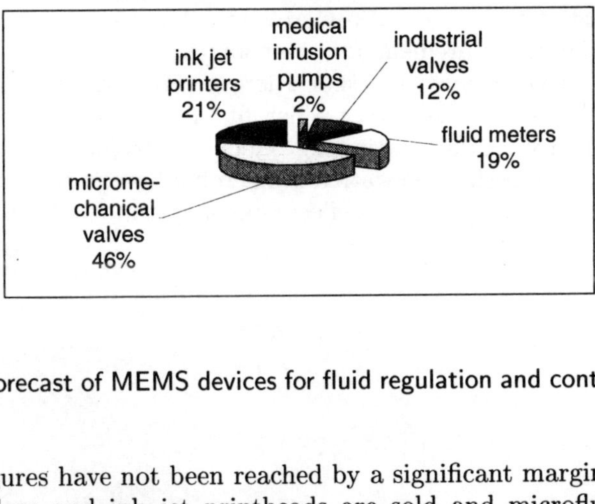

Figure 1.1: Forecast of MEMS devices for fluid regulation and control in the year 2000 [2]

These figures have not been reached by a significant margin, as only microfluidic valves and ink jet printheads are sold and microfluidic reaction systems are still in the prototype stage. However, the authors believe that the market of microfluidic devices will reach these figures as mentioned above quite soon. This will mainly be due to the "lab on a chip" vision as is necessary in molecular biology, chemistry and medicine.

With microsystems it is possible to increase the yield in certain chemical reactions dramatically and at the same time reduce the realisation of byproducts. Also, it is possible to fabricate DNA sequencers (machines to analyse DNA) working with volumes of the picolitres range and thus allow the analysis of even the smallest sample of DNA. In medicine, blood monitoring and drug delivery are also some of the key issues that are addressed with microsystems. Once these systems achieve the same reliability as their macroscopic counterparts and offer better cost/performance values, it is expected that the booming markets of biotechnology, analytical and clinical chemistry will also create a booming industry in microfluidics.

This book describes both the theory and the applications of microfluidics and microsystems. The aim is to present all the information necessary to design practical devices, and also to describe most of the research which has been published in the past 10 years. We thus include such topics as fluid theory, device simulation and fabrication technology, as well as a large number of practical device designs. The level of presentation should be suitable for both graduate students and engineers in industry.

As microfluidics is a subject that is based on many areas of physics, chemistry and engineering, the treatment of fluid theory and mechanics starts from fundamentals and progresses to the level required to perform quantitative device analysis. The techniques available for device fabrication are explained in detail and related to examples of the various devices to which they are best suited. We also consider the fast growing field of microfluidic systems, so that by the end of the book the reader should have a comprehensive understanding of what is possible nowadays with microfluidic devices.

A fundamental requirement for the commercial success of microfluidics is the need to develop mass markets where a very large number of devices can be sold. Possible market areas which fit this description include genetic diagnostics, environmental monitoring, medicine, biology and experimental chemistry. The successful application of microfluidics in these areas could be as important and dramatic as the effect that silicon microelectronics has had on electronics.

CHAPTER 2

Theory of Microfluidic Flow

In order to understand the behaviour of most microfluidic systems, a reasonable knowledge of the theory of fluidics is necessary. It is the aim of this chapter to introduce the reader to the theory of fluid mechanics and give examples where necessary. As this book will be an introduction to microfluidics, the readers will be from many different disciplines. Thus, a thorough introduction to the topic is given. Readers who have an engineering background will already be familiar with most of the aspects of fluid mechanics and can therefore skip section 2.1.

Section 2.2 considers the differences between macroscopic and microscopic fluid flow. This topic will be of interest to all readers, as it covers the situation when the laws of fluid mechanics no longer describe behaviour at the micron scale.

This chapter should enable the reader to calculate the flow parameters for micromechanical channels and tubes. It also gives the theoretical background necessary for an understanding of the chapter on simulation of microfluidic devices (chapter 4).

2.1 Fluid Mechanics

In this section we will start with simple problems such as two-dimensional nonviscous flow before we treat the more complicated viscous flow behaviour. After the introduction to laminar flow, it is necessary to consider the turbulent flow regime. Although most microfluidic systems operating with liquids remain in the laminar regime, turbulence occurs for more complex systems like pumps or for operation with gases.

To be able to simplify some more complex geometries into structures that are easier to analyse, an introduction to similitude (the theory of similar flows) is given in section 2.1.4. Here, it is intended to describe the flow through nozzles and diffusers, as both are very common components in micropumps and microvalves. Finally, an introduction is given to the theory of three-dimensional flow. Here, the Navier-Stokes equations are explained. This set of equations is essential for the understanding of simulation tools that enable the reader to simulate the flow behaviour of complex microfluidic systems.

The analytical model for macroscopic fluid mechanics is valid for most micro-channels of more than 1 μm size. Macroscopic fluid mechanics can also be applied to describe the operation of valves and flow sensors [3].

2.1.1 One-dimensional Nonviscous Flow

In ideal fluids no frictional force acts between two neighbouring fluid elements. At first, we consider the principle of conservation of mass. Let us assume that the fluid that we are interested in is incompressible. Then the amount of fluid streaming into a system is the same as the amount streaming out again:

$$v_1 \cdot A_1 = v_2 \cdot A_2 \tag{2.1}$$

with v_1, v_2 as entrance and exit velocity, respectively, and A_1, A_2 as the cross sectional area of entrance and exit, respectively.

In the steady state, the first law of thermodynamics, i.e. the conservation of energy, can be applied to a control volume (figure 2.1) of an incompressible ideal fluid.

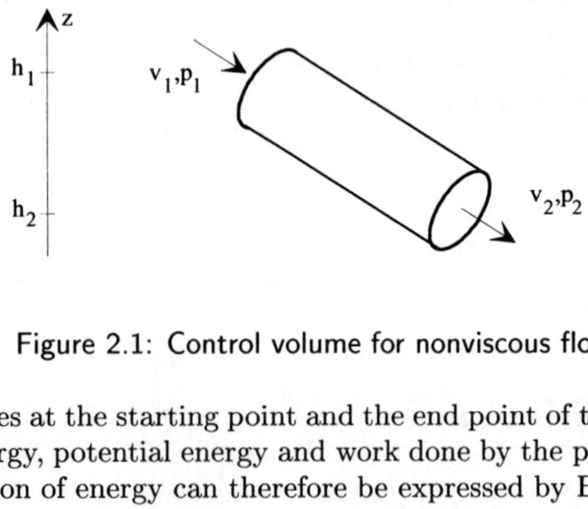

Figure 2.1: Control volume for nonviscous flow

The energies at the starting point and the end point of the control volume are kinetic energy, potential energy and work done by the pressure difference. The conservation of energy can therefore be expressed by Bernoulli's law:

$$\frac{1}{2}\rho v_1^2 + p_1 + \rho\, g\, h_1 = \frac{1}{2}\rho v_2^2 + p_2 + \rho\, g\, h_2 \tag{2.2}$$

where ρ is the fluid density, p the pressure, h the height and g the acceleration due to gravity. This equation can for example be used to calculate the velocity of the fluid ejected from an inkjet nozzle. If the pressure within the inkjet chamber is known to be p and the velocity inside the chamber is zero due to its large volume, the velocity at the exit for h constant is given by:

$$v = \sqrt{\frac{2\,p}{\rho}} \tag{2.3}$$

2.1.2 Two-dimensional Laminar Viscous Flow

Viscosity of a fluid can be explained by using two small plates on top of each other only separated by a thin layer of fluid. When the bottom plate is fixed and a dissipative force is necessary to move the top plate, the fluid is called viscous.

Let us now consider friction between two neighbouring fluid elements. Here, we want to treat only Newtonian fluids. A simple description of Newtonian fluids can be given by looking again at the two plates on top of each other only separated by a fluid. A Newtonian fluid is one where the horizontal force component required to move the top plate is proportional to the speed of the plate. In order to come up with a useful equation, we need to look at a fluid element. Then, a Newtonian fluid is obtained when the shear stress τ on an interface tangent to the direction of flow is proportional to the distance rate of change of velocity, with the differentiation taken in a direction normal to the interface. For this case:

$$\tau = \frac{F(z2) - F(z1)}{A} = \mu \frac{\mathrm{d}u}{\mathrm{d}z} \tag{2.4}$$

with μ as the viscosity of the fluid and u as velocity in x-direction (figure 2.2).

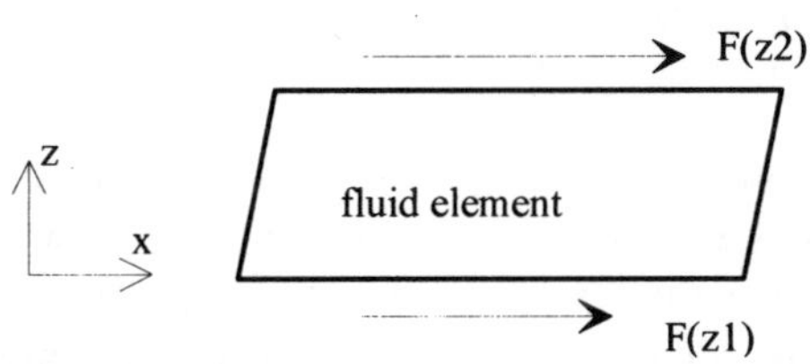

Figure 2.2: Shear stress on a fluid element

This equation can be applied to calculate the velocity distribution of a fluid flowing between two infinite parallel plates. This flow is depicted in figure 2.3. Such flow profiles occur for example in micromixers where the width of the channel is much bigger than the height of the channel (D).

Consider a rectangular element of length l, height $2z$ and width b in the y direction centered about $z = 0$, as shown in figure 2.3. Due to symmetry there is a velocity gradient $\frac{\mathrm{d}u}{\mathrm{d}z}$ and thus a frictional force acting on the upper and lower walls with a total value of $F_f = 2lb\mu\frac{\mathrm{d}u}{\mathrm{d}z}$. The frictional force is compensated by a pressure difference between the front and back of the cube: $F_p = 2zbl\frac{\mathrm{d}p}{\mathrm{d}x}$. This leads to a differential equation for the flow:

$$\frac{\mathrm{d}u}{\mathrm{d}z} = \frac{1}{\mu}\frac{\mathrm{d}p}{\mathrm{d}x} z \tag{2.5}$$

The solution to this differential equation can be found with the boundary

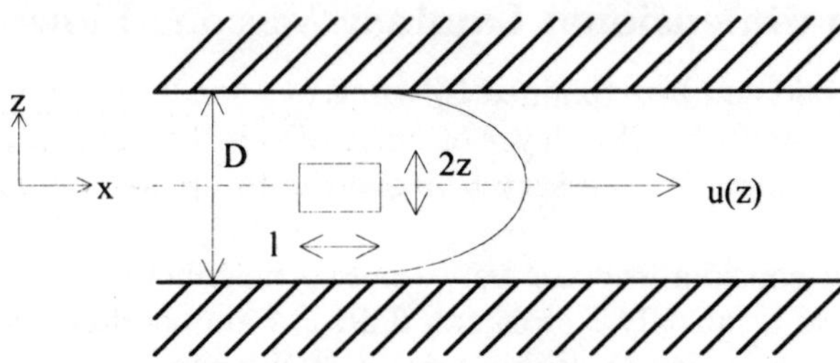

Figure 2.3: Flow between two infinite parallel planes

conditions (the velocity is zero at the tube wall ($z = 0.5D$)):

$$u(z) = \frac{1}{2\mu}\frac{\mathrm{d}p}{\mathrm{d}x}\left(\frac{D^2}{4} - z^2\right) \tag{2.6}$$

The volume flow per unit width is then calculated as:

$$Q_w = \int_{z=-0.5D}^{0.5D} u(z)\,dz = \frac{D^3}{12\mu}\frac{\mathrm{d}p}{\mathrm{d}x} \tag{2.7}$$

For flow through a pipe of radius R we consider a small cylinder of radius r and length l at the centre of the tube. In this case the frictional force on the surface of the cylinder is $F_f = 2\pi\, r\, l \mu \frac{\mathrm{d}u}{\mathrm{d}r}$ and the pressure force is $F_p = \pi\, r^2\, l\, \frac{\mathrm{d}p}{\mathrm{d}x}$. In the steady state, these two forces compensate each other and we get the differential equation for the flow through a tube:

$$\frac{\mathrm{d}u}{\mathrm{d}r} = \frac{1}{2\mu}\frac{\mathrm{d}p}{\mathrm{d}x} r \tag{2.8}$$

The solution of this equation, with the boundary condition that the velocity is zero at the tube wall, is given by:

$$u(z) = \frac{1}{4\mu}\frac{\mathrm{d}p}{\mathrm{d}x}\left(R^2 - r^2\right) \tag{2.9}$$

This flow profile is again parabolic and the volume flow through the tube is given by the Hagen-Poisseuille law as follows:

$$Q = \frac{\pi}{8\mu}\frac{\mathrm{d}p}{\mathrm{d}x} R^4 \tag{2.10}$$

The flow for other tube cross sections produced by micromachining can often be approximated by equation 2.10.

2.1.3 Dimensional Analysis

Often in fluidic problems, the variables of a problem are known but the relationship between them is not. By a procedure called dimensional analysis, a fluidic problem can be described by a set of dimensionless groups rather than the variables. The dimensionless groups are arithmetic products of the variables with the unit "1". The move to dimensionless variables often reduces the number of relationships and thus the number of experiments needed to describe the system behaviour.

To obtain the number of dimensionless variables, every variable has to be expressed in terms of the three basic dimensions: mass, length and time. We now create a matrix, where we put the powers to which the basic units are raised against the corresponding variable. For example, suppose that in a particular problem the pressure drop depends on the variables density ρ (mass per unit volume), viscosity μ, velocity v and the two lengths D and L. The dimensional matrix for this problem is as follows:

	ρ	μ	v	D	L
mass M	1	1	0	0	0
length L	-3	-1	1	1	1
time T	0	-1	-1	0	0

The next step is to calculate the rank of the dimensional matrix, which is 3 in the case of our problem. By subtracting the rank of the dimensional matrix from the number of variables, we obtain the number of independent dimensionless groups for our problem. This step is called Buckingham's Π theorem. We can thus describe our problem with only 2 dimensionless groups.

In order to form the dimensionless groups in our problem, we have to rewrite the basic relation of pressure drop and variables, i.e. $\Delta p = F(\rho, \mu, v, D, L)$ into an infinite series:

$$\Delta p = \left(K_1 \rho^{a_1} \mu^{b_1} v^{c_1} D^{d_1} L^{e_1}\right) + \left(K_2 \rho^{a_2} \mu^{b_2} v^{c_2} D^{d_2} L^{e_2}\right) + ... \tag{2.11}$$

with K_i as dimensionless coefficients. Expressing the components dimensionally we get:

$$\left[\frac{M}{LT^2}\right] \equiv \left[\frac{M}{L^3}\right]^a \left[\frac{M}{LT}\right]^b \left[\frac{L}{T}\right]^c [L]^d [L]^e \tag{2.12}$$

Equating the powers of the variables on each side of the equation gives:

$$\begin{array}{lrcl} \text{For M:} & 1 & = & a+b \\ \text{For L:} & -1 & = & -3a-b+c+d+e \\ \text{For T:} & -2 & = & -b-c \end{array} \tag{2.13}$$

Now, we have three equations for five quantities. In order to solve this, we need to choose three dependent variables and two independent variables. It is now essential to look for three variables from our initial problem that one wants in the same dimensionless group. In our case we select ρ, v and D and thus a,c and d as their corresponding exponents. Next, equation 2.13 is solved for the selected dependent variables, and equation 2.11 is rewritten:

$$\Delta p = \sum_i \left(K_i \rho^{1-b_i} \mu^{b_i} v^{2-b_i} D^{-b_i-e_i} L^{e_i} \right) \tag{2.14}$$

Collecting terms with the same exponent, equation 2.14 reduces to:

$$\Delta p = \sum_i \left(\frac{\mu}{\rho\, v\, D} \right)^{b_i} \left(\frac{L}{D} \right)^{e_i} \tag{2.15}$$

Here, the first dimensionless group $(\frac{\mu}{\rho\, v\, D})$ is the inverse of the Reynolds number and the second $(\frac{L}{D})$ is the ratio of two characteristic lengths. This simplification has big advantages for experiments where the relation of the variables is not known. In the case of microfluidics, structures are often too complicated to be treated analytically, but their behaviour can be expressed with ease in terms of the dimensionless groups of the problem.

2.1.4 Similitude

Two flows are geometrically similar when they have a constant relation between all dimensions. This implies that all angles need to be identical. The feature of similitude is often used for modelling the flow in a large system with a smaller but geometrically similar device. In microfluidics, the opposite can be quite interesting. A larger model allows simple observation of the flow pattern through a micromachined device.

Kinematic similarity describes two flows with similar sets of streamlines. As the streamlines close to the boundaries need to be similar for kinematic similarity, this property also includes geometrical similarity. A similarity of even higher order is dynamic similarity, in which the forces at corresponding points in the model and the real device have the same direction and also a constant relation between each other. This is identical to a constant relation linking the density at corresponding points.

The last similarity, i.e. dynamic similarity, is only important for lifting or dragging. This is necessary for micromachined turbines, but for most other devices geometrically similar flows are enough for modelling.

Although exact similarity is important for modelling, it is sometimes more interesting if it is possible to predict the flow for almost identical structures, i.e. structures where the important dimensions are scaled but the unimportant ones are not. Thus, the pressure - velocity relationships of the two systems to be compared are not exactly identical but almost. As an example, a diffuser with rectangular cross section is used (figure 2.4). For large opening angles, $\alpha > 25°$, the velocity-pressure relationship is almost independent of the length of the diffuser but only dependent on the opening angle α and the opening

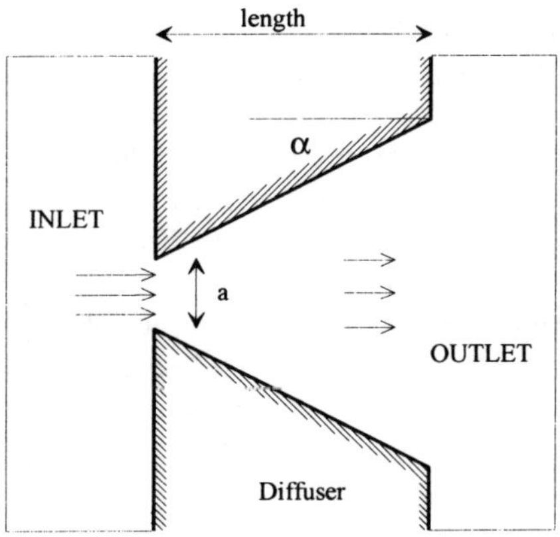

Figure 2.4: Diffuser structure

size a [4]. This implies that the basic flow characteristics of two such systems are almost identical although the diffuser lengths are not scaled.

A second interesting point along these lines is the similarity of flows when orifices of different cross sectional shape are used. The question as to whether two flows with different cross sections are almost similar has been solved by the introduction of the *hydraulic diameter*. The hydraulic diameter is defined as follows:

$$D_h = 4 \cdot \frac{\text{cross sectional area}}{\text{wetted perimeter}} \tag{2.16}$$

It has been verified experimentally that the flow-pressure relation is similar for two systems if the hydraulic diameter as well as all other dimensions are geometrically similar. For laminar flow, the approximate theory can lead to discrepancies of $\pm 40\%$ [4]. For the turbulent case, which is covered in the next section, the difference between theory and practice is better than $\pm 15\%$ [4].

In the case of the diffuser described above, this implies that two diffusers show almost identical pressure-flow relationships if the hydraulic diameter fulfils geometrical similarity and the opening angle is identical. The length of the diffuser is not very relevant in this case.

2.1.5 Turbulent Flow

Turbulent flow is obtained when the effect of a small but macroscopically fluctuating velocity perpendicular to the flow direction is superimposed on a regular flow. This behaviour makes it rather complicated to predict the velocity field for turbulent flow.

The origin of turbulence lies in the nonlinear differential equations which determine the flow pattern. In general, it is possible that nonlinear differential equations behave chaotically. The chaotic behaviour is observed by plotting all variables (pressure, velocity, location, acceleration etc.) in an N-dimensional coordinate system with the time as variable. In comparing two curves with almost identical starting points within this coordinate system, it will be obvious that laminar flow (not chaotic) is dominant if the two curves are constantly close to each other. Turbulent flow is obtained when the curves differ drastically after a short time.

In a simple pipe that is described by the laws of Bernoulli, turbulence can be added as an empirical head loss H_L. The head loss factor can be regarded as the effect of the transformation of potential into kinetic energy as well as the generation of dissipative heat. Using a coefficient K for turbulence, Bernoulli's equation is altered to [5]:

$$\begin{aligned} \frac{v_1^2}{2} + \frac{p_1}{\rho} + g\,h_1 &= \frac{v_2^2}{2} + \frac{p_2}{\rho} + g\,h_2 + H_L \\ \text{where } H_L &= K \cdot \frac{\bar{v}^2}{2} = f\frac{L}{D} \cdot \frac{\bar{v}^2}{2} \end{aligned} \tag{2.17}$$

with $\bar{v}$ as an average velocity, L and D as length and diameter of the tube. The calculation of the average velocity cannot be generalised for diffusers or nozzles but the average velocity is always the velocity at the smallest dimension. The coefficient f is the friction factor and can be obtained from books for specific flow problems or from experiments using dimensional analysis to simplify matters [5].

In general, both laminar and turbulent flows can occur in the same fluidic systems depending on the applied pressure. However, at low velocities only laminar flow is obtained, and at high velocities only turbulent flow is obtained. The transition between the two flow regimes can be obtained from the Reynolds number Re. This number is the ratio of inertial force to viscous force and can be shown to be given by [5]:

$$Re = \frac{\rho}{\mu}\,u\,D \tag{2.18}$$

where ρ, μ, u and D are the density, dynamic viscosity, velocity and a characteristic length within the system, respectively.

Table 2.1 describes the values of the critical transitional Reynolds number Re_t where the flow changes from viscous-loss dominated (laminar) to inertial-loss dominated (turbulent). The equations are also given for the calculation of the flow for some pipes. The characteristic length D in equation 2.18 is given by the hydraulic diameter D_h in table 2.1.

In the table, L is the length of the channel and Q is the volume flow at smaller and larger Reynolds numbers than the transitional Reynolds number. $D_h = 4 \cdot \text{area} \cdot (\text{circumference}^{-1})$ is the hydraulic diameter, i.e. the characteristic length of the channel. The use of the hydraulic diameter makes it possible to use the equations to calculate the flow in non-circular cross section. A is the cross sectional area of the flow, ξ is the pressure loss coefficient,

Type of flow	Definition	Volume Flow Q Q_{lam} (viscous losses dominate) Q_{inert} (inertial losses dominate)	Re_t empirical
orifice	$L/D_h < 0.5$	$Q_{lam} = A\frac{2}{C_{fr}}\frac{D_h}{\mu}\Delta p$ $Q_{inert} = A\sqrt{\frac{2}{\xi\rho}}\,\Delta p$ with $\xi \simeq 2.6$	15
short channel	$2 < L/D_h < 50$	$Q_{lam} = A\frac{2}{C_{fr}}\frac{D_h^2}{L\mu}\Delta p$ Q_{inert}: orifice with $1 < \xi < 1.5$	$30{\cdot}L/D_h$
long channel	$L/D_h > 100$	$Q_{lam} = A\frac{2}{C_{fr}}\frac{D_h^2}{L\mu}\Delta p$ $Q_{inert} = A\sqrt{\frac{2\,D_h}{L\,f_{fr}\,\rho}}\,\Delta p$	2300

Table 2.1: Transitional Reynolds number and volume flow for incompressible liquids [3]

C_{fr} is the friction coefficient ($C_{fr} = 96$ for rectangular cross section where the width is much larger than the height, $C_{fr} = 64$ for circular cross section), f_{fr} is the friction factor ($\simeq 0.14\,Re^{-0.18}$ for fully developed turbulent flow) and Δp is the pressure drop.

For the development of the above equations, the assumption that the viscosity is independent of the dimension of the flow channel has been made. This can be violated sometimes, as the measured flow through very small capillaries (0.5 to 40 μm) does not seem to agree with macroscopic theory [3].

2.1.6 Flow in a Diffuser and Nozzle

The behaviour of a diffuser has already been introduced. Here we want to examine the flow through such a device in more detail and focus also on the reverse flow through a diffuser, i.e. a nozzle (see figure 2.5). The theory developed here applies not only to diffusers and nozzles, but also to all sorts of cross section changes that occur in microfluidics.

Basically, there are two different types of behaviour for diffuser or nozzle structures. Depending on the angle of the truncated pyramid, the flow in the enlarging or contracting direction is preferred. However, the general flow behaviour is identical and will be discussed independently of the opening angle. The difference in the preferred direction depends only on the weight of each contributing factor.

All losses other than those due to normal fluid friction can be described by the head loss factor χ, which correlates with the head loss as presented earlier by $\chi = 2H_L \cdot v^{-2}$ [5]. Normally, all losses have to be determined experimentally and can then be given as the dimensionless head loss factor χ [6], which is given by

$$\chi = \Delta p \left(\frac{\rho}{2}\bar{u}^2\right)^{-1} \tag{2.19}$$

where Δp is the pressure difference across the object, ρ is the density of the fluid and $\bar{u}$ is the average velocity. Here, the head loss factor is used for comparing the individual losses occurring through contracting or expanding ducts. As the head loss is an additional loss, the head loss factor χ becomes zero when the problem can be described analytically, i.e. a straight pipe.

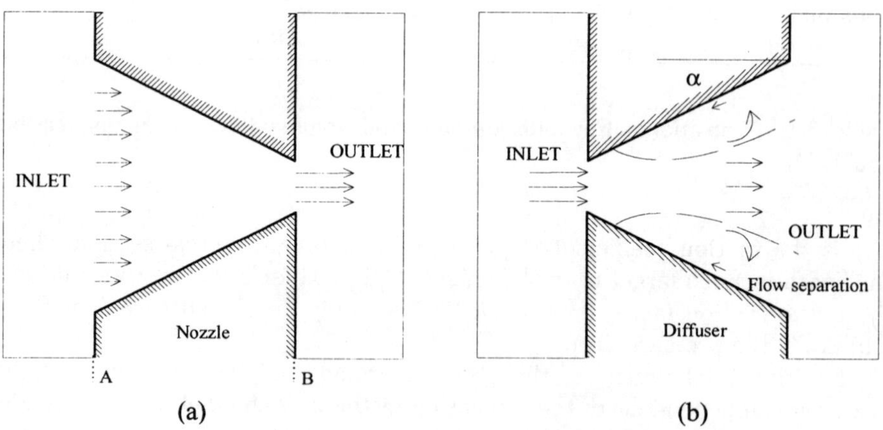

Figure 2.5: Nozzle (a) and diffuser (b) structures

The flow in the nozzle direction is shown in figure 2.5.a. The flow enters through the inlet where it is accelerated by the sudden contraction at plane A. The head loss factor for this sudden contraction is 0.44 [7]. Then the pressure drop across the duct increases the kinetic energy and thus also the flow velocity. In the case of small taper angles of 3° the head loss is less than 0.02, whereas for large taper angles of 35.3° a factor of 0.1 is reached [6]. Both values are rather small, leading to the conclusion that a large amount of the potential energy is transformed into kinetic energy without much dissipation loss. Finally, the flow exits at maximum speed into a large reservoir at plane B where the flow velocity is reduced to smaller values. This exit is referred to as sudden expansion with a head loss factor of $\chi_{se} = 1$ [7]. All head loss values are summarised in table 2.2. The factors were determined for large Re numbers and dimensions but can be applied for small dimensions as well due to fluidic similarity [6]. For a simple approximation, the total head loss factor can be calculated by summing all individual components. Here, a head loss factor of 1.1 is obtained for nozzles with a taper angle of 35.3° and 1.46 for small angles of about 2.6 degrees. For the small angles, all factors are summed, as the inlet and outlet dimensions are similar, but for the large taper angles only the factors for the small opening area and the gradual contraction are taken. The real pressure loss is dependent on the square of the average velocity. For large angles, this velocity is far greater at plane B and so the head loss at plane A is negligible. In the case of small opening angles the second diameter of the cone is slightly bigger than the first, but is regarded as identical for the summation of the head loss factors. This simplification gives an error that has to be borne in mind when discussing flow rates.

A comparison of both factor sums for different angles shows that the flow is mainly dependent on the sudden contraction and expansion rather than the gradual contraction of the nozzle. Thus, the influence of the taper angle on the total head loss factor is minor.

sudden contraction	sudden expansion	gradual contraction	gradual expansion
$\chi_{sc} = 0.44$	$\chi_{se} = 1$	$\chi_{gc1} = 0.1$ (35.3°) $\chi_{gc2} < 0.02$ (2.6°)	$\chi_{ge1} = 1$ to 1.2 (35.3°) $\chi_{ge2} = 0.25$ (2.6°)

Table 2.2: Head loss factors for contraction and expansion derived from measurements [6]

A picture of the flow within a diffuser was shown in figure 2.5.b. The flow enters through a small opening, i.e. a sudden contraction. The head loss factor for this case is 0.44 as above. Then a gradual expansion region follows, where the kinetic energy is transformed into potential energy at head loss factors of 0.25 for $\alpha = 2.6°$ and 1 – 1.2 for $\alpha = 35.3°$ [6]. The large difference in the head loss factor can be explained as the flow separates from the walls for diffuser angles of 35.3 degrees (see figure 2.5). The same phenomenon can be observed only for very large flow rates in the case of diffusers with small taper angles. In the case of the nozzle direction, no such separation from the wall is present, which leads to far lower values than in the diffuser direction. The large value of the head loss factor for diffusers of 35.3° indicates that a considerable amount of energy is lost due to dissipation. The decelerated fluid then drains into a large reservoir by passing a sudden expansion with the head loss factor of 1.0 as given above. Summing up the determining factors for the diffuser gives a total value of 1.44 – 1.64 for $\alpha = 35.3°$ and 1.69 at angles of 2.6°. Again all factors are used in the case of the small taper angle and only sudden contraction and gradual expansion need to be taken into account for large angles.

	small angles α=2.6°	large angles $\alpha = 35.3°$
nozzle	$\sum \chi_N = 1.46$	$\sum \chi_N = 1.1$
diffuser	$\sum \chi_D = 1.69$	$\sum \chi_D = 1.44 - 1.64$
$\sum(\chi_D - \chi_N)$	+0.23	-0.34 – -0.54

Table 2.3: Sum of head loss factors for small and large angles

The comparison in table 2.3 of both sums $\sum(\chi_D - \chi_N)$ for diffuser and nozzle directions shows clearly the different behaviour of the cone for large and small taper angles. The difference between the total head loss factor for diffuser and nozzle is -0.3 – -0.5 for large angles of 35.3° and therefore shows a higher flowrate in the contracting direction. In contrast, a positive value of the difference in the total head loss factor of 0.23 is obtained for small angles of 2.6° . Therefore the gradual expanding direction shows higher flowrates at small taper angles.

Some estimates have been carried out to show the real behaviour of the flow in diffusers and nozzles. In the cases of both small and large opening angles the square shape of the cone has been simplified to a circular cross section. Similarity between circular and square shaped pyramidal structures in fluid-dynamical results (pressure drop, flowrate) can be assumed if the hydraulic diameter in the smallest cross section and the opening angle α are geometrically similar. Thus the results can be directly applied to the KOH etched diffuser/nozzle structure which is described in section 6.3.1.4.

Some simplifications have to be made in order to calculate the volume flow through a nozzle or a diffuser. The simplifications and the boundary conditions necessary for the calculation of the pressure losses are as follows:

- The main pressure drop across both nozzle and diffuser is determined by the pressure losses due to viscosity as well as geometry variations at the entrance, exit and across the channel.
- For both entrance and exit behaviour the average velocity is calculated from the volume flow divided by the entrance or exit area.
- The average velocity of gradual expanding and contracting flow is given by the volume flow through the smaller opening.

The two main sources of the pressure drop can be calculated by the use of the Hagen-Poisseuille law as well as a modified Bernoulli equation. In the case of the diffuser the pressure drop for large taper angles is determined by [8]:

$$\begin{aligned}\Delta p_{\text{diff}} = & \int_0^L \frac{128 \cdot \mu \cdot \Phi_v}{\pi \left(D_0 + x \tan \alpha\right)^4} \mathrm{d}x \\ + & \left(\chi_{sc} \frac{\rho}{2} \left(\frac{4\Phi_v}{\pi D_0^2} \right)^2 + \chi_{ge1} \frac{\rho}{2} \left(\frac{4\Phi_v}{\pi D_0^2} \right)^2 \right. \\ & \left. + \chi_{se} \frac{\rho}{2} \left(\frac{4\Phi_v}{\pi \left(D_0 + L \tan \alpha\right)^2} \right)^2 \right)\end{aligned} \tag{2.20}$$

where the first part is Hagen-Poisseuille's law applied to a diverging duct with D_0 as the smaller diameter of the cone, α as the taper angle, μ as the dynamic viscosity and Φ_v as the volume flow through the device. The second part is the modified law of Bernoulli, i.e. the sum of all additional pressure losses, with χ_i as the head loss factor from table 2.2, L as the length of the duct and ρ as the density of the fluid. Solving and simplifying gives then a polynomial of second order, relating the pressure drop Δp and the volume flow Φ_v:

$$\begin{aligned}\Delta p_{\text{diff}} = & \frac{128 \cdot \mu \cdot \Phi_v}{3\pi \tan \alpha} \left(\frac{1}{D_0^3} + \frac{1}{\left(D_0 + L \tan \alpha\right)^3} \right) \\ & + \frac{8\rho\Phi_v^2}{\pi^2} \left(\chi_{sc} \frac{1}{D_0^4} + \chi_{gc1} \frac{1}{\pi D_0^4} + \chi_{se} \frac{1}{\left(D_0 + L \tan \alpha\right)^4} \right)\end{aligned} \tag{2.21}$$

In the same way, an equation for the pressure as a function of the volume flow can be found for the nozzle direction. The viscous flow for this direction is identical to that in the diffuser, but the modified Bernoulli part is responsible for a flow directing behaviour of the diffuser/nozzle structure:

$$\begin{aligned}\Delta p_{\text{nozzle}} = & \frac{128 \cdot \mu \cdot \Phi_v}{3\pi \tan \alpha} \left(\frac{1}{D_0^3} + \frac{1}{\left(D_0 + L \tan \alpha\right)^3} \right) \\ & + \frac{8\rho\Phi_v^2}{\pi^2} \left(\chi_{sc} \frac{1}{\left(D_0 + L \tan \alpha\right)^4} + \chi_{gc1} \frac{1}{\pi D_0^4} + \chi_{se} \frac{1}{D_0^4} \right)\end{aligned} \tag{2.22}$$

A graph of the pressure as a function of the volume flow can be seen in figure 2.6. The dimensions for the calculations were $L = 500\mu$m, $D_0 = 130$ μm, $\alpha = 35.3°$, $\mu = 1.002 \cdot 10^{-3}$ Pa s, $\rho = 1000$ kg m^{-3}. A closer inspection of the individual terms in equations 2.21 and 2.22 for these dimensions shows that the pressure drop due to viscous flow is small compared to the other terms. Therefore, the main losses are transition losses arising from the modified Bernoulli equation.

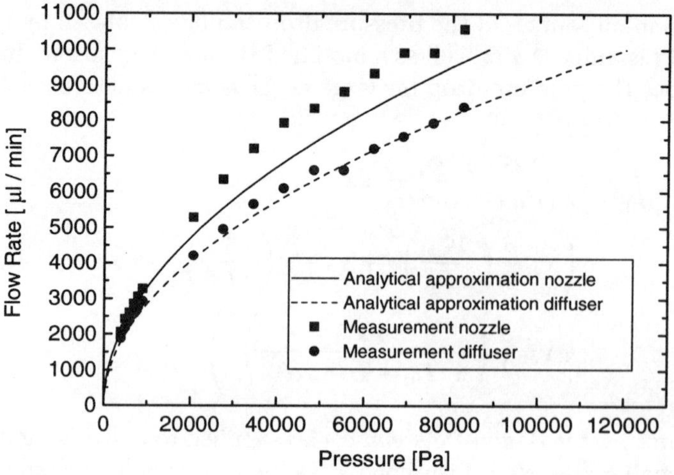

Figure 2.6: Measured and analytical results for diffuser and nozzle at a taper angle of 35.3°

According to the equations, the graphs for the flow rate against pressure of both diffuser and nozzle show a typical square root shape. The comparison with actual measurements on diffusers and nozzles shows that the diffuser is described quite well by the analytical calculation. The calculations have used the concept of the hydraulic diameter. As stated before (section 2.1.4), the error due to similitude is much smaller for turbulent flow than for laminar flow. This is clearly seen in the above figure, as the diffuser operates in the turbulent flow regime and the nozzle operates in the laminar flow regime.

These calculations provide some estimates of the efficiency of the flow rate difference between positive (nozzle) and negative (diffuser) flow through a duct with a tapered angle. The equation for the efficiency is given by the relationship of the flow rate difference and the average flow rate at a certain pressure. Using the fact that the Hagen-Poisseuille pressure loss is small and $D_0 + L\tan\alpha \gg D_0$, the efficiency ϵ can be described by:

$$\epsilon(p) = \frac{\Phi_{\text{pos}}(p) - \Phi_{\text{neg}}(p)}{\frac{1}{2}\left(\Phi_{\text{pos}}(p) + \Phi_{\text{neg}}(p)\right)} \approx 2\frac{\sqrt{\dfrac{1}{\chi_{gc1}+\chi_{se}}} - \sqrt{\dfrac{1}{\chi_{sc}+\chi_{ge1}}}}{\sqrt{\dfrac{1}{\chi_{gc1}+\chi_{se}}} + \sqrt{\dfrac{1}{\chi_{sc}+\chi_{ge1}}}} \tag{2.23}$$

Thus, ϵ is independent of the pressure difference and the size of the inlet D_0. When accounting for the head loss factors from table 2.2, the efficiency is calculated to be 0.155. This number gives the fraction of the liquid that is truly moved in one direction using a membrane pump. A closer inspection including all terms, however, reveals a small change in efficiency for a variation of D_0. For a pressure of 50 kPa, a range of opening sizes varying from 70 μm to 210 μm has been examined, where the efficiency was calculated to be 0.149 for 70 μm openings and 0.158 for 210 μm openings with a steady change between these limits.

Equations 2.21 and 2.22 are also valid for small taper angles of around 2.6° although the head loss factors for gradual contraction and expansion have to be adapted from table 2.2. By etching truly three-dimensional structures through a wafer, using eg. laser assisted etching [9], a cone with a 130 μm small opening and a taper angle of 2.6° could be fabricated through the wafer. This size is chosen in order to keep the smallest diameter constant. The flow rates of a diffuser and a nozzle with these dimensions are given in figure 2.7. Here the volume flow in the diffuser direction is slightly greater than in the nozzle direction, reversing the direction with higher flowrate compared to the one for 35.3° shown before. An estimate of the efficiency of this valve was made by assuming that the main pressure loss originates from transition losses:

$$\begin{aligned}\epsilon(p) &= \frac{\Phi_{\text{pos}}(p) - \Phi_{\text{neg}}(p)}{\frac{1}{2}\left(\Phi_{\text{pos}}(p) + \Phi_{\text{neg}}(p)\right)} \\ &\approx 2\frac{\sqrt{\dfrac{1}{m^{-4}\chi_{sc} + \chi_{gc2} + \chi_{se}}} - \sqrt{\dfrac{1}{\chi_{sc} + \chi_{gc2} + m^{-4}\chi_{se}}}}{\sqrt{\dfrac{1}{m^{-4}\chi_{sc} + \chi_{gc2} + \chi_{se}}} + \sqrt{\dfrac{1}{\chi_{sc} + \chi_{ge2} + m^{-4}\chi_{se}}}}\end{aligned} \tag{2.24}$$

with $m = 1 + \frac{L}{D_0}\tan\alpha$. For a length of $L = 500$ μm and a taper angle of α=2.6° an efficiency of -0.023 is realised, which is rather low compared with 0.155 for the cone with larger angles. This could be improved dramatically for smaller sizes of D_0. A calculation of the efficiency for $D_0 = 25$ μm and $L = 1200$ μm, i.e. the values used by Heschel [8], yields a value of -0.219 which is larger than the figure of merit obtained with a 35.3° angle.

As the ducts with small taper angle, published by Olsson [10] and Heschel [8], do not show a simple cross section, the values obtained from the calculations could not be verified by any measurements. Flow in the diffuser direction was measured by both authors as expected by the analytical approximations for small taper angles. In addition, the efficiency of the valve tested by Heschel was around -0.22 for flow rates between 200 and 400 μl min^{-1}. This value is very close to the figure of merit calculated above using similar dimensions of inlet and taper length, but the duct shape is not identical.

2.1.7 Three-dimensional Fluid Flow

For the full three dimensional analysis of fluid flow, as performed by numerical analysis programmes, it is necessary to use the Navier Stokes equations for fluid flow [5]. These will now be described. They can be transformed as required into a finite element or a finite difference scheme.

The starting point is Euler's equation of motion applied to an infinitesimal small control volume of mass dm restricted to no shear stresses with only

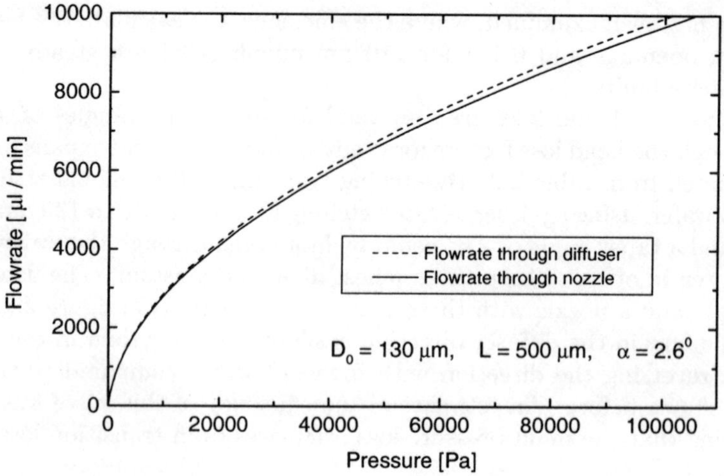

Figure 2.7: Simplified behaviour of a diffuser and nozzle structure for a taper angle of 2.6°

gravity as a body force:

$$\mathrm{d}m \begin{pmatrix} u\frac{\partial u}{\partial x} + v\frac{\partial u}{\partial y} + w\frac{\partial u}{\partial z} + \frac{\partial u}{\partial t} \\ u\frac{\partial v}{\partial x} + v\frac{\partial v}{\partial y} + w\frac{\partial v}{\partial z} + \frac{\partial v}{\partial t} \\ u\frac{\partial w}{\partial x} + v\frac{\partial w}{\partial y} + w\frac{\partial w}{\partial z} + \frac{\partial w}{\partial t} \end{pmatrix} = \mathrm{d}m \begin{pmatrix} B_x \\ B_y \\ B_z \end{pmatrix} + \begin{pmatrix} \mathrm{d}f_x \\ \mathrm{d}f_y \\ \mathrm{d}f_z \end{pmatrix} \quad (2.25)$$

with u, v, w as the velocities of the flow in x, y, z directions, $\vec{B}$ as the body force per unit mass and $\mathrm{d}\vec{f}$ as the resultant surface force, which is only due to pressure p, and thus $\mathrm{d}\vec{f} = -\frac{\mathrm{d}m}{\rho}\mathrm{grad}(p)$.
In three dimensions, the state of stress is specified by the stress tensor of a fluidic particle:

$$\begin{pmatrix} \tau_{xx} & \tau_{xy} & \tau_{xz} \\ \tau_{yx} & \tau_{yy} & \tau_{yz} \\ \tau_{zx} & \tau_{zy} & \tau_{zz} \end{pmatrix} \quad (2.26)$$

with $\tau_{ij} = \tau_{ji}$ due to rotational equilibrium and τ_{ii} describing the normal stress on the planes.
In three dimensions the resultant surface force along the y-axis (see figure 2.8) of an infinitesimally small cube is calculated as:

$$\delta f_y = \left(\frac{\partial \tau_{yy}}{\partial y} + \frac{\partial \tau_{zy}}{\partial z} + \frac{\partial \tau_{xy}}{\partial x} \right) \delta x \delta y \delta z \quad (2.27)$$

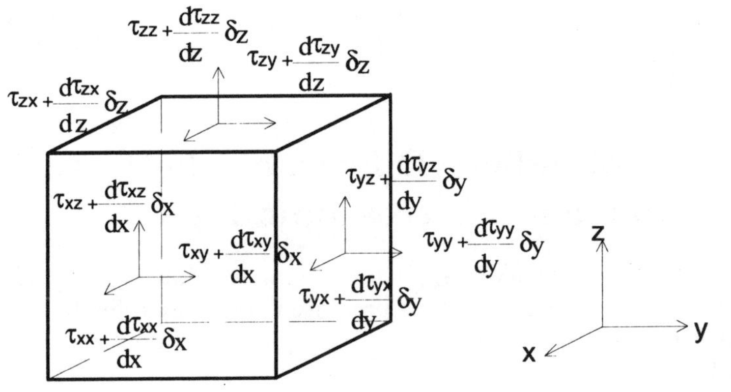

Figure 2.8: Three-dimensional stress on a cubic fluid element

The shear stress components can be expressed by the common form of Stokes's viscosity law:

$$\begin{aligned}
\tau_{xx} &= \mu\left(2\tfrac{\partial u}{\partial x} - \tfrac{2}{3}\operatorname{div}\vec{v}\right) - p; \quad & \tau_{xy} &= \mu\left(\tfrac{\partial u}{\partial y} + \tfrac{\partial v}{\partial x}\right) \\
\tau_{yy} &= \mu\left(2\tfrac{\partial v}{\partial y} - \tfrac{2}{3}\operatorname{div}\vec{v}\right) - p; \quad & \tau_{xz} &= \mu\left(\tfrac{\partial u}{\partial z} + \tfrac{\partial w}{\partial x}\right) \\
\tau_{zz} &= \mu\left(2\tfrac{\partial w}{\partial z} - \tfrac{2}{3}\operatorname{div}\vec{v}\right) - p; \quad & \tau_{yz} &= \mu\left(\tfrac{\partial v}{\partial z} + \tfrac{\partial w}{\partial y}\right)
\end{aligned} \tag{2.28}$$

with p as the pressure and μ as the coefficient of viscosity. For incompressible fluids, the expression $\operatorname{div}\vec{v}$ is zero from continuity considerations. Inserting Stokes's law into equation 2.27 and rearranging gives:

$$\delta f_y = \left(-\frac{\partial p}{\partial y} + \mu\,\nabla^2 v + \mu\left(\frac{\partial^2 u}{\partial x \partial y} + \frac{\partial^2 v}{\partial y^2} + \frac{\partial^2 w}{\partial z \partial y}\right)\right)\delta x\,\delta y\,\delta z \tag{2.29}$$

For incompressible flow the expression $\partial^2 u/(\partial x \partial y) + \partial^2 v/\partial y^2 + \partial^2 w/(\partial z \partial y)$ is zero. When the modified equation 2.29 is now inserted into Euler's equation of motion (2.25) the Navier-Stokes equation of the incompressible laminar flow for isotropic fluids is obtained:

$$\frac{\mathrm{D}\vec{v}}{\mathrm{D}t} = \vec{B} - \frac{1}{\rho}\operatorname{grad} p + \frac{\mu}{\rho}\,\Delta\,\vec{v} \tag{2.30}$$

with Δ as Laplace operator, which is, for cartesian coordinates, $\left(\frac{\partial^2}{\partial x^2} + \frac{\partial^2}{\partial y^2} + \frac{\partial^2}{\partial z^2}\right)$. Equation 2.30 can be used for solving a fluidic problem analytically. However, for complex geometries, a solution cannot be found and a numerical calculation with small elements of constant physical properties has to be performed.

It should be added that Stokes's viscosity law and the resulting Navier-Stokes equations are considered valid also for turbulent flow if the actual time-dependent variables are used.

2.2 Microfluidics: Differences between Microscopic and Macroscopic Flow

Not all of the above theory is applicable to microfluidic systems, although in most cases the calculation methods presented above are reliable. This section will deal with the cases where the macroscopic theory presented above is not applicable.

Due to the small scale in microsystems, practical problems also arise which do not occur in macroscopic systems. In this section we therefore also consider the effect of air bubbles or particles clogging the micromachined capillaries.

2.2.1 Laminar and Turbulent Flow Regimes

There are cases where macroscopic flow theory is no longer valid for micro-channels. One assumption for laminar macroscopic gas flow is that the gas velocity is much lower than the velocity of sound. This is most often the case for large structures, but in micro-channels the gas velocity can easily exceed the velocity of sound whereas the Reynolds number is at the same time far less than the critical Reynolds number (2000 to 2300) [3]. This break-down of macroscopic flow theory has been observed by Jerman [11] for gas flow in a channel of 5μm height. The second assumption is a velocity of zero (no slip) at the wall of the device. Again, this is valid at atmospheric pressure on the macroscopic scale, but for microfluidic devices the dimensions of the structure are sometimes not much larger than the mean free path for gas molecules (60 nm at 100 kPa) [3]. The violation of the no-slip condition in microfluidics leads to increased gas flows as compared with those predicted by the simple theories (table 2.1). Such an increase compared to theory has been observed by Pfahler [12] for channels of 0.5 μm height.

For liquids both the assumptions considered above are still valid. However, the length of the elements in the direction of the flow can still be shorter than the length for fully developed laminar or turbulent flow [3]. According to a study by Gravesen [3], all microfluidic devices considered (cantilever valves, diffuser/nozzle valve, micropumps, flowsensors, capillary tubes etc.) operate in regions where either laminar flow with viscous losses dominates or in the transition flow region with inertial losses dominating. Owing to the small scales no fully developed turbulent flow could be found. Therefore the transitional Reynolds number of 2300 seems to have little relevance in microfluidics [3].

Also, in macroscopic liquid flow, the viscosity is independent of the dimension of the flow channel. This can be sometimes violated in microfluidics, as the measured flow through very small capillaries (0.5 to 40 μm) seems not to agree with macroscopic theory [3].

2.2.2 Practical Problems in Liquid Flow at Small Dimensions

An important phenomenon in microfluidics is the fact that gas bubbles can block exits. Matsumoto and Colgate [13] have developed a theory for the movement of a gas bubble in a capillary. The pressure difference ΔP across a gas bubble and the pressure ΔP_b needed to transport such a bubble are given by:

$$\Delta P = \frac{2\,\gamma_{lg}\cos\theta}{r} \qquad \Delta P_b = \frac{2\gamma_f}{r} \tag{2.31}$$

where γ_{lg} is the gas-liquid surface tension, γ_f the frictional surface parameter, θ the contact angle and r the radius of the capillary. The first term only affects the situation where there is a change in the cross sectional area of the capillary as there is no net effect from this term on a gas bubble inside a channel of constant radius r. For the removal of a gas bubble from the exit of a capillary of 1 μm side length, a pressure in the order of 140 kPa, i.e. more than atmospheric pressure, is necessary for water as the test liquid [3].

The inadvertent introduction of gas bubbles into microfluidic systems often occurs during the priming procedure. One solution to this is to prime the system with high pressure, so that no bubbles can be formed or are immediately washed out again. This often helped in the experiments of the present authors with micropumps, valves and other microfluidic systems.

Other solutions to the problem of introducing gas bubbles have been described by Zengerle [14]. The first possibility is to use ethanol as the first priming liquid, as it has a solubility coefficient for air that is two decades larger than water. This will allow the air bubbles to dissolve in ethanol. After the priming, the flow can be switched from ethanol to water, guaranteeing air-bubble free channels. The solubility coefficient of CO_2 is three times that of air in water [14]. This means that an initial priming of the channels with carbon dioxide and a subsequent filling with water removes gas bubbles almost entirely.

Another problem in microfluidics is unwanted particles. For channels with dimensions less than or equal 5 μm, clogging is a critical issue. It can however be reduced if the solutions are filtered prior to use in microfluidic experiments. For tests with small channels, the present authors have used filters with 0.2μm pores quite successfully and clogging could be reduced drastically. However, this becomes more complicated if the application is sorting or counting of particles or cells. Polystyrene particles for example have the tendency to stick to each other and to silicon dioxide covered surfaces or Pyrex. The only way to remove these particles is by using a pressure shock from the outlet. Quite often, this impact pushes the particles into regions of larger cross sections.

CHAPTER 3

Mechanical Aspects of Microfluidic Devices

Many microfluidic devices include mechanical structures. This is the case for many pumps and valves fabricated by microsystem technology. Therefore this chapter gives an overview of the theory of mechanics and discusses different actuation techniques used in microfluidics.

3.1 Mechanics of Materials

We first develop the theory of elasticity in solids. This is then used to deal with deflections and vibrations in beams, bridges, shells and diaphragms. These are the structures that are used in microfluidic pumps and valves.

3.1.1 Elementary Theory of Elasticity

Let us first consider isotropic bodies, which have identical material properties in all directions. This can be for example polycrystalline silicon, or all metals deposited on silicon structures. Hooke's law relates stress $\sigma = \frac{\text{force}}{\text{area}}$ to the resultant strain $\epsilon = \frac{\text{change in length}}{\text{original length}}$ for a mechanical structure.

If a force acts along the x direction as shown in figure 3.1 it expands or contracts the material according to Hooke's law

$$\epsilon_x = E^{-1} \cdot \sigma_x \tag{3.1}$$

where E is the Young's modulus, which is material dependent. Table 3.1 lists the Young's modulus for materials of interest in microsystems.

At the same time, the material is thinned or expanded in a direction perpendicular to the applied force. With the help of the Poisson's ratio ν_P the strain in the perpendicular direction is given by:

$$\epsilon_y = -\nu_P E^{-1} \sigma_x \tag{3.2}$$

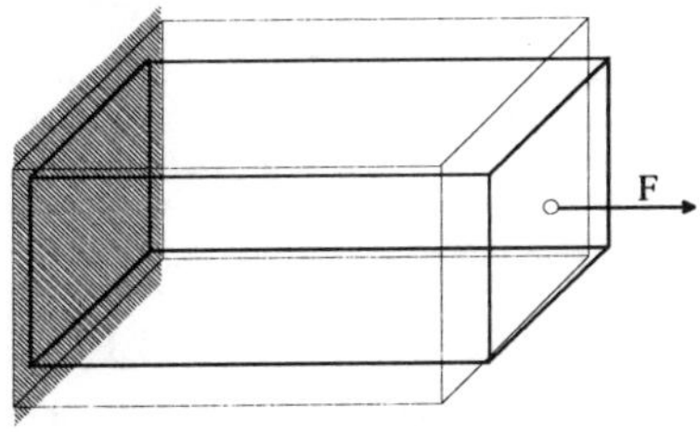

Figure 3.1: Effect of strain on an elastic material

The Poisson's ratio ν_P for bulk Al is for example 0.34, and most other materials also have a value in the range 0.2 to 0.45.

The shear stress $\tau = \frac{\text{force}}{\text{area}}$ as shown in figure 3.2 is related to the shear angle γ by

$$\tau = G \cdot \gamma \tag{3.3}$$

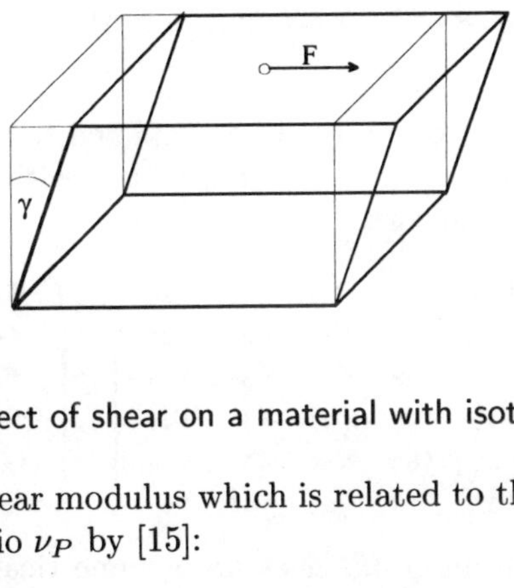

Figure 3.2: Effect of shear on a material with isotropic properties

where G is the shear modulus which is related to the Young's modulus E and the Poisson's ratio ν_P by [15]:

$$G = \frac{1}{2}E(1 + \nu_P)^{-1} \tag{3.4}$$

For silicon, the equations are slightly more complex, as it is a single crystal which shows anisotropic behaviour. Hooke's law in three dimensions is given by the following stress - strain relation for any anisotropic material:

$$\sigma = [c] \cdot \epsilon \qquad\qquad \epsilon = [s] \cdot \sigma \tag{3.5}$$

where σ and ϵ are the stress and strain tensors of second order and $[c]$ and $[s]$ are the stiffness and the compliance tensors of fourth order [16]. In order to avoid rotation of the body, a symmetry of the shear stresses is necessary: $\tau_{ij} = \tau_{ji}$. Figure 3.3 illustrates some of the components of the stress and strain

material	Young's Modulus [GPa]	comment
Al	60 – 70	sputter deposited
Ti	100 – 110	sputter deposited
Pt	170	sputter deposited
poly-Si	150 – 200	LPCVD deposited
SiO_2	70 – 75	thermally grown
SiO_2	92	sputter deposited
Si_3N_4	290	LPCVD deposited
Si_3N_4	210	PECVD deposited
Si_3N_4	130	sputter deposited
SiC	100 – 150	LPCVD deposited
Si	190	bulk, in $< 111 >$ direction
Si	130	bulk, in $< 100 >$ direction

Table 3.1: Young's modulus for some isotropic materials and for silicon in different directions (see the webpage at http://mems.isi.edu)

tensors. The tensor notation can then be simplified to a matrix-like equation with vectors of 6 dimensions σ and ϵ and a matrix with 6×6 elements for the stiffness and compliance tensors:

$$\begin{pmatrix} \sigma_x \\ \sigma_y \\ \sigma_z \\ \sigma_{yz} \\ \sigma_{zx} \\ \sigma_{xy} \end{pmatrix} = \begin{bmatrix} c_{11} & c_{12} & c_{13} & c_{14} & c_{15} & c_{16} \\ c_{21} & c_{22} & c_{23} & c_{24} & c_{25} & c_{26} \\ c_{31} & c_{32} & c_{33} & c_{34} & c_{35} & c_{36} \\ c_{41} & c_{42} & c_{43} & c_{44} & c_{45} & c_{46} \\ c_{51} & c_{52} & c_{53} & c_{54} & c_{55} & c_{56} \\ c_{61} & c_{62} & c_{63} & c_{64} & c_{65} & c_{66} \end{bmatrix} \cdot \begin{pmatrix} \epsilon_x \\ \epsilon_y \\ \epsilon_z \\ \epsilon_{yz} \\ \epsilon_{zx} \\ \epsilon_{xy} \end{pmatrix} \tag{3.6}$$

The compliance and stiffness matrices are symmetrical about the main diagonal ($c_{ij} = c_{ji}$). Thus, the number of elements is reduced to 21. Using symmetry effects for hexagonal crystals like PZT (lead zirconate titanate) or ZnO the system can be reduced further to the following equation:

$$\begin{pmatrix} \sigma_x \\ \sigma_y \\ \sigma_z \\ \sigma_{yz} \\ \sigma_{zx} \\ \sigma_{xy} \end{pmatrix} = \begin{bmatrix} c_{11} & c_{12} & c_{13} & 0 & 0 & 0 \\ c_{12} & c_{11} & c_{13} & 0 & 0 & 0 \\ c_{13} & c_{13} & c_{33} & 0 & 0 & 0 \\ 0 & 0 & 0 & c_{44} & 0 & 0 \\ 0 & 0 & 0 & 0 & c_{44} & 0 \\ 0 & 0 & 0 & 0 & 0 & 0.5(c_{11} - c_{12}) \end{bmatrix} \cdot \begin{pmatrix} \epsilon_x \\ \epsilon_y \\ \epsilon_z \\ \epsilon_{yz} \\ \epsilon_{zx} \\ \epsilon_{xy} \end{pmatrix} \tag{3.7}$$

Because of an even higher symmetry in cubic crystals like silicon, this can be

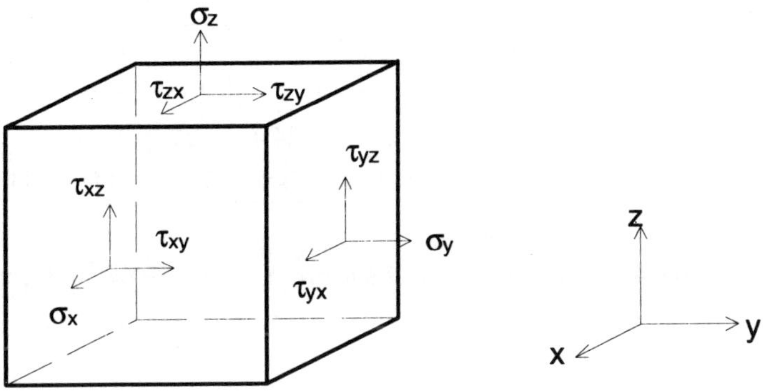

Figure 3.3: Three-dimensional stress on a mechanical element

further simplified:

$$\begin{pmatrix} \sigma_x \\ \sigma_y \\ \sigma_z \\ \sigma_{yz} \\ \sigma_{zx} \\ \sigma_{xy} \end{pmatrix} = \begin{bmatrix} c_{11} & c_{12} & c_{12} & 0 & 0 & 0 \\ c_{12} & c_{11} & c_{12} & 0 & 0 & 0 \\ c_{12} & c_{12} & c_{11} & 0 & 0 & 0 \\ 0 & 0 & 0 & c_{44} & 0 & 0 \\ 0 & 0 & 0 & 0 & c_{44} & 0 \\ 0 & 0 & 0 & 0 & 0 & c_{44} \end{bmatrix} \cdot \begin{pmatrix} \epsilon_x \\ \epsilon_y \\ \epsilon_z \\ \epsilon_{yz} \\ \epsilon_{zx} \\ \epsilon_{xy} \end{pmatrix} \tag{3.8}$$

In isotropic materials the same number of coefficients is used as for cubic crystals. These coefficients are related to the Young's modulus E, Poisson's ratio ν_P and shear modulus G:

$$c_{11} = E\frac{1-\nu_P}{(1+\nu_P)(1-2\nu_P)} \qquad c_{12} = E\frac{\nu_P}{(1+\nu_P)(1-2\nu_P)}$$

$$G = c_{44} = \frac{c_{11}-c_{12}}{2} = \frac{E}{2(1+\nu_P)}$$

The coefficients for silicon, α quartz (triclinic crystal), PZT and ZnO are given in table 3.2.

3.1.2 Bending of Beams and Bridges

A quite important aspect of mechanical engineering in microfluidics is the bending of beams and bridges. Complex structures like an actuating membrane or a cantilever within a passive valve can often be approximated by such a two dimensional structure.

material [unit]	c_{11}	c_{12}	c_{13}	c_{33}	c_{44}	c_{14}
silicon [10^9 Pa]	165.8	63.9	-	-	79.6	-
α quartz [10^9 Pa]	87	7	12	106	58	-18
PZT 5H [10^9 Pa]	126.0	79.5	84.1	117	23.0	-
ZnO [10^9 Pa]	194	102	94	217	44	44

Table 3.2: Mechanical characteristics of silicon, α quartz, PZT and ZnO [16, 17]

3.1.2.1 Basic Mechanics

The sum of the forces F and the bending moments M on a beam need to be zero so that no rotation or translational movement of the beam is obtained:

$$\sum F = 0 \qquad\qquad \sum M = 0 \tag{3.9}$$

Let us assume a beam resting on two points below ($F_{left.1}$ and $F_{right.2}$) with two forces applied from above ($F_{left.2}$ and $F_{right.1}$). Then there will be reaction forces at the resting points to hold the beam in place. At an imaginary cross section (x=0 in figure 3.4) of this beam the forces and moments balance to give:

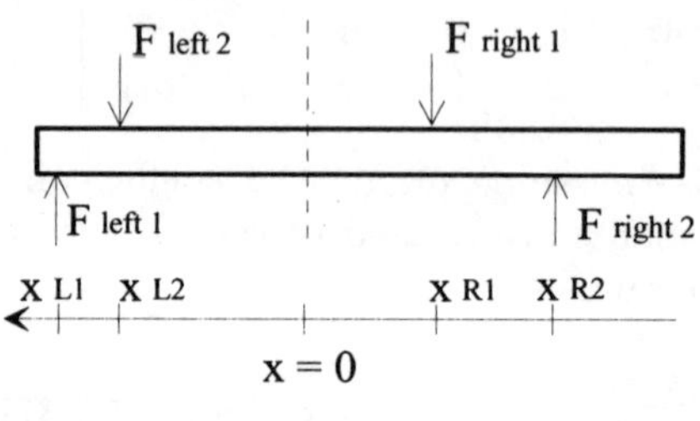

Figure 3.4: Forces and bending moments on a beam

$$\begin{aligned} -\sum F_{left.i} &= \sum F_{right.i} \\ \sum (F_{left.i} \cdot x_{L.i}) &= \sum (F_{right.i} \cdot x_{R.i}) \end{aligned} \tag{3.10}$$

Let us now consider an infinitesimally short section of a beam with constant width w as depicted in figure 3.5. The applied force per unit length along the section shall be called $f(x)$, being a constant f due to the short dimension $\mathrm{d}x$. Thus we obtain the following equations from the equation of continuity for forces and bending moments (equation 3.9):

$$\begin{aligned} f \cdot \mathrm{d}x + F + \mathrm{d}F - F &= 0 \\ M + f \cdot \mathrm{d}x \cdot \frac{\mathrm{d}x}{2} + (F + \mathrm{d}F)\mathrm{d}x - (M + \mathrm{d}M) &= 0 \end{aligned} \tag{3.11}$$

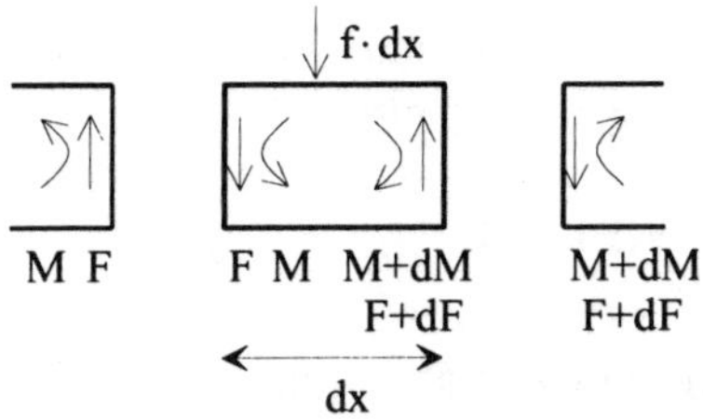

Figure 3.5: Forces and bending moments on an infinitesimally short section of a beam

Simplification and the assumption that $\mathrm{d}F \cdot \mathrm{d}x \ll F \cdot \mathrm{d}x + \mathrm{d}M$ lead to the differential equations of a simple beam:

$$\frac{\mathrm{d}F}{\mathrm{d}x} = -f(x) \qquad \frac{\mathrm{d}M}{\mathrm{d}x} = -F \tag{3.12}$$

The simple bending of a beam is depicted in figure 3.6. For small bending angles θ, the strain at a distance y from the neutral axis (zero strain) is calculated by:

$$\epsilon_{ef} = \frac{(R+y)\theta - R\theta}{R\theta} = \frac{y}{R} \tag{3.13}$$

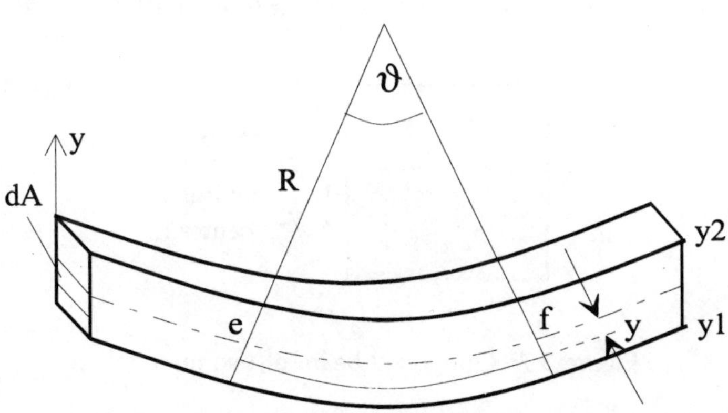

Figure 3.6: Simple bending of a beam

The strain can be used in Hooke's law (equation 3.1) to yield the stress of the beam: $\sigma = E\frac{y}{R}$. The force necessary for the bending is calculated by $F = E\frac{y}{R}\mathrm{d}A$.

As the shear is zero at any cross section through the beam, the sum of the forces from $y = y_1$ to $y = y_2$ is also zero:

$$\sum F = \frac{E}{R} \int_{y=y_1}^{y=y_2} y \, \mathrm{d}A = 0 \tag{3.14}$$

Also, the whole bending moment at a cross section is calculated by:

$$M_{ges} = \frac{E}{R} \int_{y=y_1}^{y=y_2} y^2 \mathrm{d}A = \frac{E}{R} I \tag{3.15}$$

with I as the second moment of area. For a beam consisting of only one material and having a cross section symmetric through the centre (i.e. rectangular with width w and height h or circular cross section with radius R), the second moment of area is calculated as follows:

$$\begin{aligned} I_{\text{rectangular}} &= \int_{-h/2}^{h/2} y^2 \cdot w \, \mathrm{d}y = \frac{w\,h^3}{12} \\ I_{\text{circular}} &= \int_{-R}^{R} 2y^2 \sqrt{R^2 - y^2} \mathrm{d}y = \frac{\pi\,R^4}{4} \end{aligned} \tag{3.16}$$

In order to calculate the product EI for a beam composed of layers of several materials with rectangular cross section, where the individual components do not show any kind of sliding at the material interfaces, the theorem of parallel axes is used. Here, the product EI is defined as the sum for all components: $EI = \sum E_i I_i$. The individual second moments of area are then calculated by using equation 3.17.

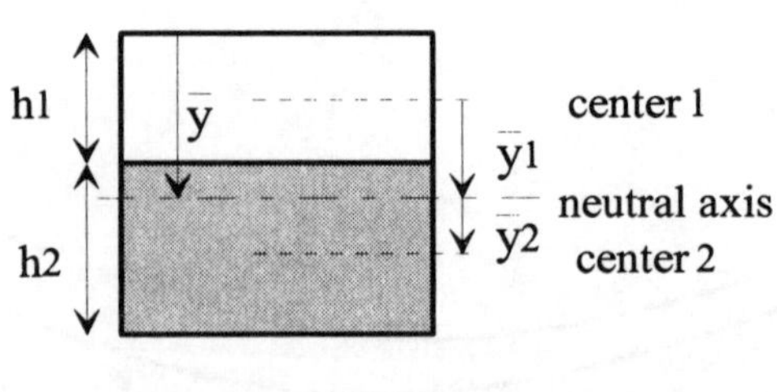

Figure 3.7: Compound beam of two materials

$$I_i = \frac{w_i\,h_i^3}{12} + h_i\,w_i\,\bar{y}_i^2 \tag{3.17}$$

with w_i and h_i as the width and height of the component i and $\bar{y}_i$ as the distance of the centre of the component i from the neutral axis. To calculate this distance, the forces F_i at the individual centre are first obtained by $F_i = \frac{1}{R} E_i\,A_i\,\bar{y}_i$, where $A_i = w_i \cdot h_i$ is the area of each component. The sum of the forces for all components must be zero and thus the following equation needs to be satisfied:

$$\sum E_i\, h_i\, w_i\, \bar{y}_i = 0 \tag{3.18}$$

For the example shown in figure 3.7, the two distances $\bar{y}_i$ can be related to the general distance $\bar{y}$ of the neutral axis from the top of the structure:

$$\bar{y}_1 = -\left(\bar{y} - \frac{1}{2}h_1\right) \qquad \bar{y}_2 = \left(h_1 + \frac{1}{2}h_2 - \bar{y}\right) \tag{3.19}$$

Using the fact that all components have the same width w and employing the three equations above we obtain a formula for calculating $\bar{y}$:

$$\bar{y} = \frac{(h_1 + h_2)^2 + h_1^2\left(\frac{E_1}{E_2} - 1\right)}{2\left(h_2 + h_1\frac{E_1}{E_2}\right)} \tag{3.20}$$

And finally we obtain for the combination of two materials with rectangular cross sections of uniform width the equation for the product EI:

$$EI = \frac{1}{12}\frac{h_1^4E_1^2 + E_1E_2\left(4h_1^3h_2 + 6h_1^2h_2^2 + 4h_1h_2^3\right) + h_2^4E_2^2}{h_1E_1 + h_2E_2} \tag{3.21}$$

3.1.2.2 Equations for Cantilevers and Clamped Beams

As a first example we calculate the bending of a cantilever that is clamped at one end (see figure 3.8). Such structures are often used in passive valves and many other micromechanical devices (cf. section 6.3). For simplification, a force per unit length $(p \cdot w)$ in the z-direction is set to be constant over the whole length of the beam.

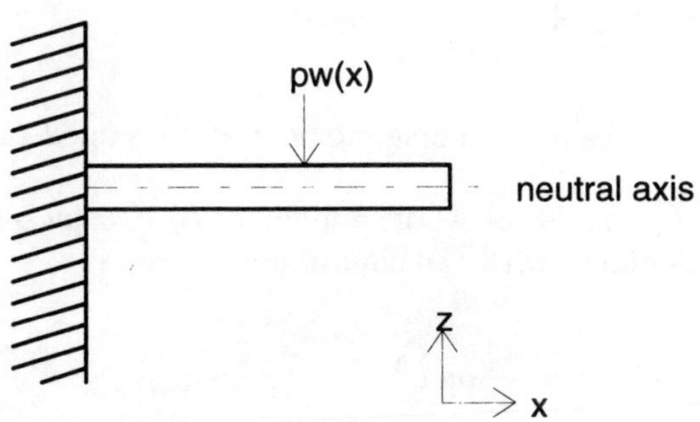

Figure 3.8: Force per unit length on a cantilever beam

The bending moment M for small deflections has already been given in equation 3.15. By replacing the inverse bending radius with the second derivative of the deflection δ, this becomes:

$$M = -EI\frac{d^2\delta}{dx^2} \tag{3.22}$$

As the material is assumed uniform through the whole beam, the neutral axis lies in the centre (see Figure 3.8). Therefore, I is given by $\frac{wh^3}{12}$ where h is the thickness of the beam. As the differentiation with respect to x of the bending moment equals the force, a further differentiation leads to the force per unit length:

$$\frac{\mathrm{d}^2M}{\mathrm{d}x^2} = \frac{\mathrm{d}F}{\mathrm{d}x} = -pw \tag{3.23}$$

Therefore, equation 3.22 can be rewritten as:

$$pw = EI\frac{\mathrm{d}^4\delta}{\mathrm{d}x^4} \tag{3.24}$$

This is the differential equation for the bending of a cantilever. As the cantilever of length L is clamped horizontally at one end, a reaction force F_1 and a bending moment M_1 exist at the support (see Figure 3.9).

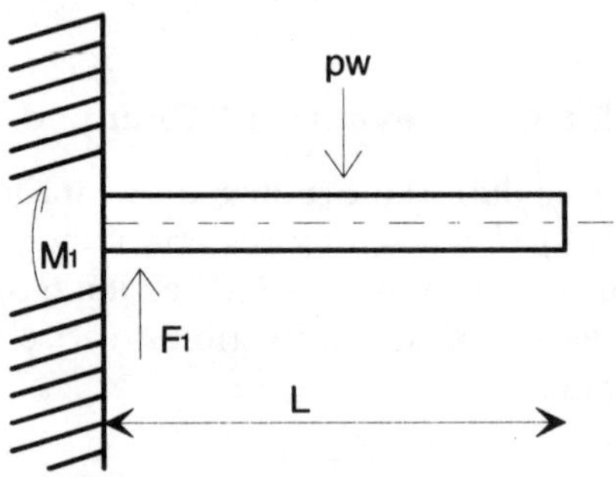

Figure 3.9: Force and bending moment at a clamped cantilever

Furthermore, F_1 equals the whole applied force $F = pwL$ (see Figure 3.8) and M_1 can be calculated with the equation:

$$M_1 = -\int_{x=0}^{L} pwx\, dx = -\frac{1}{2}pwL^2 \tag{3.25}$$

This means that the moment M_1 is equal and opposite to the sum of the bending moments due to the force per unit length pw. The differential equation 3.24 can then be integrated subject to the boundary conditions of a clamped beam ($\delta(x=0)=0$, $\frac{\mathrm{d}}{\mathrm{d}x}\delta(x=0)=0$, $\frac{\mathrm{d}^2}{\mathrm{d}x^2}\delta(x=0)=-\frac{1}{EI}M_1$, $\frac{\mathrm{d}^3}{\mathrm{d}x^3}\delta(x=0)=-\frac{1}{EI}F_1$) together with equation 3.25 to give the deflection along the cantilever:

$$\delta(x)=\frac{1}{EI}\left(\frac{pwx^4}{24}-\frac{pwLx^3}{6}+\frac{pwL^2x^2}{4}\right) \tag{3.26}$$

This equation is used for evaluating the maximum deflection of the beam at the free end ($x=L$):

$$\delta_{max}=\frac{pwl^4}{8EI} \tag{3.27}$$

Inserting the value of I for a rectangular beam (equation 3.16) leads to the final equation of the maximum deflection as a function of the applied pressure p:

$$\delta_{max}=\frac{3}{2E}\frac{L^4}{h^3}p \tag{3.28}$$

As a second example we consider a beam that is clamped at both ends (figure 3.10). This structure is a simplified model of a circular membrane as used in micropumps (cf. section 6.4). Also, it can be employed to calculate the deflection of a rectangular membrane, if the length L is much greater than its width w.

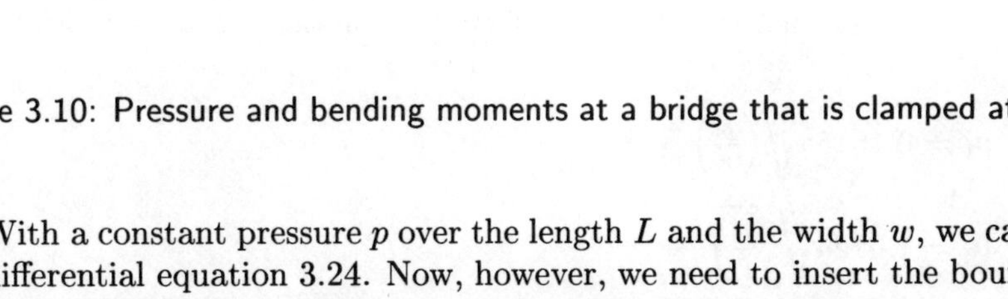

Figure 3.10: Pressure and bending moments at a bridge that is clamped at both ends

With a constant pressure p over the length L and the width w, we can use the differential equation 3.24. Now, however, we need to insert the boundary conditions differently, as the beam is clamped at both ends. Integrating the differential equation four times gives:

$$\delta(x)=\frac{1}{24}\frac{pw}{EI}x^4+\frac{1}{6}Ax^3+\frac{1}{2}Bx^2+Cx+D \tag{3.29}$$

with A, B, C, D as integration variables. The boundary conditions for a beam clamped at both ends are: zero deflection at both ends of the beam ($\delta(x=0)=0$, $\delta(x=L)=0$) and zero curvature at both ends of the beam

$\frac{\mathrm{d}}{\mathrm{d}x}\delta(x=0)=0$, $\frac{\mathrm{d}}{\mathrm{d}x}\delta(x=L)=0$. This gives the following equation for the deflection of the bridge:

$$\delta(x) = \frac{1}{24}\frac{pw}{EI}\left(x^4 - Lx^3 + L^2x^2\right) \tag{3.30}$$

with the maximum deflection at the centre of the beam: $\delta_{max} = \frac{1}{112}\frac{pw}{EI}L^4$

3.1.2.3 Natural Frequencies

The natural frequencies of a cantilever are quite important for microfluidic devices like micropumps. The wave equation for the flexural wave motion of a uniform cantilever is calculated from equations 3.22 and 3.23 [18]:

$$\begin{aligned} \frac{\mathrm{d}^2 M}{\mathrm{d}x^2} = \frac{\mathrm{d}F}{\mathrm{d}x} = \frac{\mathrm{d}}{\mathrm{d}x}\left(m\frac{\mathrm{d}^2\delta}{\mathrm{d}t^2}\right) = \frac{\mathrm{d}}{\mathrm{d}x}\left(\rho A \mathrm{d}x\frac{\mathrm{d}^2\delta}{\mathrm{d}t^2}\right) \\ \Rightarrow EI\cdot\frac{\partial^4\delta}{\partial x^4} + \rho A\frac{\partial^2\delta}{\partial t^2} = 0 \end{aligned} \tag{3.31}$$

where $A = wh$ is the area of the cross section, and ρ is the density of the material. Using Laplace transformation, the time independent solution for harmonic oscillation can be found:

$$\delta(x) = a\cos(\alpha x) + b\sin(\alpha x) + c\cosh(\alpha x) + d\sinh(\alpha x) \tag{3.32}$$

with $\alpha^4 = \frac{\omega^2\rho A}{EI}$, $I = \frac{1}{12}wh^3$ and a, b, c, d being integration constants. From the boundary conditions of a cantilever with one end clamped ($\delta(x=0)=0$, $\frac{\mathrm{d}}{\mathrm{d}x}\delta(x=0)=0 \longrightarrow c=-a,\ d=-b$) and one end free (no bending moment and no forces: $\frac{\mathrm{d}^2}{\mathrm{d}x^2}\delta(x=L)=0$, $\frac{\mathrm{d}^3}{\mathrm{d}x^3}\delta(x=L)=0$), we obtain the condition for valid resonance frequencies as $\cos(\alpha L)\cosh(\alpha L) = -1$. The first three natural frequencies ν_i can then be determined from the solutions to this equation [18]:

$$\begin{aligned} \nu_1 &= 0.16154\sqrt{\left(\frac{E}{\rho}\right)}\frac{h}{L^2} \\ \nu_2 &= 1.0124\sqrt{\left(\frac{E}{\rho}\right)}\frac{h}{L^2} \\ \nu_3 &= 2.8346\sqrt{\left(\frac{E}{\rho}\right)}\frac{h}{L^2} \end{aligned} \tag{3.33}$$

The same approach can be used for a bridge clamped at both ends. The boundary conditions for this problem are $\delta(x=0)=0$, $\frac{\mathrm{d}}{\mathrm{d}x}\delta(x=0)=0$ $\delta(x=L)=0$, $\frac{\mathrm{d}}{\mathrm{d}x}\delta(x=L)=0$. This leads to the conditional equation for valid frequencies: $\cos(\alpha L)\cosh(\alpha L) = 1$. The first three natural frequencies are given by [19]:

$$\begin{aligned} \nu_1 &= 1.028\sqrt{\left(\tfrac{E}{\rho}\right)}\tfrac{h}{L^2} \\ \nu_2 &= 2.833\sqrt{\left(\tfrac{E}{\rho}\right)}\tfrac{h}{L^2} \\ \nu_3 &= 5.556\sqrt{\left(\tfrac{E}{\rho}\right)}\tfrac{h}{L^2} \end{aligned} \tag{3.34}$$

In microfluidics, cantilevers often operate in liquids with high viscosity. Therefore, damping occurs which decreases the resonance frequency. This effect can be mathematically described with an 'added' mass m_a [19]. Since the natural frequency of an elastic structure is inversely proportional to the square root of its mass, the effect of the added mass on the natural frequency can be expressed as [19]:

$$\nu_i(\text{fluid}) = \left(1 + \frac{m_a}{m_0}\right)^{-0.5} \nu_i(\text{vacuum}) \tag{3.35}$$

where m_0 equals the original mass of the beam. For a cantilever submerged in a fluid, the added mass for the first resonance frequency predicted from strip theory [19] is 0.5 times the volume enclosed by rotating the plate of length L and width w about the longer symmetric axis [19]:

$$m_a = \frac{\pi}{8} \cdot \rho_{\text{fluid}}\, L\, w^2 \tag{3.36}$$

The above two equations allow us to calculate the first resonance frequency of a cantilever in water and these are therefore useful in predicting and interpreting the frequency behaviour of pumps utilising flap valves.

3.1.3 Mechanics of Plates

The thickness of a plate is generally small compared to its lateral dimensions. As a consequence, the bending depends very much on the thickness rather than the lateral size. The theory presented here is for out-of-plane displacements up to half of the plate thickness. When the displacement gets larger, second order effects that are omitted here become dominant. These equations, derived for deflection and resonant frequency, are used when considering diaphragm pump displacement and dynamic operation.

3.1.3.1 Fundamental Differential Equation

The derivation of the equations for a plate is more complicated than for a simple beam and shall therefore not be presented in this book. The equations are given in detail in references [20, 21] and are only summarised here. With the help of energy methods, it is possible to derive a partial differential equation for the out-of-plane displacement δ with respect to pressure p application at one side of an isotropic plate [20]:

$$\nabla^4 \delta = \frac{p}{D} \tag{3.37}$$

with ∇^4 as the gradient operator $\frac{\mathrm{d}^4\delta}{\mathrm{d}x^4} + 2\frac{\mathrm{d}^4\delta}{\mathrm{d}x^2 y^2} + \frac{\mathrm{d}^4\delta}{\mathrm{d}y^4}$ and D as the flexural rigidity of the plate:

$$D = \frac{Eh^3}{12\left(1-\nu^2\right)} \tag{3.38}$$

where E is the Young's modulus, h is the thickness of the plate and ν is the Poisson's ratio.

For the derivation of equation 3.37 it is assumed that the plate is in a state of plane stress relative to the (x,y)-plane, i.e. $\sigma_{xz} = \sigma_{yz} = \sigma_{zz} = 0$, and the material is isotropic. A very important problem in plate theory is to find a solution to the differential equation 3.37 for a variety of boundary conditions.

3.1.3.2 Rectangular Plates

Let us consider a simply supported rectangular plate (figure 3.11) of length a and width b. The boundary conditions at $y = 0$ and $y = b$ are $\delta = 0$ and the second derivative in y-direction of the bending moment is zero, i.e. $M_{yy} = -D\left(\frac{\mathrm{d}^2\delta}{\mathrm{d}y^2} + v\frac{\mathrm{d}^2\delta}{\mathrm{d}x^2}\right) = 0$, and are fulfilled with the function [20]:

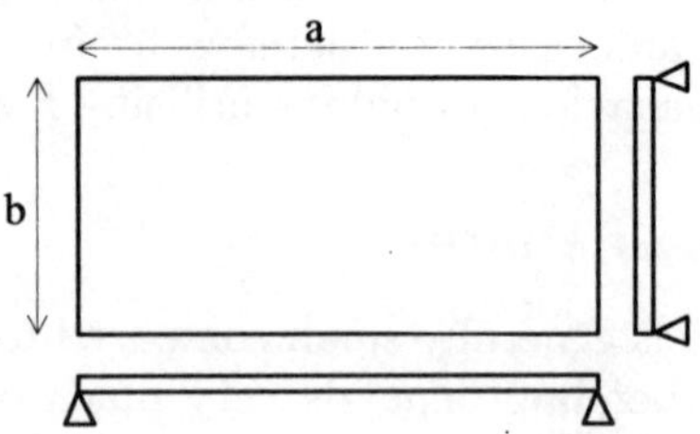

Figure 3.11: Simply supported rectangular plate

$$\delta(x,y) = \sum_{n=1}^{\infty} X_n(x) \sin\frac{n\pi y}{b} \tag{3.39}$$

with u, v, w as the displacements at any point on the plate in x, y, z directions. Also, equation 3.39 fulfils the differential equation 3.37 with

$$\begin{aligned} X_n(x) = \; & \frac{p_n}{D}\left(\frac{b}{n\pi}\right)^4 \left\{1 - \cosh\frac{n\pi x}{b} + \frac{n\pi x}{b}\sinh\frac{n\pi x}{b}\right. \\ & + \frac{1}{1+\cosh\frac{n\pi a}{b}}\left[\left(sinh\frac{n\pi a}{b} - \frac{n\pi a}{b}\right)\sinh\frac{n\pi x}{b}\right. \\ & \left.\left. - \frac{n\pi a}{b}\sinh\frac{n\pi a}{b}\cosh\frac{n\pi x}{b}\right]\right\} \end{aligned} \tag{3.40}$$

and $p(x,y) = f(x)\sum_{n=1}^{\infty} p_n \sin\frac{n\pi y}{b}$. Here, p_n is given by $\frac{2p_0}{b}\int_0^b g(y)\sin\frac{n\pi y}{b}\,dy$. This is the general solution for a plate with simply supported boundaries.

An exact solution for the problem is often not possible. From computations of a uniform pressure load p on rectangular plates [20], the maximum deflection at the centre of a rectangular plate is obtained as:

$$\delta_{max} = C(1-\nu^2)\left(\frac{pb^4}{Eh^3}\right) \tag{3.41}$$

where C is a constant. This constant is given in table 3.3 for various boundary conditions. The dimensions of the plate are length a, width b and thickness h.

boundary condition	C
four edges simply supported	$\frac{0.16}{1+2.4\alpha^3}$
span b fixed, span a simply supported	$\frac{0.032}{1+0.4\alpha^3}$
span a fixed, span b simply supported	$\frac{0.16}{1+\alpha^2+5\alpha^4}$
all edges fixed	$\frac{0.032}{1+\alpha^4}$

Table 3.3: Coefficient C for the calculation of the maximum deflection of a rectangular plate at uniform pressure load [20]; $\alpha = \frac{b}{a}$

A first approximation for the deflection of a rectangular plate with all sides fixed has been given by Heuberger [16]:

$$\delta(x,y) = \frac{\delta_{max}}{4}\left(1-\cos\frac{2\pi x}{a}\right)\left(1-\cos\frac{2\pi y}{b}\right) \tag{3.42}$$

with a maximum deflection at the centre of $\delta_{max} = \frac{p_0}{\pi^4 D}\left(\frac{3}{a^4}+\frac{2}{a^2b^2}+\frac{3}{b^4}\right)^{-1}$. The shape of the deflection is depicted in figure 3.12.

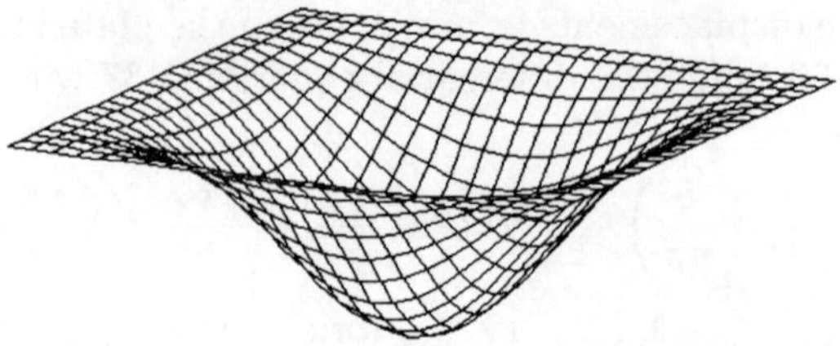

Figure 3.12: Deflection shape of a plate due to a uniform pressure load

3.1.3.3 Circular Plates

A solution of the differential equation 3.37 for the circular plate is simpler. Here, polar coordinates are used to simplify the calculation. The general solution for a circular plate under uniform pressure is given by [20]:

$$\delta = \frac{p_0 r^4}{64D} + A_1 + A_2 \ln r + B_1 r^2 + B_2 r^2 \ln r \tag{3.43}$$

with A_1, A_2, B_1 and B_2 as constants of integration, which are determined by the boundary conditions at $r = a$, i.e. the border of the circular plate. For a circular plate of radius a with simply supported edges, we get

$$\delta = \frac{p_0 a^4}{64D} \left[1 - \left(\frac{r}{a}\right)^2\right] \left[\frac{5+\nu}{1+\nu} - \left(\frac{r}{a}\right)^2\right] \tag{3.44}$$

The solution for a circular plate of radius a with fixed edges is given by

$$\delta = \frac{p_0 a^4}{64D} \left[1 - \left(\frac{r}{a}\right)^2\right]^2 \tag{3.45}$$

3.1.3.4 Natural Frequencies

The resonance frequency for a uniform rectangular plate with all sides clamped is given by Blevins [19]:

$$\nu_1 = \frac{\lambda_{11}}{2\,\pi\,L^2} \sqrt{\frac{E\,h^2}{12 \cdot \rho(1-\nu^2)}} \tag{3.46}$$

with $\lambda_{11} = 103$ as a numerical coefficient [19] for the first resonance frequency.

In order to calculate the shift in the natural frequencies for rectangular plates with one side exposed to a fluid, equation 3.35 may be employed. The added mass in this problem can be determined by [19]:

$$m_a = \alpha_{11}\,\beta\,\rho_{\text{fluid}}\,a\,b^2 \tag{3.47}$$

with $\alpha_{11} = 0.3452$ and $\beta = 0.72$ being geometrical coefficients for the first mode [19] and a, b as the length and width of the plate.

3.2 Actuation Techniques for Microfluidic Devices

Actuators are needed in a variety of microfluidic devices, for example inkjet printers, control valves and micropumps. There are several different actuation techniques which can be used. Their use in practice depends on both economic and technical constraints - e.g. biocompatibility. In order to give the reader the possibility to judge these constraints and to know about advantages and disadvantages of each actuator type, a summary is given in this section.

3.2.1 The Piezoelectric Effect

The piezoelectric effect is one of the actuation methods used in micromachining. Here, we develop the theory of the piezoelectric effect and discuss its application to microactuators.

3.2.1.1 Theory of Piezoelectricity

When a dielectric material is put between two capacitor plates, the centre of positive and negative charges within an atom become separated from their common point, leading to a tiny dipole. For a linear material, the generated dipole moment per unit volume, i.e. the polarisation P, and the electric field E can be linked by the dielectric susceptibility χ:

$$P = \chi \cdot E = (\varepsilon_r - 1)\varepsilon_0 \cdot E = (\varepsilon_r - 1)\varepsilon_0 \cdot \frac{E_0}{\varepsilon_r} \tag{3.48}$$

with ε_r as the relative dielectric constant of the material and $\varepsilon_0 = 8.85 \cdot 10^{-12}\mathrm{F\,m^{-1}}$ as the dielectric constant of vacuum. The electric displacement D within a dielectric material is given by

$$D = \varepsilon_0 \cdot E + P = \varepsilon_r \varepsilon_0 \cdot E = \varepsilon \cdot E \tag{3.49}$$

In vector notation, for a linear cubic crystal like silicon the electric displacement is written as

$$\begin{pmatrix} D_x \\ D_y \\ D_z \end{pmatrix} = \begin{bmatrix} \varepsilon_{11} & 0 & 0 \\ 0 & \varepsilon_{11} & 0 \\ 0 & 0 & \varepsilon_{11} \end{bmatrix} \cdot \begin{pmatrix} E_x \\ E_y \\ E_z \end{pmatrix} \tag{3.50}$$

For ferroelectric materials, however, the relationship between P and E is more complicated. Here, the polarisation is a non-linear function of the electric field as given in figure 3.13, implying that the material is capable of sustaining a permanent dipole moment even when E is zero. A ferroelectric material can therefore be defined as a polar crystal with permanent dipole moment whose direction can be altered by an applied electric field.

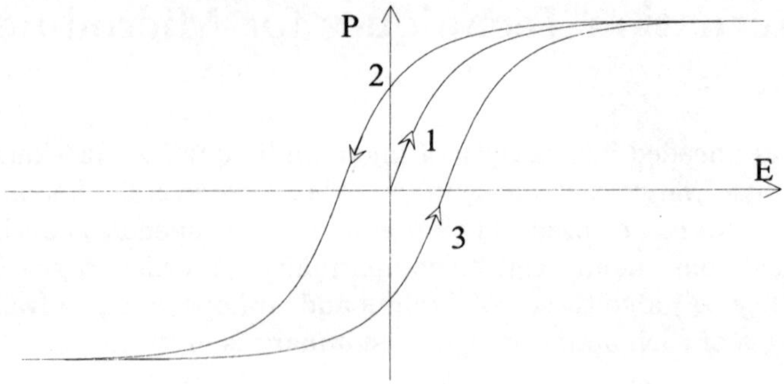

Figure 3.13: Polarisation and electric field for ferroelectric materials

If the ferroelectric material has not been exposed to a strong electric field, it is called unpoled. Thus, when a weak field is applied, the susceptibility χ is the gradient of the curve marked "1" in the graph. If the field gets stronger (preferably at elevated temperatures), a strong hysteresis is obtained in the curve. This leads to a remanent polarisation P when the electric field gets to zero (near point "2"). Now, the material is called poled. If the temperature gets too hot (beyond the so-called Curie temperature, which is a material characteristic), the remanent polarisation disappears.

A material is called piezoelectric when the application of an electric field causes a mechanical deformation of the material. This is the case in all materials which show a spontaneous electric polarisation without the application of an electric field, and so all ferroelectrics are also piezoelectric. Other crystalline materials which are not ferroelectric may also be piezoelectric, for example quartz, ZnO and CdS. Thus, a necessary condition for piezoelectricity is that the crystalline structure lacks a centre of symmetry. If there is an asymmetry of positive and negative charged ions within the crystal structure, there will be distortion of the crystal structure when an electric field is applied and the whole crystal will be changed in length, as shown in figure 3.14.

In microfabrication, the deposition of a piezoelectric material is usually polycrystalline, and therefore all crystallites would be expected to have a different direction of their polarisation resulting in zero overall polarisation. However, in the case of the hexagonal crystals ZnO and CdS, the crystallites exhibit a preferred orientation even when deposited on an amorphous substrate, such that the c axis tends to be normal to the substrate. Since the c axis is the polar axis in these materials, such deposited layers therefore show piezoelectric properties [22].

If the material is ferroelectric, then even if there is no preferred crystallite orientation, the deposited layer can be made piezoelectric by poling in the manner described earlier.

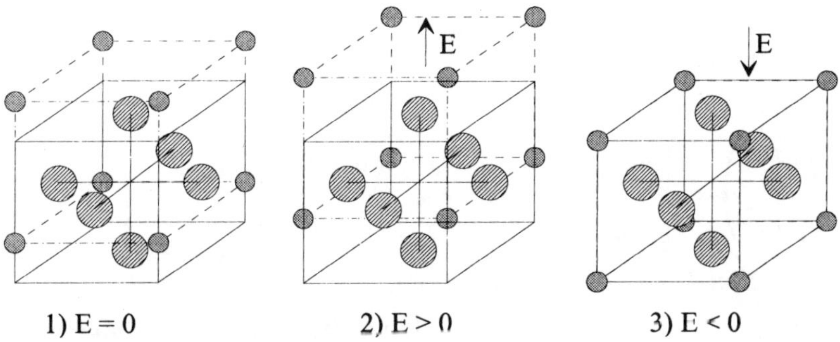

Figure 3.14: Piezoelectric behaviour of an elementary cell at rest (1) and when an external electric field is applied (2 and 3)

To summarise: For piezoelectric interaction to exist it is necessary that one or more axes of the material intrinsically possess a polarity. The polarity needed to impart piezoelectric properties can be given to an originally isotropic polycrystalline ceramic more or less permanently by temporary application of a strong electric field (poling). Poling is the application of a high voltage sufficient to reverse and align electric moments of spontaneously polarised regions in the ceramic.

For a hexagonal crystal like PZT (lead zirconate titanate) which shows ferroelectric and piezoelectric behaviour, the electric displacement in the linear region may be written as:

$$\begin{pmatrix} D_x \\ D_y \\ D_z \end{pmatrix} = \begin{bmatrix} \varepsilon_{11} & 0 & 0 \\ 0 & \varepsilon_{11} & 0 \\ 0 & 0 & \varepsilon_{33} \end{bmatrix} \cdot \begin{pmatrix} E_x \\ E_y \\ E_z \end{pmatrix} \tag{3.51}$$

The values of the dielectric constant ε for PZT and ZnO, two common piezoelectric materials, are given in table 3.4.

material	ε_{11}	ε_{33}	unit
PZT-5A	1730	1700	ε_0
PZT-5H	3130	3400	ε_0
ZnO	8.3	8.8	ε_0

Table 3.4: Dielectric constant for two different compositions of PZT and ZnO [17]

The direct piezoelectric effect is obtained when a strain ϵ causes a polarisation P in a piezoelectric material:

$$\begin{pmatrix} P_x \\ P_y \\ P_z \end{pmatrix} = \begin{bmatrix} e_{11} & e_{12} & e_{13} & e_{14} & e_{15} & e_{16} \\ e_{21} & e_{22} & e_{23} & e_{24} & e_{25} & e_{26} \\ e_{31} & e_{32} & e_{33} & e_{34} & e_{35} & e_{36} \end{bmatrix} \cdot \begin{pmatrix} \epsilon_x \\ \epsilon_y \\ \epsilon_z \\ \epsilon_{yz} \\ \epsilon_{zx} \\ \epsilon_{xy} \end{pmatrix} \tag{3.52}$$

The matrix [e] is the piezoelectric modulus. It has the same units as polarisation which is electric charge per area [$\mathrm{C\,m^{-2}}$]. Often, the stress σ is given instead of the strain ϵ. The polarisation is then given with the help of the matrix of the piezoelectric coefficients [d] which has the unit $\mathrm{C\,N^{-1}}$:

$$\begin{pmatrix} P_x \\ P_y \\ P_z \end{pmatrix} = \begin{bmatrix} d_{11} & d_{12} & d_{13} & d_{14} & d_{15} & d_{16} \\ d_{21} & d_{22} & d_{23} & d_{24} & d_{25} & d_{26} \\ d_{31} & d_{32} & d_{33} & d_{34} & d_{35} & d_{36} \end{bmatrix} \cdot \begin{pmatrix} \sigma_x \\ \sigma_y \\ \sigma_z \\ \sigma_{yz} \\ \sigma_{zx} \\ \sigma_{xy} \end{pmatrix} \tag{3.53}$$

The matrices [e] for the piezoelectric modulus and [d] for the piezoelectric coefficient can be converted into each other with the matrix [s] for the compliance and [c] for the stiffness:

$$e_{rs} = \sum_{t=1}^{6} d_{rt} \cdot c_{ts} \qquad d_{rs} = \sum_{t=1}^{6} e_{rt} \cdot s_{ts} \qquad r = 1-3,\ s = 1-6 \tag{3.54}$$

The converse piezoelectric effect exists as well. Here, an electric field E generates mechanical stress σ or strain ϵ and thus performs mechanical work. The matrix relationships use the same constants e_{ij} and d_{ij} as in equations 3.52 and 3.53 and are given as follows:

$$\begin{pmatrix} \sigma_x \\ \sigma_y \\ \sigma_z \\ \sigma_{yz} \\ \sigma_{zx} \\ \sigma_{xy} \end{pmatrix} = \begin{bmatrix} e_{11} & e_{21} & e_{31} \\ e_{12} & e_{22} & e_{23} \\ e_{13} & e_{23} & e_{33} \\ e_{14} & e_{24} & e_{34} \\ e_{15} & e_{25} & e_{35} \\ e_{16} & e_{26} & e_{36} \end{bmatrix} \cdot \begin{pmatrix} E_x \\ E_y \\ E_z \end{pmatrix} \tag{3.55}$$

$$\begin{pmatrix} \epsilon_x \\ \epsilon_y \\ \epsilon_z \\ \epsilon_{yz} \\ \epsilon_{zx} \\ \epsilon_{xy} \end{pmatrix} = \begin{bmatrix} d_{11} & d_{21} & d_{31} \\ d_{12} & d_{22} & d_{23} \\ d_{13} & d_{23} & d_{33} \\ d_{14} & d_{24} & d_{34} \\ d_{15} & d_{25} & d_{35} \\ d_{16} & d_{26} & d_{36} \end{bmatrix} \cdot \begin{pmatrix} E_x \\ E_y \\ E_z \end{pmatrix} \tag{3.56}$$

For hexagonal crystals like PZT, the piezoelectric moduli [e] and the piezoelectric coefficients [d] are simplified due to symmetry:

$$\begin{pmatrix} \sigma_x \\ \sigma_y \\ \sigma_z \\ \sigma_{yz} \\ \sigma_{zx} \\ \sigma_{xy} \end{pmatrix} = \begin{bmatrix} 0 & 0 & e_{31} \\ 0 & 0 & e_{31} \\ 0 & 0 & e_{33} \\ 0 & e_{15} & 0 \\ e_{15} & 0 & 0 \\ 0 & 0 & 0 \end{bmatrix} \cdot \begin{pmatrix} E_x \\ E_y \\ E_z \end{pmatrix} \tag{3.57}$$

$$\begin{pmatrix} \epsilon_x \\ \epsilon_y \\ \epsilon_z \\ \epsilon_{yz} \\ \epsilon_{zx} \\ \epsilon_{xy} \end{pmatrix} = \begin{bmatrix} 0 & 0 & d_{31} \\ 0 & 0 & d_{31} \\ 0 & 0 & d_{33} \\ 0 & d_{15} & 0 \\ d_{15} & 0 & 0 \\ 0 & 0 & 0 \end{bmatrix} \cdot \begin{pmatrix} E_x \\ E_y \\ E_z \end{pmatrix} \tag{3.58}$$

For example, PZT 5H has the following values for the piezoelectric moduli [17]:
e_{31} = -6.5 C m^{-2} e_{33} = 23.3 C m^{-2} e_{15} = 17 C m^{-2}

3.2.1.2 Applications in Micromechanics

The piezoelectric actuator illustrated in figure 3.15 is a dielectric device usually driven by voltage and so no ohmic current should flow. In practice, high voltages (100-1000 V) are required at almost zero current. This means that the actuation depends not only on the material properties, but also on the stability of the electric domain.

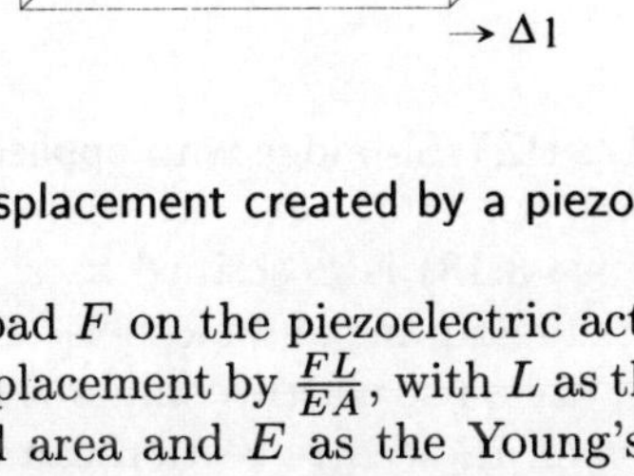

Figure 3.15: Displacement created by a piezoelectric actuator

The influence of a load F on the piezoelectric actuation can be estimated by a reduction of the displacement by $\frac{FL}{EA}$, with L as the length of the actuator, A as the cross sectional area and E as the Young's modulus. Therefore, a minimal length and maximal area will give best performance.

Besides the simple actuator, a membrane or bimorph actuator is often used in microfluidics (figure 3.16). In this actuator type, the piezoelectric material is mounted on top of a bridge or membrane, and a voltage is applied perpendicular to the membrane generating the electric field E. This leads to a lateral expansion of the piezoelectric crystal and hence, due to the connection to the membrane, to bending. Considerable mechanical amplification of the piezoelectric effect is obtained in this way.

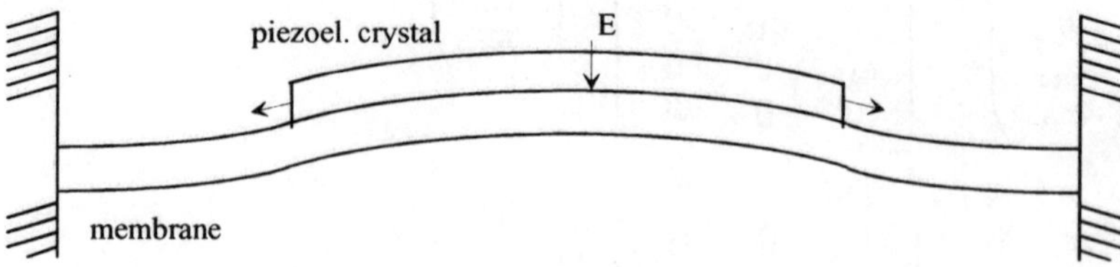

Figure 3.16: PZT-Si-bridge with applied voltage

We now calculate the static behaviour of a piezoelectric crystal covering part of a membrane with fixed ends (see figure 3.17). In order to simplify the problem, we assume that the extent of the membrane is infinite in the plane normal to the diagram, so that the problem reduces to a two-dimensional one with the behaviour determined by the lengths L and s. Another assumption is necessary for the analytical modelling: Owing to a non-uniform beam geometry the full bridge has to be divided into three sections showing symmetry about the midpoint of the centre section.

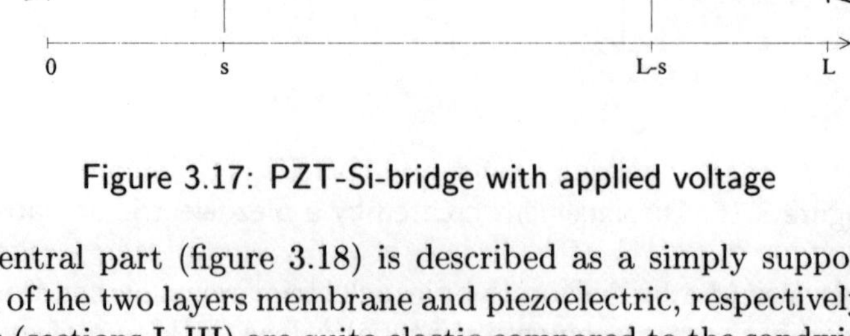

Figure 3.17: PZT-Si-bridge with applied voltage

The central part (figure 3.18) is described as a simply supported beam consisting of the two layers membrane and piezoelectric, respectively. Because the beams (sections I, III) are quite elastic compared to the sandwich (section II), the longitudinal stress on the centre section can be neglected.

In the following paragraphs we derive equations to predict the behaviour of the bridge with respect to piezoelectric actuation and back pressure.

The first actuating force is the stress in the piezoelectric layer caused by the

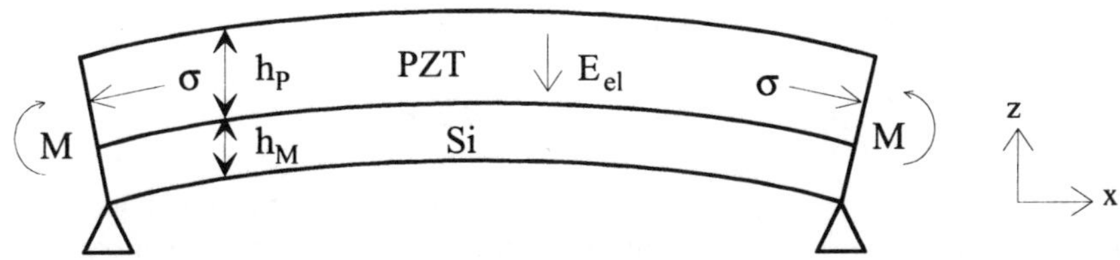

Figure 3.18: Central section of the PZT-Si-bridge

piezoelectric charge constant for the transverse effect d_{31}. This generates a stress in the plane perpendicular to the applied electrical field. A moment M, applied at both ends, is needed to give a smooth transition from the other sections. The same, but opposite moment M is applied at the end of the cantilevers of section one and three as shown in figure 3.17. For a continuous differentiable deflection, slope and displacement have to be the same at both boundaries.

Using similar calculations to the bimetal problem [23], the inverse bending radius, i.e. the second derivative of the deflection, per unit width in section II for piezoelectric actuation is given by:

$$\kappa = \frac{\partial^2 \delta}{\partial x^2} = \frac{-d_{31} \cdot E_{el}}{\frac{h}{2} + \frac{2}{h} \cdot \left(\frac{1}{E_P h_P} + \frac{1}{E_M h_M}\right)(E_M I_M + E_P I_P)} \tag{3.59}$$

with E_P as the Young's modulus of the piezoelectric material and E_M as the Young's modulus of the membrane, respectively. The second moment of area for unit width is calculated from $I_i = \frac{h_i^3}{12}$ with h_i as the height of the individual layer, and $h = h_M + h_P$. The electrical field in z-direction is given by E_{el}.

Using the principle of superposition, the moment M can be added to the bending equation 3.59 to obtain the complete second derivative of the deflection for this section:

$$\delta''_{II} = \kappa + \frac{M}{EI} \qquad \text{for} \qquad s < x < L - s \tag{3.60}$$

where EI, the Young's modulus times the second moment of area for the compound beam, is calculated from the theorem of parallel axes (equation 3.21):

$$EI = \frac{1}{12} \frac{h_P^4 E_P^2 + E_P E_M \left(4h_P^3 h_M + 6h_P^2 h_M^2 + 4h_P h_M^3\right) + h_M^4 E_M^2}{h_P E_P + h_M E_M} \tag{3.61}$$

The application of the moment M to the cantilevers of sections I and III leads to:

$$\delta_I'' = \frac{M}{E_M I_M} \quad \text{for} \quad 0 < x < s \tag{3.62}$$

$$\delta_{III}'' = \frac{M}{E_M I_M} \quad \text{for} \quad L - s < x < L \tag{3.63}$$

Applying the boundary conditions (fixed beam at both ends as well as identical deflection and slope at the intersections), the unknown moment M can be determined as:

$$M = \frac{1 - \frac{L}{2s}}{\frac{1}{E_M I_M} + \frac{1}{EI}\left(\frac{L}{2s} - 1\right)} \cdot \kappa \tag{3.64}$$

When M is inserted into equations 3.60, 3.62 and 3.63 and the equations are integrated, the following deflection function is obtained:

$$\delta(x) = \begin{cases} \frac{1}{2}\frac{M}{E_M I_M}x^2 & \text{for } 0 < x < s \\ \left(\kappa + \frac{M}{EI}\right)\left(\frac{1}{2}x^2 - \frac{L}{2}x + \frac{1}{2}Ls - \frac{1}{2}s^2\right) & \\ +\frac{1}{2}\frac{M}{E_M I_M}s^2 & \text{for } s < x < L - s \\ \frac{M}{E_M I_M}\left(\frac{1}{2}x^2 - Lx + \frac{1}{2}L^2\right) & \text{for } L - s < x < L \end{cases} \tag{3.65}$$

The maximum deflection at the centre ($x = \frac{L}{2}$) is thus:

$$\delta_{\max} = \left(\kappa + \frac{M}{EI}\right)\left(-\frac{1}{8}L^2 + \frac{s}{2}L - \frac{1}{2}s^2\right) + \frac{1}{2}\frac{M}{E_M I_M}s^2 \tag{3.66}$$

Another factor which can cause deflection of the membrane is its reaction to an applied pressure (see figure 3.19). Again, the membrane can be simplified as a bridge with fixed ends, where the length of the bridge equals the smaller dimension of the membrane.

For a constant pressure p over the full length L of a bridge, the bending moment M_p along the bridge is given by:

$$M_p(x) = M_1 + \frac{pL}{2}x - \frac{1}{2}px^2 \tag{3.67}$$

with M_1 signifying the moment at the fixed end, to fulfil the boundary condition of no slope at the edge.

The moment-area method [24] states that a complete integration of the inverse bending radius over the full length of a bridge is zero for clamped ends. The general moment area method [24] relates the slope of the deflection

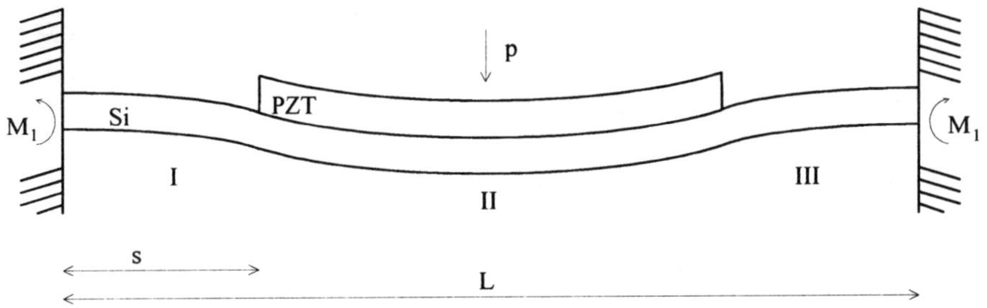

Figure 3.19: PZT-Si-bridge with applied pressure

δ of a beam at the locations $x = a$ and $x = b$ as follows:

$$\begin{aligned} \frac{\partial \delta}{\partial x}\Big|_{x=b} - \frac{\partial \delta}{\partial x}\Big|_{x=a} &= -\int_a^b \frac{M(x)}{E(x)I(x)}\,dx \\ \left(b * \frac{\partial \delta}{\partial x}\Big|_{x=b} - \delta(b)\right) - \left(a * \frac{\partial \delta}{\partial x}\Big|_{x=a} - \delta(a)\right) &= -\int_a^b \frac{M(x)}{E(x)I(x)}\,x\,dx \end{aligned} \tag{3.68}$$

Using both theorems for calculating the deflection of the bridge then gives:

$$\delta(x) = \begin{cases} \displaystyle\int_0^x \frac{M_p(x)\cdot x}{E_M I_M}\,dx - x\cdot\int_0^x \frac{M_p(x)}{E_M I_M}\,dx & \text{for } 0 < x < s \\ \displaystyle\int_0^s \frac{M_p(x)\cdot x}{E_M I_M}\,dx + \int_s^x \frac{M_p(x)\cdot x}{EI}\,dx & \\ \displaystyle -x\cdot\left(\int_0^s \frac{M_p(x)}{E_M I_M}\,dx + \int_s^x \frac{M_p(x)}{EI}\,dx\right) & \text{for } s < x < L-s \\ \displaystyle\int_0^s \frac{M_p(x)\cdot x}{E_M I_M}\,dx + \int_s^{L-s} \frac{M_p(x)\cdot x}{EI}\,dx & \\ \displaystyle +\int_{L-s}^x \frac{M_p(x)\cdot x}{E_M I_M}\,dx - x\cdot\left(\int_0^s \frac{M_p(x)}{E_M I_M}\,dx\right. & \\ \displaystyle \left. +\int_s^{L-s} \frac{M_p(x)}{EI}\,dx + \int_{L-s}^x \frac{M_p(x)}{E_M I_M}\,dx\right) & \text{for } L-s < x < L \end{cases} \tag{3.69}$$

From the solution at $x = \frac{L}{2}$, the maximum deflection can be determined as:

$$\begin{aligned}\delta_{\max} = & \frac{p}{384\left(-2sEI - LE_M I_M + 2sE_M I_M\right)} \\ & \left(\frac{EI}{E_M I_M}\left(32s^5 - 32s^4 L\right) + 40s^2 L^3 + 112s^4 L - 64s^5 \right. \\ & -80s^3 L^2 - 10L^4 s + \frac{E_M I_M}{EI}\left(10L^4 s - L^5 + 80L^2 s^3\right. \\ & \left.\left. -40L^3 s^2 - 80s^4 L + 32s^5\right)\right)\end{aligned} \tag{3.70}$$

Assuming $E_M = 168{\cdot}10^9$ Pa for Si in <110> direction [16], $E_P = C_{11} = 126{\cdot}10^9$ Pa and $d_{31} = -274 \cdot 10^{-12}$ C N^{-1} for PZT-5H material [17], the behaviour of the maximum static deflection versus PZT distance from the membrane edge can be determined from equation 3.66 (figure 3.20).

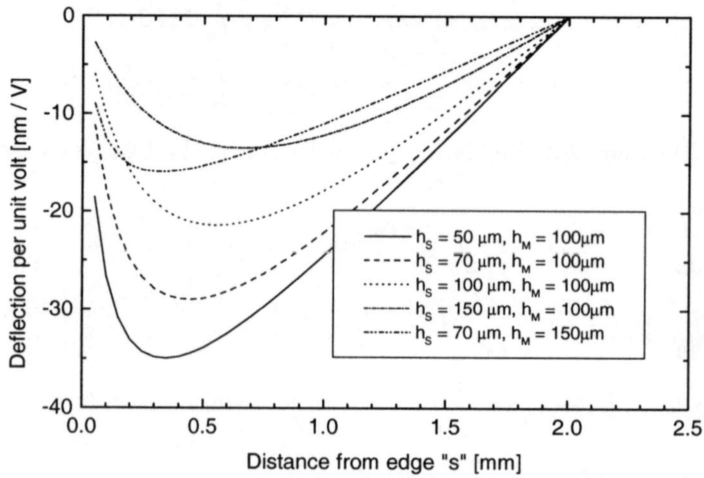

Figure 3.20: Achievable deflection per unit volt actuation with respect to the distance of the PZT layer edge from the silicon membrane edge (cf. figure 3.17). Silicon membrane: 8 × 4 mm; actuation voltage = 100 V

The maximum deflection and therefore the maximum volume stroke is achieved with a 50 μm-thick silicon membrane and a 100 μm-thick piezoelectric layer at approximately 300 μm distance from the fixed edge. Thicker silicon membranes reduce the maximum deflection with a shift of the distance s to slightly larger values. An increase of the piezoelectric film thickness, however, decreases the maximum deflection value. For a silicon thickness of 70 μm and a PZT thickness of 250 μm, the static peak-to-peak deflections for 3 to 3.5 mm wide PZT layers are around 1.2 μm at 100 V actuation voltage.

As a validation of the theory, some examples for micromachined membrane actuators of 8 × 4 mm sidelength are given. The tests were done with surface mounted piezoelectric devices. Bulk PZT plates of 250 μm thickness and different lateral dimensions were glued onto a 70 μm-thick silicon membrane of 8 × 4 mm. The peak-to-peak displacement values for different sizes and actuation frequencies can be seen in figure 3.21. The actuation voltage was kept constant at 100 V_{pp}.

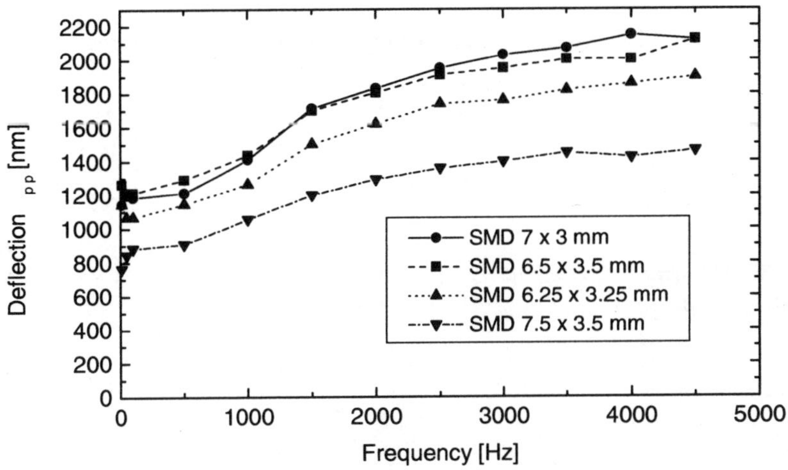

Figure 3.21: Peak to peak deflection of a piezoelectric actuator. SMD = surface mounted PZT actuator of 250 μm thickness on a 8 × 4 × 0.07 mm sized silicon membrane. Actuation voltage = 100 V

The peak-to-peak deflections in figure 3.21 at low frequencies for the devices 7 × 3 mm, 6.5 × 3.5 mm and 6.25 × 3.25 mm are comparable with the theoretical estimations of figure 3.20. Only the device with 7.5 × 3.5 mm disagrees with theory, but this can be explained due to misalignment of the surface mounted actuator to the membrane.

3.2.2 Bimetallic Actuation Technique

Bimetallic actuation is obtained if a beam, bridge or plate consists of two materials of different thermal expansion. When the temperature changes the different thermal expansion of the two materials creates stress in the compound structure and leads to bending of the structure (figure 3.22).

The thermal expansion coefficients as well as the Young's modulus of several metals and silicon are given in table 3.5.

Some basic calculations of bimetallic actuation [23] lead to an equation for the second derivative of the deflection η_x of a simple supported bimorph beam which can be used as a first approximation for the optimisation of materials

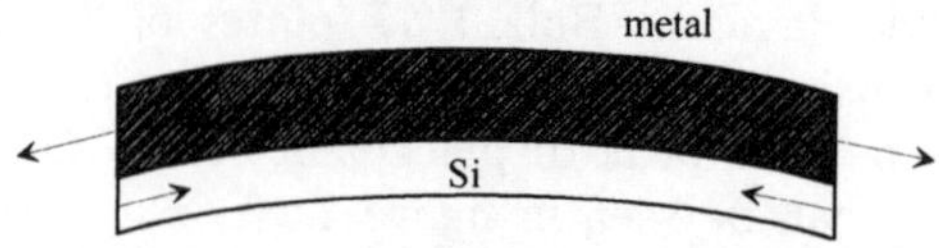

Figure 3.22: Bimetallic actuation at elevated temperatures

Material	Si	SiO_2	Si_3N_4	Al	Au	Ni	Cu
TCE $[10^{-6}\ K^{-1}]$	2.33	0.55	0.8	23.8	14.3	12.8	16.8
E [GPa]	$130_{<100>}$	73	385	71	78	210	123

Table 3.5: Thermal coefficient of expansion TCE and Young's modulus at T=300K [17, 25]

and thickness ratio:

$$\frac{\mathrm{d}^2\eta_x}{\mathrm{d}x^2} = \frac{-6\Delta T(\alpha_M - \alpha_S)(h_M + h_S)\,h_M\,h_S}{4\,h_S\,h_M^3 + \frac{E_S}{E_M}h_S^4 + \frac{E_M}{E_S}h_M^4 + 6\,h_M^2\,h_S^2 + 4\,h_M\,h_S^3} \tag{3.71}$$

where ΔT is the change in temperature, E the Young's modulus, h the layer thickness and α the thermal expansion coefficient for silicon S and metal M, respectively. The second derivative of the deflection is also the inverse of the radius of curvature, i.e. $\frac{1}{r}$. Assuming the ratio of the Young's moduli $E_M/E_S = r_E$ and a variable thickness ratio of $h_M\ /\ h_S = r_t$, equation 3.71 can be simplified to:

$$\frac{\mathrm{d}^2\eta_x}{\mathrm{d}x^2} = \frac{1}{h_S}\frac{-6\Delta T\,r_t\,r_E(\alpha_M - \alpha_S)(r_T + 1)}{4r_E\,r_t^3 + 1 + 6r_t^2r_E + r_E^2r_t^4 + 4r_E\,r_t} \tag{3.72}$$

From this equation the maximum bending can be derived for different metals and thickness ratios. A comparison for a silicon thickness of 30 μm and a temperature difference ΔT of 20°C is given in figure 3.23. The figure shows that aluminium is the best metal for actuation of a silicon bridge and the optimum thickness ratio is approximately $r_t = 0.652$, implying that, for maximum performance of the actuator, the thickness of the metal has to be 0.652 times 30 μm. Other materials like copper, nickel and gold show less bending but allow thinner layers for their optimum deflection. Note that the thickness ratio is not at all critical and can vary over a range of about two-to-one about optimum without reducing the deflection by more than 10% .

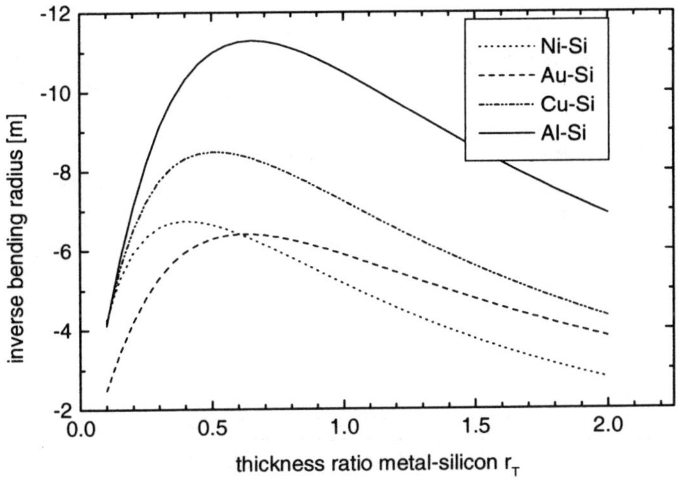

Figure 3.23: Inverse radius of curvature for different metals on a silicon bridge of 30 μm thickness at 20 °C

3.2.3 Shape-memory Effect in Micromechanical Applications

The shape-memory device is similar to the bimetallic actuator. In shape-memory alloys (usually Ti-Ni) the material is bent elastically at a high temperature (figure 3.24) and then allowed to cool, when it will return to its original shape. The device returns to the deformed shape when it is heated to the temperature where it was bent originally. Such actuators are capable of large displacements.

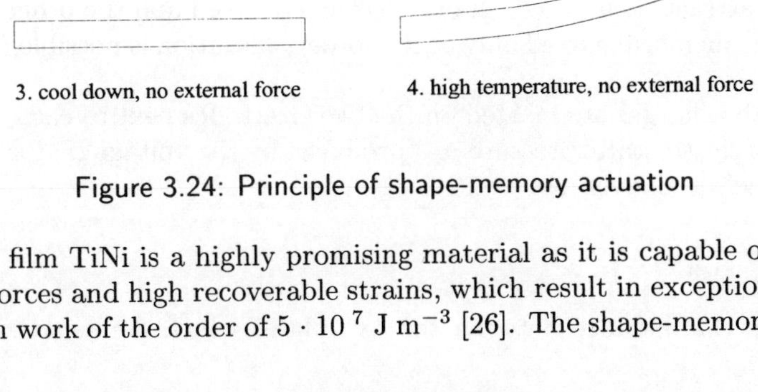

Figure 3.24: Principle of shape-memory actuation

Thin film TiNi is a highly promising material as it is capable of high actuation forces and high recoverable strains, which result in exceptionally high actuation work of the order of $5 \cdot 10^{7}$ J m^{-3} [26]. The shape-memory effect in

TiNi is the result of a reversible solid state phase transition at around 65°C. At low temperatures the phase of TiNi is martensite and at high temperature austenite. The shape stored in the austenite phase is always attained when the material is heated from the martensite shape.

In microfabrication, the materials are deposited at high temperatures, and therefore the shape at deposition is the parent shape. If the material is deposited as a straight beam, and stressed by a spring at room temperature, the material tries to return to the straight shape when it is reheated (see figure 3.25).

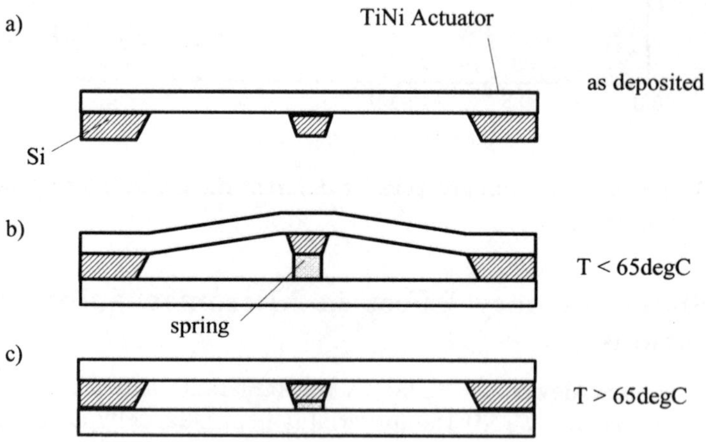

Figure 3.25: Shape-memory actuation: a) material as deposited in austenite phase, b) material at room temperature pre-stressed, c) deformation work in the spring done by changing to austenite phase through heating [27]

3.2.4 Electrostatic Actuation Technique

Electrostatic actuation uses the force developed between the plates of a capacitor when a voltage is applied. Due to electric charges, the two plates of a capacitor attract each other . If one electrode is fixed and the other electrode is a flexible membrane fixed only at its border, actuation is possible (see figure 3.26).

If all the charges are located on the two electrodes next to each other, the equivalent electrostatic pressure p_{el} (produced by the voltage U) for a parallel plate capacitor is given by:

$$p_{el} = \frac{\varepsilon_0}{2s^2} U^2 \tag{3.73}$$

where s is the distance between the two plates which are separated by air

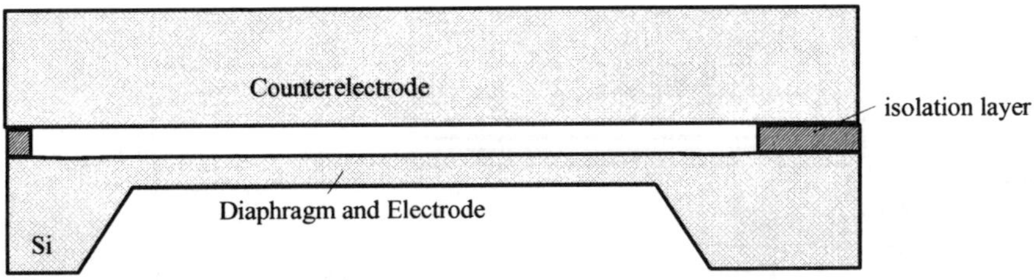

Figure 3.26: Electrostatic actuation of a membrane [28]

or vacuum. With the help of equation 3.42, we can calculate the displaced volume V for a rectangular diaphragm with dimensions $a \times b$ as:

$$V = \int_a \int_b w(x,y) \mathrm{d}x \mathrm{d}y = \frac{w_{max}}{2} a \cdot b \tag{3.74}$$

where the maximum deflection is:

$$w_{max} = \frac{12\, p_{el} \left(1 - \nu^2\right)}{\pi^4\, E\, h^3} \left(\frac{3}{a^4} + \frac{2}{a^2 b^2} + \frac{3}{b^4} \right)^{-1} \tag{3.75}$$

In order to get an idea about the power of the device, let us assume a membrane of 8×4 mm size and 25 μm thickness and 25μm gap to the fixed electrode. For silicon, we use the elastic modulus in the (100) direction, 130 GPa, and an average Poisson ratio of 0.2. An actuation voltage of 1000 V gives an electrical pressure of around 7000 Pa leading to a volume deflection of approx. 500 nl.

An electrical short can occur if the two electrodes touch each other. To prevent this, it is useful to cover one electrode with silicon dioxide. The electric pressure is then calculated by [28]:

$$p_{el} = \frac{\varepsilon_0}{2} \left(\frac{\varepsilon_{ox}}{\varepsilon_{air} s_{ox} + \varepsilon_{ox} s_{air}} \right)^2 \varepsilon_{air} U^2 \tag{3.76}$$

with the dielectric constants ε_i and the gap heights s_i for air and silicon dioxide (ox).

3.2.5 Thermopneumatic Actuators

A membrane can also be bent by using thermopneumatic actuation. A cavity is heated and the air inside the cavity tries to expand. This creates a pressure that bends the diaphragm (figure 3.27).

As the volume stays almost constant, we can derive a formula for the thermopneumatic actuation from the equation of state for an ideal gas ($\frac{p \cdot V}{T} =$ const.) and equation 3.42:

$$p_1 = (\frac{T_2}{T_1} - 1) \cdot p_0 \tag{3.77}$$

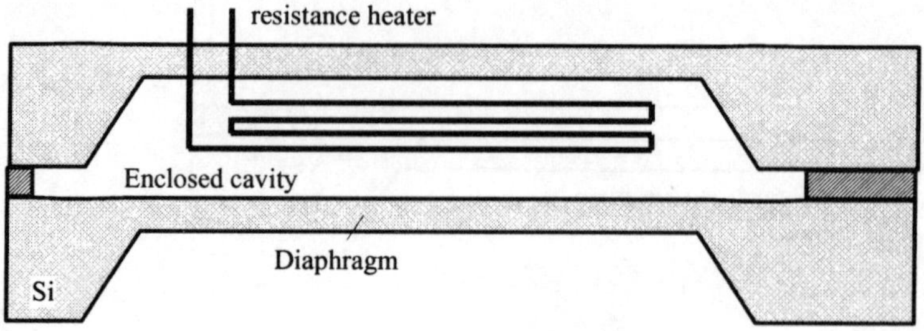

Figure 3.27: Thermopneumatic actuation of a membrane [29]

$$V = \frac{1}{2} a \cdot b \frac{12\left(\frac{T_2}{T_1} - 1\right) \cdot p_0 \left(1 - \nu^2\right)}{\pi^4 \, E \, h^3} \left(\frac{3}{a^4} + \frac{2}{a^2 b^2} + \frac{3}{b^4}\right)^{-1} \tag{3.78}$$

where V is the displaced volume, p_0 is the atmospheric pressure and p_1 is the pressure inside the cavity after changing the temperature from room temperature T_1 to the operating temperature T_2. The membrane dimensions are $a \times b$ in lateral direction and h is the thickness.

The temperature can be approximated by assuming a volume of gas that is at the same temperature as the heater. The power generated by a resistive heater is then $j_{T.el} = \frac{U^2}{R}$ where U is the applied voltage and R is the resistance of the heater. The heat is conducted through the silicon of area A_{Si} and length l_{Si}. The borders of the silicon chips are connected to an infinitely large volume at room temperature (T1).

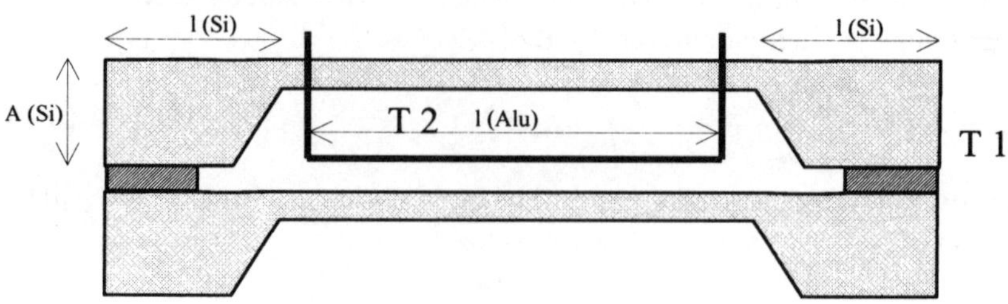

Figure 3.28: Thermopneumatic actuation of a membrane: estimation of the heat generation

This simple approximation gives the following equation for calculating the temperature T_2 inside the chamber:

$$T_2 = T_1 + \frac{U^2}{R} \frac{l_{Si}}{\lambda A_{Si}} \tag{3.79}$$

We take as an example a resistive heater of aluminium with a length of 2 mm, a thickness of 0.5 μm and a width of 40 μm. The electrical resistance of aluminium is 0.027 10^{-6} Ω m. The applied voltage is 25 V and the silicon to the left and right of the chamber (figure 3.28) is 2 mm long, 4 mm wide and 0.5 mm deep. The silicon to the front and back is also 2mm long and 0.5 mm thick, but 8 mm wide. Therefore, A_{Si} is 0.5·2·(8+4) mm. The thermal conductivity of silicon is 170 W m^{-1} K^{-1}. The thermopneumatic pressure obtained is thus 7700 Pa, which is comparable with the electrostatic actuation and leads to a volume displacement of approx. 500 nl of a 25 μm-thick membrane.

3.2.6 Magnetic Microactuators

The last actuation technique to be discussed in this book is the principle of magnetic actuation. For this, a permanent magnet is moved within a varying magnetic field [30]. An example of a magnetic actuator is given in figure 3.29. The permanent magnet can be surface mounted and the material for the coils can be sputter deposited.

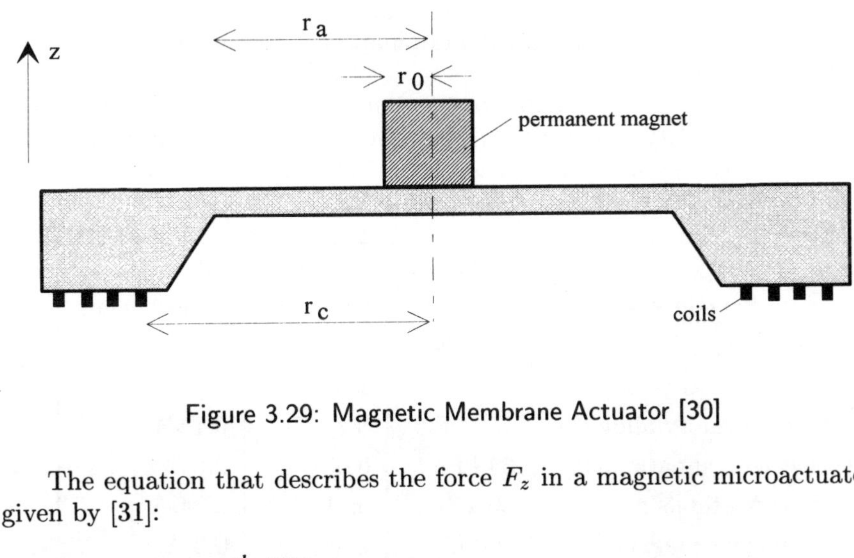

Figure 3.29: Magnetic Membrane Actuator [30]

The equation that describes the force F_z in a magnetic microactuator is given by [31]:

$$F_z = B_r \cdot A \int_0^h \frac{\mathrm{d}Hz}{\mathrm{d}z} \mathrm{d}z \tag{3.80}$$

where B_r is the remanence of the magnetic material (i.e. the magnetic field B after magnetising the material) and A and h are the area and height of the permanent magnet. H is the variable magnetic field which can be determined for a single turn of a coil by Biot Savart's law:

$$H(z) = \frac{I\, r_c^2}{2\,(r_c^2 + z^2)^{\frac{3}{2}}} \tag{3.81}$$

where r_c is the radius of the coil and I is the current. Superposition can be used if the magnetic field is generated by a coil with several windings.

If the force F_z on the permanent magnet is assumed to be uniform over the whole membrane, it is possible to calculate the maximum deflection at the centre of a circular membrane of radius r_a which is fixed at the edge and loaded by the pressure p between $r = 0$ and $r = r_0$ [30]:

$$w_{max} = \frac{3(1-\nu^2)r_0^2 r_a^2}{4Et^3}p \tag{3.82}$$

where ν is the Poisson's ratio, E the Young's modulus and t the thickness of the plate. With equation 3.80, p can be determined as:

$$p = F_z \cdot (\pi r_0^2)^{-1} \tag{3.83}$$

3.2.7 Evaluation of the Actuation Techniques for Microfluidics

Finally, we want to compare the different actuation techniques and evaluate them for use in microfluidics. Generally, the actuating components in microfluidics are used for active valves and pumps so only membrane actuators are considered here.

Table 3.6 below gives an approximate comparison of the force and power required for each actuation scheme. A membrane of 9 × 1 mm and 25 μm thickness was employed for the calculations. In the case of electrostatics, the distance between the two electrodes was also 25 μm.

actuation concept	actuation strength	actuation time (s)	input requirements
piezoelectric [32]	25 kPa	m.d.	1000 V
bimetallic strip [33]	50 kPa	1.0	0.5 A
shape-memory alloy [34]	150 kPa	0.2	0.12 A
thermopneumatic [35]	34 kPa	0.03	2.5 W
electrostatic [36]	50 kPa	m.d.	2700 V
permanent magnet [31]	300 kPa	m.d.	0.3 A

Table 3.6: Summary of actuation driver analysis [37]; m.d.: membrane dominated and much faster than thermopneumatic actuation

It is obvious that no actuation mechanism is optimum. Therefore the actuator has to be chosen according to the individual application of the device. In an environment where high voltage is forbidden, a micropump can be driven by bimetal or shape-memory actuation. Magnetic actuation is also possible, but mass fabrication is not as far advanced as for the other two drives.

However, the generated pressure is very large compared to the bimetallic actuation.

In places where high voltage can be tolerated, electrostatic and piezoelectric devices are favourable as both require only a small amount of electric energy compared to the other devices. Also, both techniques show a fast response time that is only limited by the mechanics rather than the drive mechanism.

Table 3.7 shows the effect of scaling on the force/unit area produced by various actuation mechanisms. By scaling we mean an increase of all three linear dimensions by a factor of k. If the resultant k equals 1, this implies that the force/unit area is independent of k. In the case of the electrically actuated schemes we can also distinguish between different voltage and current scaling conditions. For example, for voltage actuation, as k is varied we can either keep the current constant or we can keep the current density (current per unit area) constant. The results shown in the table are derived from the equations noted and given previously in this chapter.

actuation scheme	relevant equations	const. voltage	const. current	const. el. field	const current density
electrostatic	3.73	k^{-2}	/	1	/
electromagnetic	3.80, 3.83	/	k^{-1}	/	k
piezoelectric	3.65, 3.69	k^{-1}	/	1	/
bimetallic (*)	3.72, 3.69	-	-	-	-
thermopneum. (*)	3.77	-	-	-	-

Table 3.7: Effect of scaling on force per unit area. (* independent of k for constant actuation temperature)

The techniques which are favoured for micromachining are those for which the power of k is negative, implying that the force per unit area gets larger as the dimensions are reduced. This is the case for both electrostatic and piezoelectric actuation at constant voltage, and for electromagnetic at constant current. Whilst constant voltage scaling is usually practical, this is not the case for constant current scaling, where the ohmic heat produced becomes prohibitive as k is reduced. For this reason the voltage operated schemes are usually preferred. As can be seen from the table, the thermally operated mechanisms (bimetallic and thermopneumatic) are neutral as regards scaling at constant temperature.

Table 3.8 gives more general information about the advantages and disadvantages of the various techniques.

The issue of integration possibility with electronics and other microflu-

actuation technique	displacement	generated force	applied voltage	power consumption
piezoelectric (disc type)	medium (0)	small (-)	large (-)	low (+)
piezoelectric (stack type)	small (-)	large (+)	large (-)	low (+)
electrostatic	small (-)	small (-)	large (-)	low (+)
electromagnetic	large (+)	large (+)	small (+)	high (-)
bimetallic	small (-)	small (-)	small (+)	medium (0)
shape-memory	large (+)	large (+)	small (+)	medium (0)
thermo-pneumatic	small (-)	small (-)	small (+)	high (-)

Table 3.8: Advantages and disadvantages of the different actuation techniques

idic devices is more complicated. It very much depends on the technology used. For example, piezoelectrics is not integratable with electronics if screen printing is used but has no such problem if the piezoelectric device is surface mounted. Bimetallic actuators are easily integrated, as it is possible to use silicon and aluminium, two custom materials in electronics. Shape-memory and electromagnetic devices are harder to integrate, as they require special materials. Thermopneumatic devices are also harder to integrate, as they require several layers of silicon. In contrast, electrostatic devices only require two layers, do not use any odd materials and are therefore easy to integrate.

Chapter 4

Modelling Techniques

Modelling techniques are very important for the design of a microfluidic system. Problems in mechanics and fluidics can often be expressed with differential equations but it is, even for many simple cases, impossible to solve these equations analytically. Sometimes, however, it is not possible to describe a whole system using a single differential equation. In this case it is usually possible to divide the system into a number of smaller elements and the differential equations for the individual elements can then be solved numerically.

4.1 The Finite Element Method: an Introduction

In order to understand the working principle of finite element programs, it is the intention of this chapter to introduce the methods used to solve differential equations. Most mechanical and fluidic problems can be described by differential equations like:

$$\begin{aligned} -\left(\frac{\partial^2 u}{\partial x^2} + \frac{\partial^2 u}{\partial y^2} + \frac{\partial^2 u}{\partial z^2}\right) &= F(x,y,z,t) \\ \alpha \frac{\partial^2 v}{\partial t^2} - \left(\frac{\partial^2 v}{\partial x^2} + \frac{\partial^2 v}{\partial y^2} + \frac{\partial^2 v}{\partial z^2}\right) &= p(x,y,z,t) \end{aligned} \qquad (4.1)$$

where u is the displacement, F is the force on the element, v is the fluid velocity in a pipe and p is the pressure distribution.

After the numerical part, we will discuss the generation of mechanical and fluidic elements and other requirements for any simulation.

4.1.1 Numerical Integration of Differential Equations

In order to discuss the numerical integration of a differential equation, we first look at a differential equation of the kind:

$$y'(x) = f(x, y(x)) \qquad (4.2)$$

The simplest and best known method of solution is the use of Euler's method. Here, the function $y(x)$ can be approximated by a Taylor series in the neighbourhood of x_0:

$$y(x_0 + \Delta x) = y(x_0) + \Delta x y'(x_0) + \frac{\Delta x^2}{2} y''(x_0) + ... + \frac{\Delta x_0^n}{n!} y^{(n)}(x_0) \quad (4.3)$$

Euler's method only uses the linear term of the differentials. As this linear approach is not very efficient it is better to employ the Runge-Kutta method which uses the higher differentials up to the 4th order. A simple algorithm for solving differential equations with the Runge-Kutta method of 4th order is:

$$\begin{aligned} y_{n+1} &= y_n + \tfrac{1}{6}(k_1 + 2k_2 + 2k_3 + k_4) \\ k_1 &= hf(x_n, y_n) \\ k_2 &= hf\left(x_n + \tfrac{h}{2}, y_n + \tfrac{1}{2}k_1\right) \\ k_3 &= hf\left(x_n + \tfrac{h}{2}, y_n + \tfrac{1}{2}k_2\right) \\ k_4 &= hf(x_n + h, y_n + k_3) \end{aligned} \quad (4.4)$$

where y_n is the known value of the function, y_{n+1} is the value of the function at $x_n + h$ and k_i are variables.

A more complex solution of the differential equation above is the method of Ritz. Here, we approximate the unknown solution of the differential equation by:

$$y(x) = h_0(x) + a_1 h_1(x) + a_2 h_2(x) + ... + a_n h_n(x) \quad (4.5)$$

with a_i being constants and $h_i(x)$ being functions of the variable x. Often, the above approximation is written with the use of polynomials up to the n-th order. The differential equation 4.2 can thus be rewritten by:

$$f(x, y) - y' = 0 \quad (4.6)$$

Using the approximation above, this can be modified into:

$$f\left(x, h_0(x) + \sum_{i=1}^{n} a_i h_i(x)\right) - \left\{h_0'(x) + \sum_{i=1}^{n} a_i h_i'(x)\right\} = 0 \quad (4.7)$$

For this, we can calculate the square of the error of the approximation as a function of the variables a_i within the boundaries of x (set to x_0 and x_1):

$$\begin{aligned} &F(a_1, a_2, ..., a_n) = \\ &\int_{x_0}^{x_1} \left(f\left(x, h_0(x) + \sum_{i=1}^{n} a_i h_i(x)\right) - \left\{h_0'(x) + \sum_{i=1}^{n} a_i h_i'(x)\right\}\right)^2 dx \end{aligned} \quad (4.8)$$

The minimum of the error function F is then given by setting the derivatives to zero for i from 1 to n:

$$\frac{\partial F}{\partial a_i} = 0 \tag{4.9}$$

This leads to an equation system that can be solved numerically. The whole technique shall now be demonstrated for the simple differential equation $y'(x) = y(x)$ with $y(0) = 1$ and the boundaries of x lying between 0 and 1. As an approximation, we use a polynomial of second order:

$$y(x) = a_2 h_2(x) + a_1 h_1(x) + h_0(x) = ax^2 + bx + 1 \tag{4.10}$$

with its derivative $y'(x) = 2ax + b$. The square of the error of the approximation is then calculated as:

$$\begin{aligned} F(a,b) &= \int_0^1 \left((ax^2 + bx + 1) - (2ax + b)\right)^2 \mathrm{d}x \\ &= 1 - \frac{4a}{3} + \frac{8a^2}{15} + \frac{ab}{2} - b + \frac{b^2}{3} \end{aligned} \tag{4.11}$$

The equation system is finally obtained by calculating the minimum of the error function:

$$\begin{aligned} \frac{\partial}{\partial a} F(a,b) &= \frac{b}{2} + \frac{16a}{15} - \frac{4}{3} := 0 \\ \frac{\partial}{\partial b} F(a,b) &= \frac{a}{2} + \frac{2b}{3} - 1 := 0 \end{aligned} \tag{4.12}$$

From the equation system we obtain the values for a, b and thus the approximate solution:

$$y(x) = \frac{70}{83}x^2 + \frac{72}{83}x + 1 \tag{4.13}$$

Another approach is the Galerkin method. Here, we use the same approximation as above (equation 4.5), but calculate the error function $r(x)$ of the approximation:

$$r(x) = f\left(x, h_0(x) + \sum_{i=1}^{n} a_i h_i(x)\right) - \left\{h_0'(x) + \sum_{i=1}^{n} a_i h_i'(x)\right\} \tag{4.14}$$

Instead of minimising the total error (see equation 4.9), we minimise the error R_k of the weighted residuals:

$$R_k(a_1, .., a_n) = \int_{x_0}^{x_1} h_k(x) r(x) \mathrm{d}x \tag{4.15}$$

This leads to a symmetric equation system which helps to reduce the computer memory usage during calculation:

$$R_k(a_1, .., a_n) = 0 \tag{4.16}$$

Both the Ritz and the Galerkin methods are often used in finite element programs for solving the differential equations of the elements.

4.2 One-dimensional Mechanical Finite Element Method

When the mechanical behaviour of a structure is determined with the finite element method, the differential equations of the structure have to be solved. Therefore, a complex structure is divided into small parts, called elements. The elements should consist of locally constant physical properties. A link between the elements is represented by nodes. The equations necessary for the numerical solution of a mechanical problem will be discussed for the one-dimensional case only.

A finite element with Young's modulus E and length l is coupled via the cross sectional area A to its neighbour elements. In the steady state condition, the sum of the internal forces acting on an element equals zero:

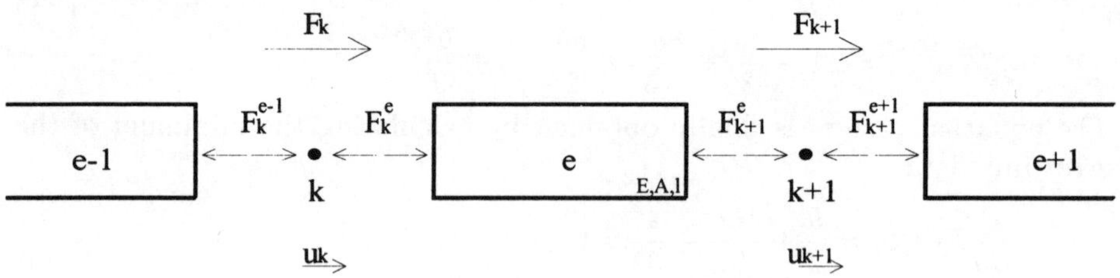

Figure 4.1: Linear finite elements for mechanical simulations

$$F_k^e + F_{k+1}^e = 0 \tag{4.17}$$

Using Hooke's law, the change in length for an element is calculated as follows:

$$\left(\frac{EA}{l}\right)_e \cdot (u_{k+1} - u_k) = F_k^e = -F_{k+1}^e \tag{4.18}$$

Combining the two equations leads to the local finite element equation:

$$\begin{pmatrix} F_k^e \\ F_{k+1}^e \end{pmatrix} = \left(\frac{EA}{l}\right)_e \cdot \begin{bmatrix} 1 & -1 \\ -1 & 1 \end{bmatrix} \cdot \begin{pmatrix} u_{k+1} \\ u_k \end{pmatrix} \tag{4.19}$$

where $\frac{EA}{l}$ expresses the stiffness of the element. The steady state condition for nodes is that the sum of the internal forces equals the applied force at the respective node, whereas this sum equals zero where no force is applied:

$$F_k^{e-1} + F_k^e = \begin{cases} 0 \\ F_k \end{cases} \tag{4.20}$$

Now the finite element equation for the whole structure can be drawn from the items above:

$$\vec{F} = K \cdot \vec{u} \tag{4.21}$$

with K as the system matrix, $\vec{F}$ as the vector of external forces and $\vec{u}$ as the displacement vector. As the force vector is known, the displacement vector can be calculated from equation 4.21.

Numerical solution methods for more complex problems than the given example most often use energy principles. Here, the variation in work of the external forces can be set equal to the variation of internal energy. The elements for this approach are then created with the simplification of a linear variation of the nodal displacement over the element and the finite element equation is solved numerically.

As an example of energy methods, let us again look at the one-dimensional problem considered previously. The elastic energy in each element is given by:

$$W_e = \frac{1}{2} E_e \left(\frac{u_{k+1} - u_k}{l} \right)^2 V_e = \frac{1}{2} E_e \varepsilon_e^2 V_e \tag{4.22}$$

where V and l are the volume and length of the element and ε is the strain within the element. In matrix form, the energy stored in the element is given by:

$$W_e = \frac{1}{2} (u_k, u_{k+1}) \left(\frac{EA}{l} \right)_e \cdot \begin{bmatrix} 1 & -1 \\ -1 & 1 \end{bmatrix} \cdot \begin{pmatrix} u_k \\ u_{k+1} \end{pmatrix} \tag{4.23}$$

From external forces $\vec{F}$, we obtain the externally applied work $W_{e.\text{ext}}$. This is equal to:

$$W_{e.\text{ext}} = \frac{1}{2} (u_k, u_{k+1}) \begin{pmatrix} F_k \\ F_{k+1} \end{pmatrix} \tag{4.24}$$

Combining both these equations leads again to the local finite element equation 4.19. Although this approach is more complicated for the one-dimensional case, it is an easier way to obtain the element stiffness matrix K_e for the two- and three-dimensional cases:

$$K_e = \left(\frac{EA}{l} \right)_e \cdot \begin{bmatrix} 1 & -1 \\ -1 & 1 \end{bmatrix} \tag{4.25}$$

4.3 Computational Fluid Mechanics

Computational Fluid Mechanics can be treated in a similar way to static simulations of mechanical devices. Besides the finite element method, fluid mechanics often uses a finite difference method. Both methods shall be explained in this section, so that the future user can apply both methods.

4.3.1 Finite Difference Method

Let us first consider the differential equation for two-dimensional ideal steady flow. In ideal flow, Laplace's equation needs to be valid:

$$\frac{\partial^2 U}{\partial x^2} + \frac{\partial^2 U}{\partial y^2} = 0 \tag{4.26}$$

where U is the velocity potential ($\vec{v}$ = -grad U) and x, y are the spatial coordinates. For a two-dimensional grid as in figure 4.2, this differential equation can be translated into a numeric equation:

$$\frac{U_{n-1,m} - 2U_{n,m} + U_{n+1,m}}{(\Delta x)^2} + \frac{U_{n,m-1} - 2U_{n,m} + U_{n,m+1}}{(\Delta y)^2} = 0 \tag{4.27}$$

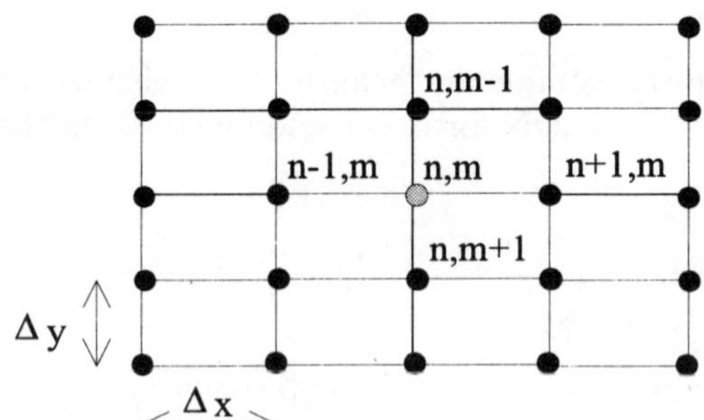

Figure 4.2: Finite Difference grid for ideal flow in computational fluid mechanics

Finally, we reach an equation for the node n, m by rearranging the equation above and using $\alpha = \Delta y/\Delta x$ as dimensional relation:

$$U_{n,m} = \frac{\alpha^2}{2(1+\alpha^2)}\left(U_{n-1,m} + U_{n+1,m}\right) + \frac{1}{2(1+\alpha^2)}\left(U_{n,m-1} + U_{n,m+1}\right) \tag{4.28}$$

Starting from the known boundary conditions, several iterations are necessary through the whole velocity array, in order to reach convergence with the finite difference method. As a very common convergence criterion, the relative error ε is taken:

$$\varepsilon = \frac{U_{n,m}^{\text{iteration } k} - U_{n,m}^{\text{iteration } k-1}}{U_{n,m}^{\text{iteration } k-1}} \tag{4.29}$$

The iteration is finished when ε drops below a given value. The velocity can finally be calculated from the velocity potential.

For viscous steady state flow without gravity, the Navier-Stokes equation is used. For example, the equation for the y direction using the simplified Navier-Stokes equations becomes:

$$\mu\left(\frac{\partial^2 v}{\partial x^2}+\frac{\partial^2 v}{\partial y^2}+\frac{\partial^2 v}{\partial z^2}\right)-\frac{\partial p}{\partial y}=0 \tag{4.30}$$

The numerical grid for a partial difference solution is assumed to have a spacing of $\Delta x, \Delta y, \Delta z$ in x, y, z directions. Linearisation and indexing the grid points for pressure and flow velocity by i, j, k in x, y, z directions gives the following numerical approach:

$$\begin{aligned}\mu\Bigg(&\frac{v_{i-1,j,k}-2v_{i,j,k}+v_{i+1,j,k}}{(\Delta x)^2}+\frac{v_{i,j-1,k}-2v_{i,j,k}+v_{i,j+1,k}}{(\Delta y)^2}+\\&+\frac{v_{i,j,k-1}-2v_{i,j,k}+v_{i,j,k+1}}{(\Delta z)^2}\Bigg)-\frac{p_{i,j+1,k}-p_{i,j-1,k}}{(2\Delta y)}=0\end{aligned} \tag{4.31}$$

This is the continuity equation for the whole grid that can be solved numerically. Boundary conditions such as no-slip at the wall are necessary for the solution.

4.3.2 Finite Element Method (FEM)

The same equations are taken in finite element programs. However, it is necessary to define the differential equation of the elements. Here, we use a pipe of radius r_0 and a fluid of viscosity μ. From the theory section, we know that the laminar viscous flow within a pipe element of length l is described by the law of Hagen-Poisseuille:

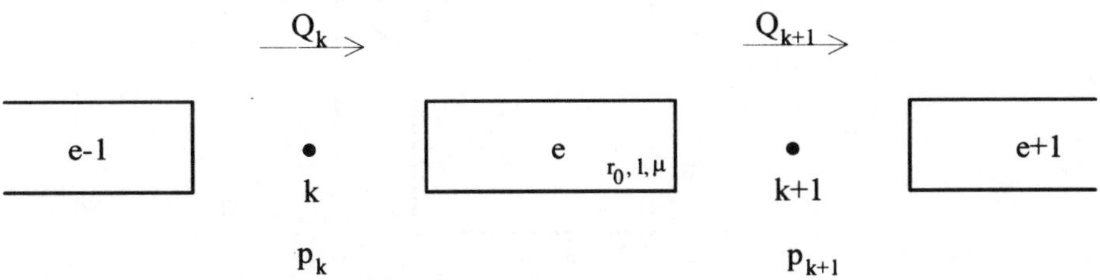

Figure 4.3: Linear finite elements for the simulations of flow in pipes

$$Q_k=-Q_{k+1}=\frac{\pi r_0^4}{8\mu\, l}\left(p_{k+1}-p_k\right) \tag{4.32}$$

where Q_k is the volume flow into the pipe with positive sign and Q_{k+1} is the volume flow out of the pipe with negative sign and p is the pressure. From the conservation of mass, i.e. that no fluid is lost within the pipe, it is obvious that the flows at the nodes k and $k+1$ are equal. We can thus form the local finite element equation:

$$\begin{pmatrix} Q_k^e \\ Q_{k+1}^e \end{pmatrix} = \left(\frac{\pi r_0^4}{8\mu l}\right)_e \cdot \begin{bmatrix} 1 & -1 \\ -1 & 1 \end{bmatrix} \cdot \begin{pmatrix} p_{k+1} \\ p_k \end{pmatrix} \tag{4.33}$$

From this, the full finite element equation is obtained as above:

$$\vec{Q} = K \cdot \vec{p} \tag{4.34}$$

which can be solved numerically with the help of computers.

4.4 Use of FEM Programs

With modern FEM programs it is not necessary to set up the equation systems for a given problem. Here, you can enter the geometry and the physical properties of the materials and the computer does the rest. However, it is still necessary to understand the background to the finite element method in order to interpret the results correctly. The computer can only give proper results if the user has fed in the right information.

One of the most important things in finite element modelling is the generation of the grid, i.e. the substructure of the whole analysis. For both microfluidic and mechanical problems, it is necessary to consider the following design rules:

- The aspect ratios of the side length of any element should be smaller than 10:1.

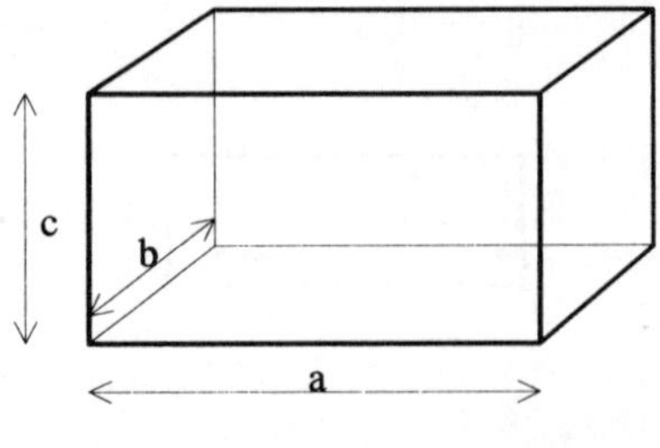

Figure 4.4: Modelling requirement: 1:10 < a:b, a:c, b:c < 10:1

- Internal angles ought to be between 30 and 150 degrees.

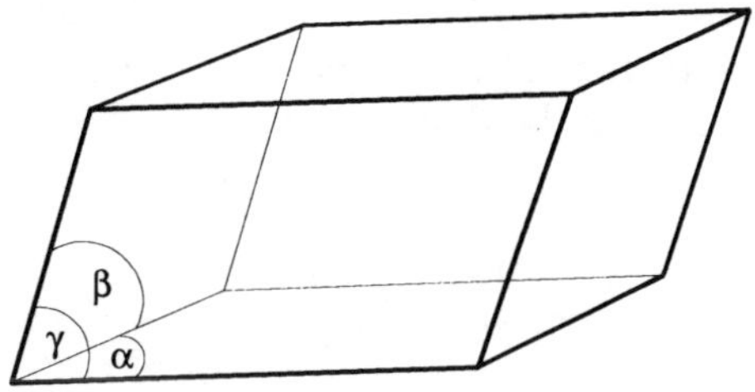

Figure 4.5: Modelling requirement: $30° < \alpha, \beta, \gamma < 150°$

- In the region of high gradients (fixing points of membranes etc.), a finer mesh should be used.

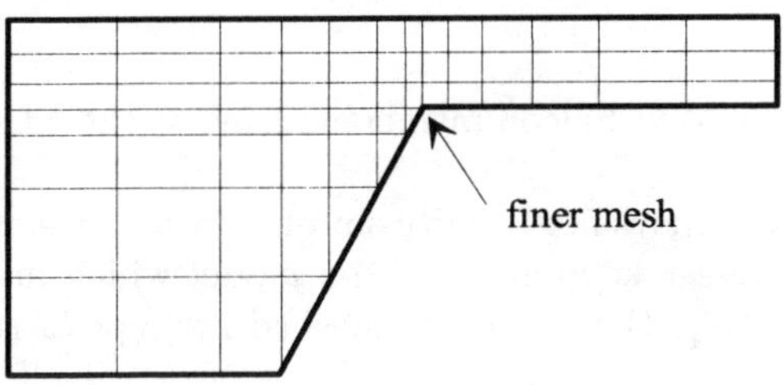

Figure 4.6: Modelling requirement: mesh size has to be adapted at fixing points

- Symmetry is to be exploited for reduction of the calculation time. For example, a rectangular membrane suspended on all edges can be modelled with only one quarter when the nodes at the symmetry lines are constrained in the displacement perpendicular to the line.

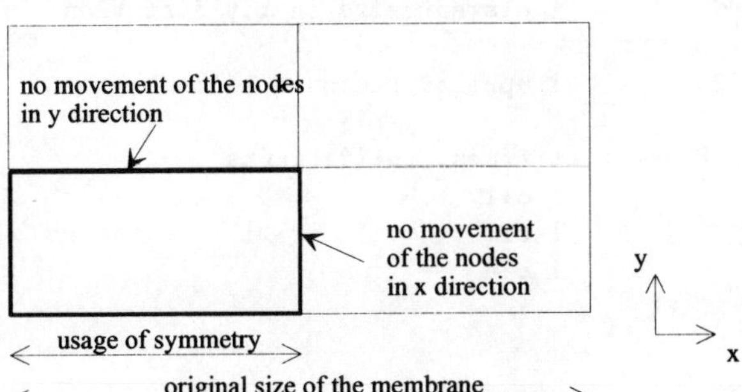

Figure 4.7: Modelling requirement: symmetry can be used for membranes of uniform thickness

4.4.1 Examples for mechanical FEM

As an example for a mechanical FEM model, we choose a silicon membrane fixed at the edges. A picture of the membrane is given in figure 4.8. Using symmetry, the modelling can be done using a quarter of the membrane.

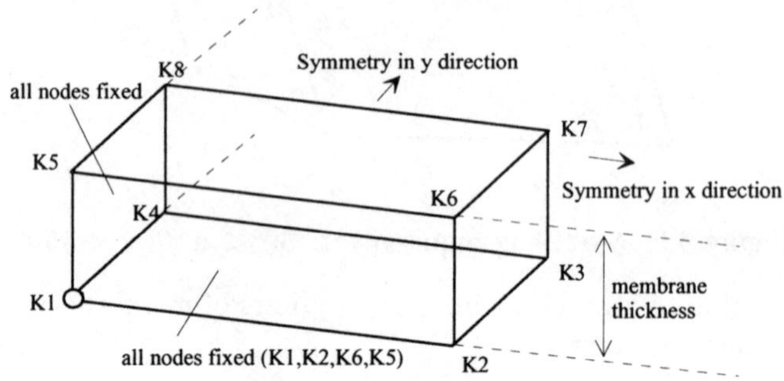

Figure 4.8: Silicon membrane fixed at the edges

The program for Ansys 5.4 is quite simple: At first we define the material properties. Then we set keypoints for the model which are the 8 endpoints of the block. After this, the volume is meshed in steps of *elsize*. Finally, the boundaries and the loads are applied. The program itself is given below so that anyone can start modelling immediately.

```
/COM, Si membrane modelled with ANSYS 5.4

sheight = 70e-6          ! height of the Si-membrane
length = 8e-3            ! length of the Si-membrane
width = 8e-3             ! width of the Si-membrane

elsize = 250e-6          ! element size in x,y direction

pressure = 10000         ! applied pressure to membrane

/COM, Silicon: Elastic stiffness coefficients
caa=165e9                ! c11
cab=64e9                 ! c12
cdd=79.2e9               ! c44

/PREP7

/COM, Silicon material definitions
MAT,1                    ! Material number
ET,1,64                  ! Element type
TYPE,1                   ! Material type number
TB,anel,1                ! Definition of the elastic stiffness matrix
```

```
TBDATA,1,caa,cab,cab,0,0,0
TBDATA,,caa,cab,0,0,0,caa
TBDATA,,0,0,0,cdd,0,0
TBDATA,,cdd,0,cdd
MP,dens,si,2.33e3          ! Density of silicon


/COM, Blocks and Meshing

K,1,0,0,0                          ! keypoint 1 at x =0, y=0, z=0
K,2,length/2,0,0                   ! keypoint 2
K,3,length/2, width/2, 0           ! keypoint 3
K,4,0,width/2,0                    ! keypoint 4

K,5,0,0,sheight                    ! keypoint 5
K,6,length/2,0,sheight             ! keypoint 6
K,7,length/2, width/2, sheight     ! keypoint 7
K,8,0,width/2,sheight              ! keypoint 8

V,1,2,3,4,5,6,7,8                  ! volume of the keypoints

VSEL,all                 ! select all volumes
VATT,1,,1                ! apply Si-material

ESIZE,elsize             ! set meshing size
VMESH,all                ! mesh the whole grid

/COM, boundaries at membrane edge
NSEL,s,loc,x,0           ! select all nodes with x=0
D,all,ux,,,,,uy,uz       ! constrain any displacement at these nodes
NSEL,s,loc,y,0           ! select all nodes with y=0
D,all,ux,,,,,uy,uz       ! constrain any displacement

/COM, symmetry-boundaries at symmetry lines
/COM, right side only x-constraint displacement
NSEL,x,length/2          ! select all nodes with x=length/2
D,all,ux                 ! constrain displacement in x direction
/COM, top side only y-constraint displacement
NSEL,y,width/2           ! select all nodes with y=width/2
D,all,uy                 ! constrain displacement in y direction

EALL                     ! select all elements
NALL                     ! select all nodes

/COM,Solution
FINISH
/SOLUTION
ANTYPE,Static            ! Static simulation
```

```
/COM, pressure application
NSEL,s,loc,z,0            ! select nodes with z=0
SF,all,pres,0             ! apply pressure p=0
NSEL,s,loc,z,sheight      ! select nodes with z=sheight
SF,all,pres,pressure      ! apply pressure (defined above)

EALL
NALL
SOLVE
FINI

/POST1                    ! Post processor
/VIEW,,1,1,1              ! Viewing Vector
PLDISP,1                  ! Plot displaced membrane
```

The program text is almost self explanatory with the help of the comments following the exclamation marks. After the start of Ansys, just type

```
/INP,filename,extension
```

in order to run the simulation. This program enables the reader to do some simple FEM with Ansys 5.4 without any further knowledge and view the results immediately. Although this program is quite simple, it can be used to solve many problems in microfluidics when modified slightly. To re-run the program with different geometries or pressures, just type

```
FINI
```

followed by

```
/CLEAR
```

Most problems that depend on membranes or cantilevers can be analysed with the program above. The results are depicted in figure 4.9. The bending of the membrane is clearly visible, with the maximum displacement at the centre (symmetry point). In this example, the maximum displacement of a 8×8 mm membrane of 70μm thickness is 11.8 μm for an applied pressure difference of 10 kPa.

4.4.2 Examples for Computational Fluid Dynamics (CFD)

In a similar way it is possible to simulate the flow of fluid through a pipe of rectangular cross section. The whole pipe with inlet, outlet and walls is depicted in figure 4.10.

The simulation program for this device is given below. Again, Ansys 5.4 is used for the simulations. The standard Ansys 5.4 package, however, has to be expanded with the Ansys 5.4 CFD package to be able to perform the simulation.

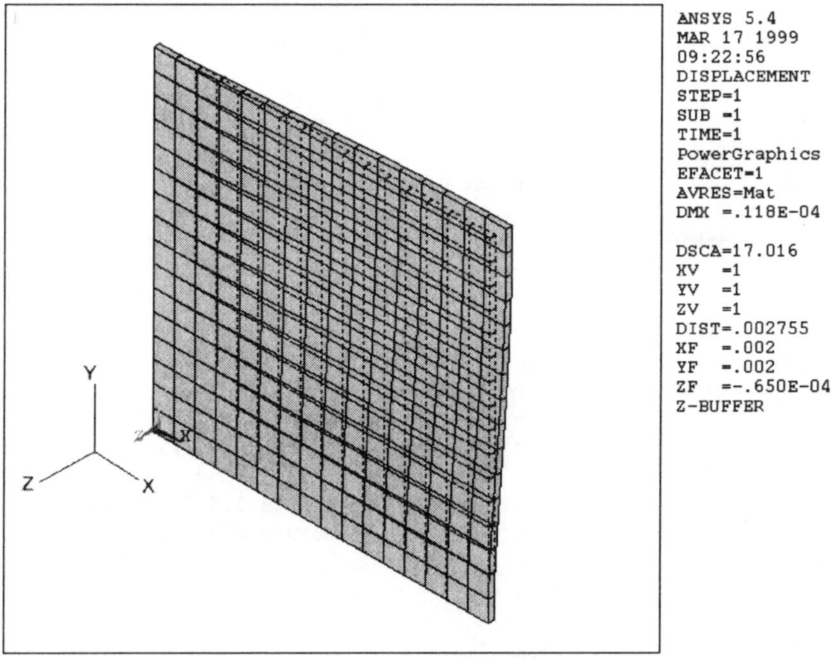

Figure 4.9: Displacement of a silicon membrane computed with Ansys 5.4. The maximum z-displacement is given by the variable DMX. All other variables are of minor importance in this example.

```
/COM, Modelling of microfluidic flow in a duct

/COM, Definition of the variables
```

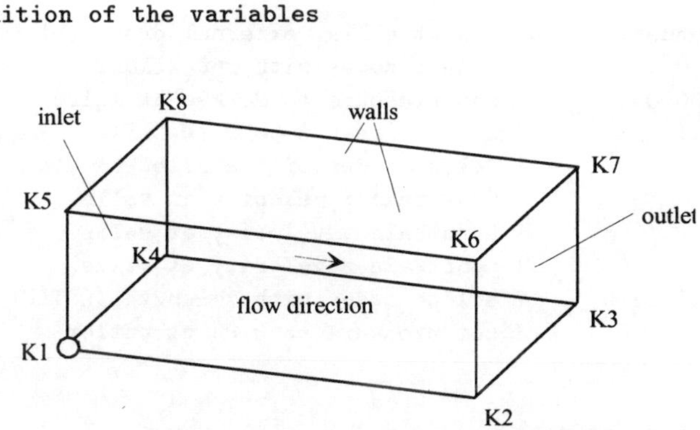

Figure 4.10: Rectangular duct for fluidic simulations

```
width=100e-6              ! duct size in y and z direction
length=500e-6             ! length of duct in x direction

/PREP7

/COM, Definition of Keypoints
K,1,0,0,0                 ! keypoint 1 at x=0, y=0, z=0
K,2,length,0,0            ! keypoint 2
K,3,length,width,0        ! keypoint 3
K,4,0,width,0             ! keypoint 4
K,5,0,0,width             ! keypoint 5
K,6,length,0,width        ! keypoint 6
K,7,length,width,width    ! keypoint 7
K,8,0,width,width         ! keypoint 8
L,1,2                     ! Line 1 between keypoint 1 and 2
L,4,1                     ! Line 2 between keypoint 4 and 1
L,1,5                     ! Line 3 between keypoint 1 and 5
A,1,4,8,5                 ! Area 1 between keypoints 1,4,8,5 - INLET
A,2,3,7,6                 ! Area 2 between keypoints 2,3,7,6 - OUTLET
A,1,2,3,4                 ! WALL
A,1,2,6,5                 ! WALL
A,3,4,8,7                 ! WALL
A,5,6,7,8                 ! WALL
V,1,2,3,4,5,6,7,8         ! Volume between keypoints 1,2,3,4,5,6,7,8
/TRIAD,OFF

LESIZE,1,,,6              ! Number of nodes on line 1 is 6
LESIZE,2,,,6              ! Number of nodes on line 2 is 6
LESIZE,3,,,20             ! Number of nodes on line 3 is 20

ET,1,142          ! 3-D Fluid Element
VMESH,1                   ! Generate gridpoints and elements

/COM, Set Boundaries (no slip at walls, external pressure)
NSEL,s,loc,x,0            ! select nodes with x=0 (INLET)
D,all,pres,20000          ! set pressure to 20 kPa at inlet
ASEL,S,,,3,6,1            ! Select areas 3 to 6 (WALLS)
NSLA,S,1                  ! Select nodes of the selected areas
D,ALL,VX,0                ! Constrain x velocity at walls
D,ALL,VY,0                ! Constrain y velocity at walls
D,all,vz,0                ! Constrain z velocity at walls
NSEL,s,loc,x,length       ! select nodes with x=length (OUTLET)
D,ALL,PRES,0              ! set pressure to 0 Pa at outlet
ALLS

/COM, Define Simulation
FLDATA2,iter,exec,50              ! maximum 50 global iterations
FLDATA6,conv,outp,LAND            ! controls output
```

```
FLDATA6,conv,iter,10             ! number of iterations for convergency
FLDATA1,solu,flow,t              ! set laminar flow
FLDATA7,prot,visc,CONSTANT       ! property viscosity constant
FLDATA7,prot,dens,CONSTANT       ! property density constant
FLDATA8,nomi,dens,1.0e3          ! Density of the fluid
FLDATA8,nomi,visc,1.0e-3         ! Viscosity of the fluid
FLDATA21,conv,pres,1e-8          ! Convergence Control

/COM, Start solution and view results
/SOLU                            ! Start Solution
SAVE
SOLVE
FINISH
/POST1                           ! Start Viewer
SET,last                         ! Use last iteration results for display
/TYPE,,1                         ! Show only cross sections
/VIEW,,0,1,0                     ! Viewing Vector
/CPLANE,0                        ! Cutting plane normal to viewing vector
PLVEC,v,z                        ! Plot velocity vector
```

Again, the program listing is quite simple and can be adapted for a variety of problems in microfluidics. The result for this simulation is depicted in figure 4.11. It is quite clear that a Hagen-Poisseuille flow profile is obtained.

These program listings only give an introduction to the use of Ansys 5.4 for microfluidic problems. For more complicated tasks, please refer to the manual supplied with Ansys 5.4 [38].

4.5 Coupled FEM and CFD Simulators in Microfluidics

A very interesting problem in microfluidics is the coupling between mechanical and fluidic elements. This is for example the case for cantilever valves, where a cantilever controls the flow through a duct, or in membrane-actuated pumps, where the membrane displaces the fluid. There are two possible ways to solve such problems. The first method is to solve each problem individually and feed the results into the other model. This has to be done iteratively until both solutions converge [39,40]. Yet, there can be problems that will not converge. In this case, a full integrated mechanical/fluidic analysis is necessary [41]. However, as most commercially available FEM programs do not offer the possibility of solving mechanical and fluidic problems at the same time, we now describe a method for coupling fluidic and mechanical FEM programs.

As an example of a coupled FEM/CFD simulation, a cantilever valve is taken. Cantilever valves (see figure 4.12) are passive one-way valves allowing the flow of a liquid or a gas in one direction and preventing it in the other. This type of valve is either fabricated macroscopically with cutting tools or

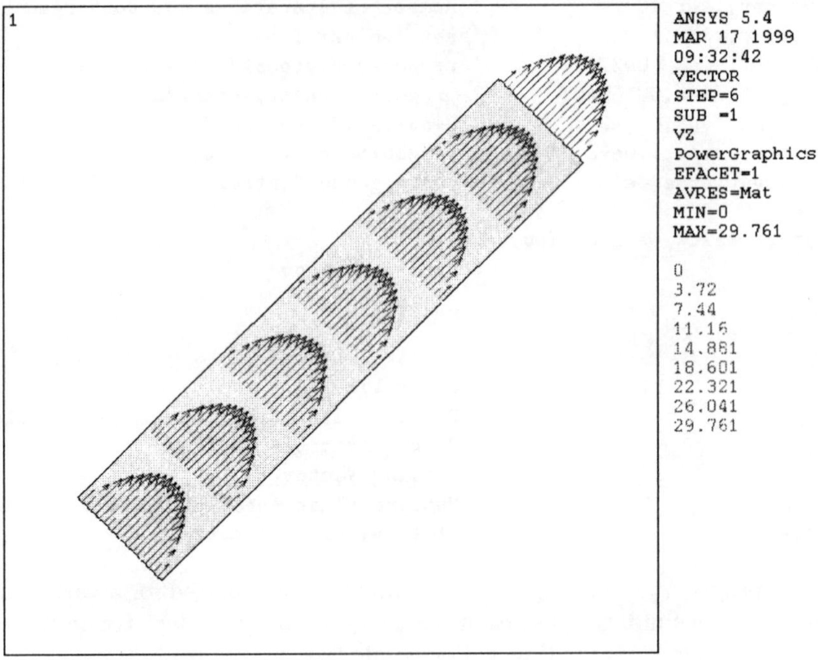

Figure 4.11: Modelling results for a rectangular duct: Velocity vector

micromachined. Both techniques give a similar device with an identical working principle. However, micromachining allows the creation of small-scale devices in the sub-millimetre and even micrometre range as well as low-cost mass production.

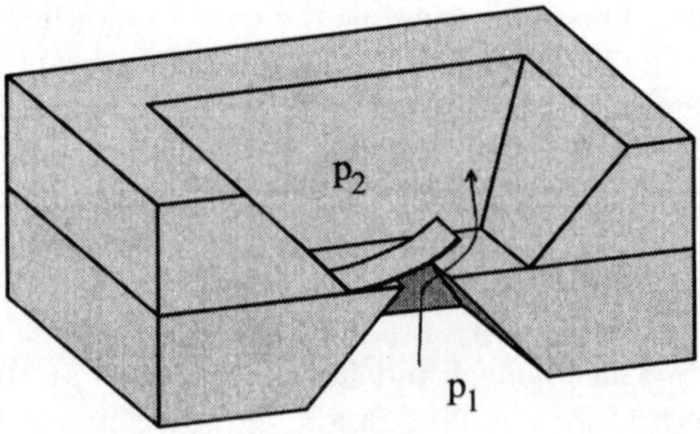

Figure 4.12: Principle of the passive cantilever valve

A cantilever valve generally consists of a thin flap sitting on top of a duct.

The flap controls the flow through the valve-system. In the case of a decrease in pressure towards the "free" side ($p_2 < p_1$), the cantilever bends into an open state and allows the fluid to pass through. For an increasing pressure towards the free side ($p_2 > p_1$), the flap is pressed against the valve seat and the passage is sealed.

4.5.1 Description of the Simulation Method

The two key issues for the modelling, i.e. the mechanical bending and the fluid flow, cannot be handled independently. Therefore, a coupling of two computational solvers is required [40, 42].In the case described here, the continuum mechanics of the cantilever is computed with the FEM package Ansys 5.4[1], whereas the fluid flow is solved with the CFD software program CFX Flow3D 4.1[2]. The coupling of FEM and CFD simulators can be achieved with an entirely computer controlled macro-program that couples Ansys 5.4 (FEM) and CFX Flow3D 4.1 (CFD). The control program (figure 4.15) converts the output of one simulator to a suitable format for use by the other. It is started with the input of the desired pressure difference across the valve and the geometry of the cantilever. This includes the possibility of entering an initial deflection of the cantilever, eg. originating from stress due to non-uniform boron concentrations. Two grids are used for the modelling. The Ansys 5.4 grid (figure 4.13) contains a cuboid beam consisting of nearly cubic elements. To reduce the size of the problem and thus simulation times, the nodes at the root end of the cantilever are fixed. Owing to the thin cantilever (5μm) and the thick bulk silicon (500 μm), the resulting error will be negligible.

Initial fluidic simulations of the whole valve as seen in figure 4.12, i.e. the duct and the area around the cantilever, have shown that the main pressure drop occurs in the region between cantilever and valve seat. Consequently the CFX Flow3D 4.1 grid was simplified to the duct linked with the area below the cantilever (see figure 4.14).

The first task of the control program is to create an initial pressure array containing the pressure below the cantilever at each node of the Ansys 5.4 grid. The pressure is assumed to be the maximum applied value in the duct area and zero elsewhere. With the pressure as input, Ansys 5.4 then calculates and outputs the deflection at each node of the cantilever grid. The file containing the deflection array is now converted to the CFX Flow3D 4.1 CFD grid with a 2-dimensional cubic spline interpolation. For the simulation itself, a laminar flow behaviour is assumed. This has been confirmed by previously computing the same fluidic problem with both laminar and turbulent behaviour where approximately the same fluid flow was calculated.

The output of the CFX Flow3D 4.1 simulation is a logfile, containing pressure and flow velocities for all nodes. Pressure values directly beneath the cantilever are then extracted and converted from the CFX Flow3D 4.1

[1] Ansys Inc., USA
[2] AEA Technology, USA

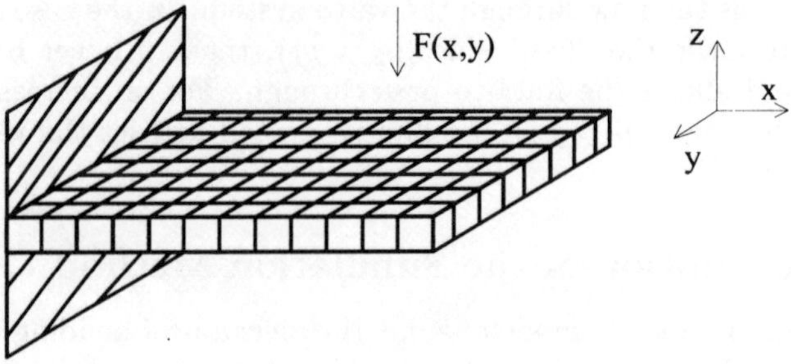

Figure 4.13: Grid for mechanical simulations

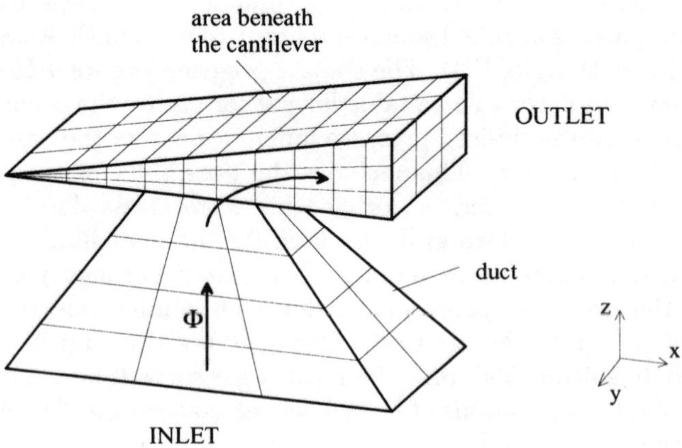

Figure 4.14: Grid for flow simulations

logfile into an Ansys 5.4-compatible form. Again, cubic spline fitting is used for conversion. After this, Ansys 5.4 is started with the new pressure file as input, resulting in the new deflection for this pressure condition.

Finally, a test is done to evaluate whether the results converge. The old and new pressure files as well as the old and new deflection files are compared and the program finishes if the average and maximum difference is less than a given percentage value. In our case, 5% and 10% respectively were taken. In order to accelerate the convergence, non-linear successive overrelaxation (using the mean value between previous and current simulation) was used for the input of the next CFX Flow3D 4.1 simulation. The simulation time for a 700 μm $\times$ 700 μm cantilever for one iteration was approximately 20 minutes on a SPARC 1000 computer. On average, the simulation was finished after three to five iterations.

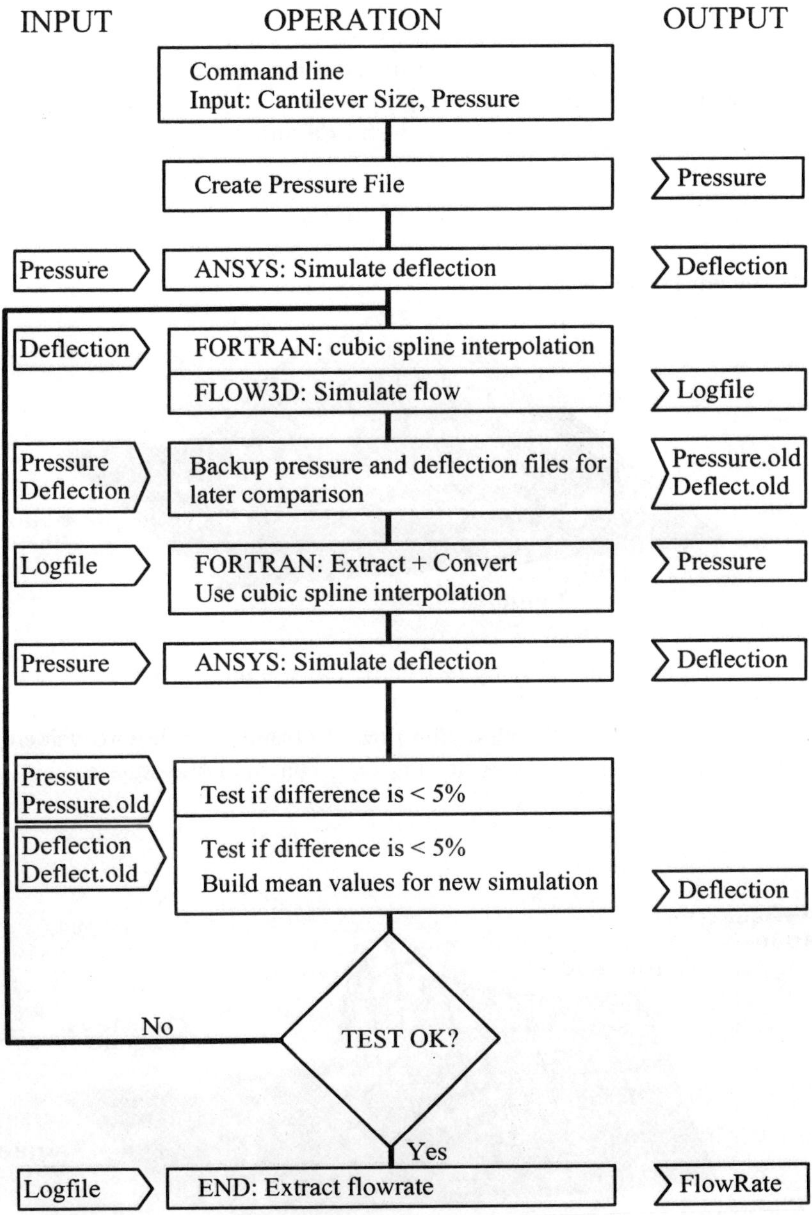

Figure 4.15: Flow chart for the macro-program

4.5.2 Simulation Results and Comparison with Measurements

At first, let us have a look at the velocity distribution of the flow at the inlet of the valve. Figure 4.16 shows this distribution along the 700 × 700 μm^2 entrance, which is labelled *inlet* in figure 4.14. For this case the inlet and outlet pressures p_1 and p_2 were taken as 8 kPa and zero, respectively.

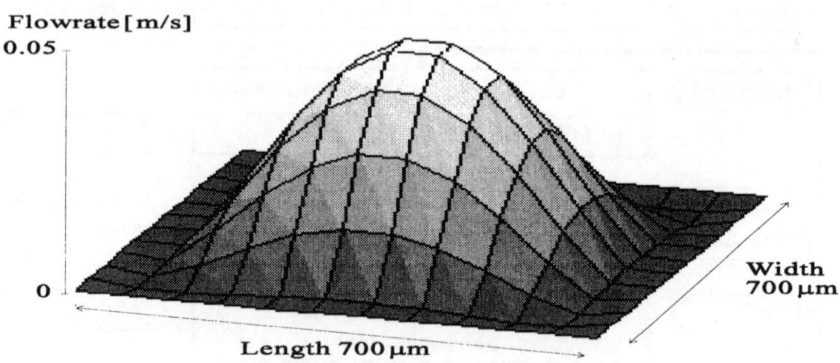

Figure 4.16: Velocity distribution at the inlet (8 kPa)

As expected, a kind of Poisseuille flow is obtained with zero velocity at the walls and the maximum velocity near the centre of the square.

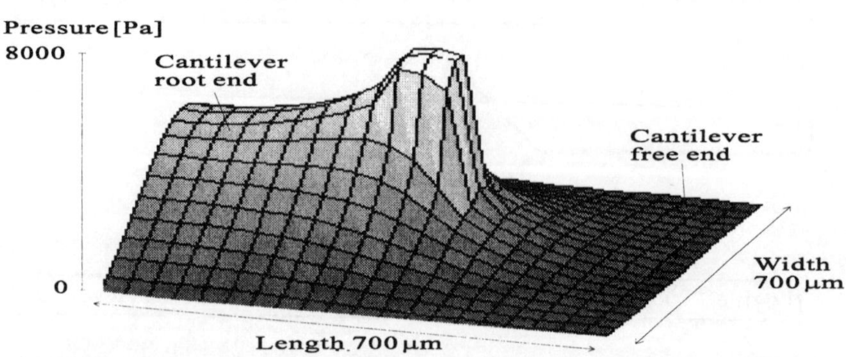

Figure 4.17: Pressure distribution beneath the cantilever at 8 kPa

In order to verify the assumption that the main pressure difference occurs beneath the cantilever, figure 4.17 depicts the resulting pressure distribution for a simulation with a flow grid which includes the whole valve at an inlet pressure of $p_1 = 8$ kPa.

As discussed previously, the maximum pressure can be found directly over the orifice. This value is close to the inlet pressure. The pressure at the edges of the cantilever drops to zero, except at the root, demonstrating that modelling only the region of the cantilevers is adequate.

In figure 4.18 the measured and simulated curves for three different cantilever valves are depicted. The error bars were determined by varying the thickness and the maximum initial deflection of the cantilever by 0.2 μm. It is obvious that both simulation and measurement agree very well.

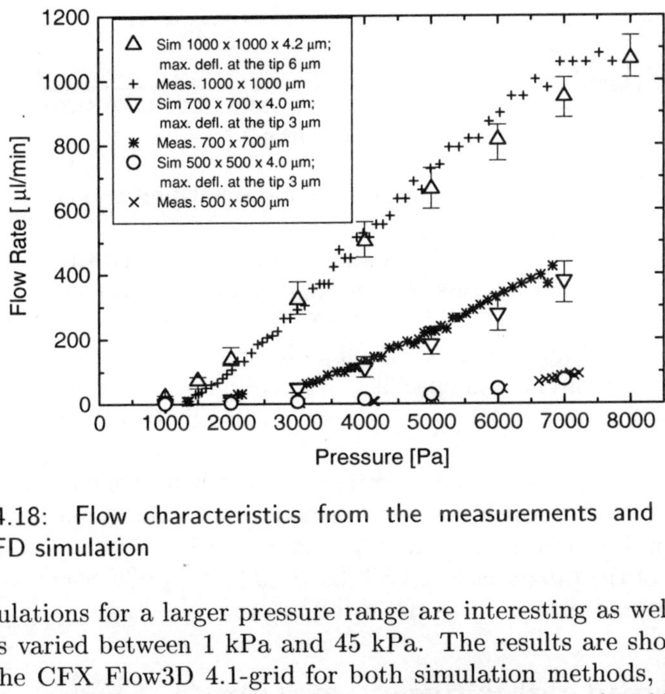

Figure 4.18: Flow characteristics from the measurements and the coupled FEM/CFD simulation

Simulations for a larger pressure range are interesting as well. The pressure was varied between 1 kPa and 45 kPa. The results are shown in figure 4.19. The CFX Flow3D 4.1-grid for both simulation methods, laminar and turbulent, can use the same grid spacing up to pressures of 20 kPa. In order to achieve convergence with the laminar model for higher pressures a finer grid had to be used which increases the calculation time considerably. However, since the results of both laminar and turbulent simulations are nearly identical, the faster turbulent model can be used for the simulations at high pressures. An estimate of the Reynold's number for this problem gives approximately 50 for a maximum cantilever deflection of 50 μm and a volume flow of 2000 μl min^{-1}, being well below the critical Reynold's number 2300 [5] for smooth tubes. The Reynold's number together with the simulation results both suggest that the inertia terms in the Navier Stokes equations are playing a minor role except in improving the convergence. The flow can therefore be assumed laminar over the whole pressure range.

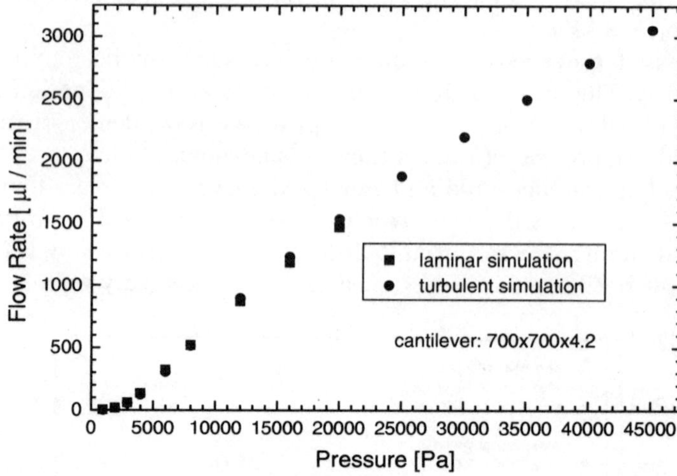

Figure 4.19: Turbulent and laminar simulation results

It is obvious from the discussion above that the coupled FEM/CFD simulation is a useful tool to solve simple fluid-structural problems. This can be done by combining Ansys 5.4 and CFX Flow3D 4.1 [39] or using the two packages Ansys 5.4 and Ansys CFD [40]. With such tools the development time of cantilever valves with specified characterisations can be dramatically reduced, avoiding the time consuming need for batch processing in microfabrication facilities.

More complex problems, like simulating a whole micropump, have to be solved with a system which couples mechanics and fluidics mathematically. Such a system has been presented at Transducers'97, Chicago [41]. The system consists of the programs FEM STRESS and CFD-ACE which are linked internally, and not externally as discussed before.

4.6 Analytical Approximations

Another possible way to model complicated structures in microfluidics is by using differential equations to describe the device approximately. This technique was used by Zengerle [43] when he analysed a micropump. In this section we will show how the technique can be used to model valves, micropumps and flow sensors.

4.6.1 Micromachined Cantilever Valves

We will first describe how to set up the differential equation for a microvalve and a micropump. However, quite often the solution to such approximate equations is not very accurate. Therefore it is necessary that some initial modelling be done with finite element programs. In the case of the microvalve we will demonstrate how to perform this task.

4.6.1.1 Simple Analytical Approach for the Volume Flow

A simple approach to estimate the fluid flow through a cantilever valve was used by Tiren [44]. Here, we have slightly modified this approximation with the assumption that the pressure for the cantilever valve is constant in the region between root end and duct end (L_{D2} in figure 4.20) and zero elsewhere. The flow is assumed to be laminar in a slit of uniform height (δ_{av}), i.e. the average between duct end (L_{D2}) and free end (L) of the cantilever.

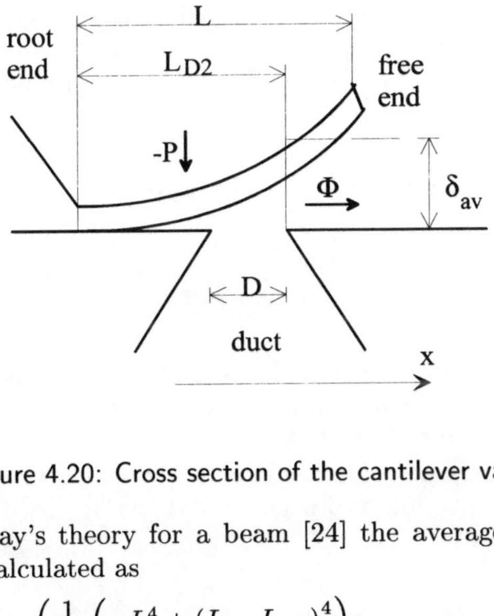

Figure 4.20: Cross section of the cantilever valve

Using Macaulay's theory for a beam [24] the average deflection as described above is calculated as

$$\delta_{av} = \frac{p\,W}{2\,E\,I} \left(\frac{1}{24}\left(-L^4 + (L - L_{D2})^4\right) + \frac{1}{6}L^3\,L_{D2} - \frac{1}{24}L_{D2}^4 \right) \tag{4.35}$$

where E is Young's modulus and $I = \frac{1}{12}W \cdot t_B^3$ is the second moment of cross sectional area about the neutral axis. L, W and t_B are equal to the length, width and thickness of the beam and p is the applied pressure. The fluid flow Φ_v of an incompressible liquid through a rectangular slit [5] defined by the height equal to the average deflection of the beam is

$$\Phi_v = -\frac{\delta_{av}^3\,W}{12\,\mu}\frac{\partial p}{\partial x} \tag{4.36}$$

where δ_{av} and W are the height and width of the slit, μ is the dynamic viscosity and $\partial p/\partial x$ is the pressure gradient. Because of the small dimensions the flow is assumed laminar, the width W of the flow is estimated to be 3 times the duct width ($W_D = 100\ \mu$m) and the pressure gradient for the flow is simplified to $\frac{p}{L-L_{D2}}$. Combining both equations and solving for the fluid flow Φ_v gives:

$$\Phi_v = -\frac{\left(\frac{1}{24}\left(-L^4 + (L - L_{D2})^4\right) + \frac{1}{6}L^3\,L_{D2} - \frac{1}{24}L_{D2}^4\right)^3}{96\,E^3\,I^3\,\mu} \cdot \frac{W^4\,p^4}{L - L_{D2}} \tag{4.37}$$

Using Young's modulus for (100) silicon in <110> direction $E = 168$ GPa, the dynamic viscosity of water $\mu = 1.002 \times 10^{-3}$ Pa s and the dimensions of the cantilever $L = 700\ \mu$m, $L_{D2} = 400\ \mu$m, $W = 300\ \mu$m and $t_B = 4.2\ \mu$m, the fluid flow for a pressure of 3000 Pa is calculated to be 131 μl min^{-1} which is more than 2 times larger than the measured valve (see figure 4.22). This can be explained by the fact that this approach does not couple deflection and flow equations properly. Therefore, equation 4.37 is not suitable for calculating the flow rate.

4.6.1.2 Coupled Analytical Approximation

A more exact analytical approach was undertaken by Park [45] and Ulrich [40]. Both derived two coupled differential equations for the deflection of the cantilever and the volume flow and solved the resulting equations numerically. This approach is used here as well. To simplify the three-dimensional problem, a two-dimensional model as seen in figure 4.21 is used. The equations of Park are adapted to the problem of a cantilever valve and solved with iterative coupling as adopted by Ulrich.

The starting point for the bending of the cantilever is the differential equation relating deflection δ of a beam and the applied force $f_W(x)$ on an element of width W:

$$E\,I\,\frac{\mathrm{d}^4\delta}{\mathrm{d}x^4} = f_W(x) \tag{4.38}$$

where I is the second moment of cross sectional area about the neutral axis. For a uniform beam of rectangular cross sectional area this is $I = \frac{1}{12}W \cdot t_b^3$. Integrating equation 4.38 four times and applying the boundary conditions of a cantilever beam, i.e.

$$\frac{\mathrm{d}\delta(x=0)}{\mathrm{d}x} = 0; \quad \delta(x=0) = 0; \quad M(x=0) = -\int_0^L f_W(x)\cdot x\,\mathrm{d}x; \quad F_W(x=0) = -\int_0^L f_W(x)\mathrm{d}x \tag{4.39}$$

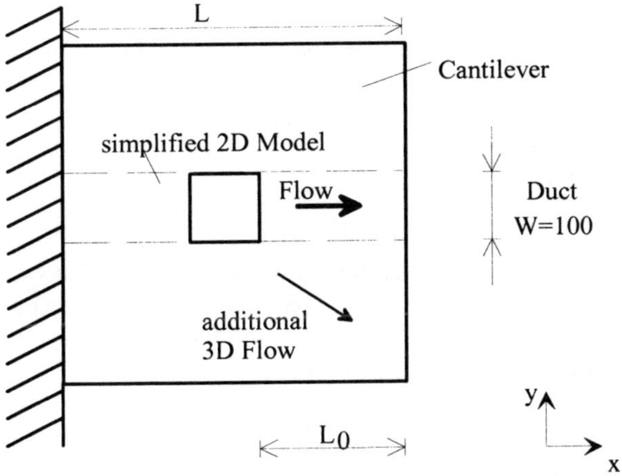

Figure 4.21: Top view of the cantilever sitting on a duct

for deflection, bending moment M and force $F_W = p(x) \cdot W$ on an element of width W gives the first of the two equations relating the pressure $p(x)$ and the deflection $\delta(x)$:

$$\begin{aligned}\delta(x) &= \frac{1}{EI} \cdot \left(\int_0^x \int_0^\varsigma \int_0^\varphi \int_0^\zeta f_W(\xi)\, \mathrm{d}\xi\, \mathrm{d}\zeta\, \mathrm{d}\varphi\, \mathrm{d}\varsigma \right. \\ &\left. -\frac{1}{6}x^3 \int_0^L f_W(\xi)\, \mathrm{d}\xi + \frac{1}{2}x^2 \int_0^L f_W(\xi)\xi\, \mathrm{d}\xi \right)\end{aligned} \tag{4.40}$$

To reach a coupled solution method an equation for the fluid flow beneath the cantilever is necessary. Assuming steady, laminar, incompressible flow at a low Reynold's number the derivation of such an equation is started with Newton's viscosity law:

$$\frac{\partial p}{\partial x} = \mu \frac{\partial^2 u}{\partial z^2} \tag{4.41}$$

which relates the pressure gradient of p in x-direction and the flow velocity u in x-direction, with μ as the dynamic viscosity. Integration and employment of the boundary conditions ($u_{z=0} = 0;\ \ u_{z=\delta(x)} = 0$) yields the velocity field:

$$u(x,z) = \frac{1}{2\mu} \frac{\partial p}{\partial x} \left(z^2 - \delta(x) z\right) \tag{4.42}$$

Two further integrations along the y and z axes give the flow rate Φ_v below the cantilever for a given height distribution at a specific location x:

$$\Phi_v = \int_0^W \int_0^{\delta(x)} u(x,z)\, \mathrm{d}z\, \mathrm{d}y = \frac{W}{12} \left(-\frac{\mathrm{d}p}{\mathrm{d}x} \delta^3(x) \right) \tag{4.43}$$

As the equation of continuity of the flowrate has to be fulfilled, Φ_v given by equation 4.43 is true for all x. Rearranging, integrating with respect to x and inserting the boundary condition ($p_{x=L} = 0$) yields the pressure as a function of the x-direction:

$$p(x) = \frac{12\,\mu\,\Phi_v}{W} \int_x^L \frac{1}{\delta^3(x)}\,\mathrm{d}x \tag{4.44}$$

Equation 4.44 is the second equation necessary for the coupled solution method. However, as mentioned before, this equation is only valid for the x-coordinates from the edge of the duct to the free end of the cantilever. Therefore, some additional assumptions have to be made:

- The pressure in the whole duct area is equal to the maximum pressure at $x = L - L_0$ from equation 4.44.
- From CFD simulations of a three-dimensional cantilever the pressure distribution between the root end of the cantilever and the duct was found to be approximately $\frac{2}{3}$ of the maximum pressure in the duct area for a cantilever of 700 μm sidelength and about $\frac{1}{3}$ for the 1000 μm-long cantilever.

Unlike the problem of Ulrich [40] and Park [45], some additional three-dimensional fluid flow occurs apart from the main liquid stream under the centre of the valve (see additional 3D flow in figure 4.21). This is accounted for by the multiplication of the pressure field and the amount of fluid flow by two constants:

- The pressure over the width is not constant but drops quickly to zero. In order to achieve this behaviour with the beam-model (equation 4.40) some FEM simulations with Ansys 5.4 have been made and a linear behaviour between the correct pressure drop and the constant pressure model over the width has been established. For the cantilever of 700 μm (1000 μm) width, the pressure in the beam model has to be multiplied by a constant factor of 0.56 (0.25) to achieve the same deflection behaviour as the three-dimensional finite element beam.
- A more complicated consideration is the amount of flow which travels under the central part of the cantilever valve, i.e. only over a width equal to that of the duct. From flow simulations with a CFD program a nearly linear behaviour of the percentage of flow that travels through this part could be determined for maximum deflections greater than 10 μm. The factor for the flow rate through the whole valve can be described by:

  $$\Phi_{\text{centre}} = \Phi_{\text{full}} \cdot \left(-0.02 + 0.0071 \cdot \left(\delta_{\text{max}} \cdot 10^{-6}\text{m}^{-1}\right)\right)$$

  with the maximum deflection δ_{max} at the tip of the cantilever.

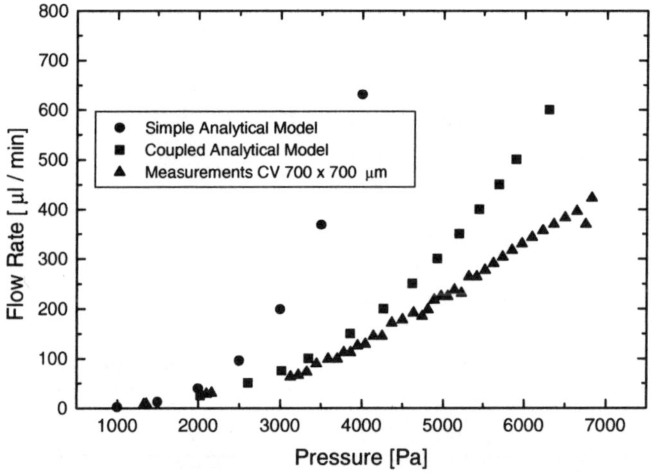

Figure 4.22: Flow characteristics from measurements, simple analytical model and coupled analytical model

In our case, the solution of equations 4.40 and 4.44 with the above simplifications have been obtained with a coupled numerical approach written in the C programming language. The integrals are solved with the Riemann Sum [46] and the coupling is achieved by subdividing the problem into finite elements. Two arrays for deflection and pressure at all nodes are defined by the program and the volume flow Φ_v has to be specified by the user. In a similar way to the flow diagram of figure 4.15, the pressure array is calculated from equation 4.44 and used to get the deflection array from equation 4.40. This process is iterated until convergence is reached. The break-off condition was implemented by comparing the previous and new arrays. If the maximum relative error of this comparison is below a certain percentage (here: 5%) the iteration is stopped and the maximum pressure at $x = L - L_0$ printed. This maximum pressure was taken as the pressure difference applied at the cantilever valve. The results for the 700 μm-long cantilever are plotted in figure 4.22. A comparison of measurements and analytical approach shows that this model gives a reasonable description of the behaviour of the cantilever valve. However, due to the two-dimensional simplifications the error between measurements and calculations is still up to 50% and increases for higher pressure values.

4.6.2 Membrane-Actuated Pumps

4.6.2.1 Continuity Equation and Quasi-Static Model

The next problem to be considered here is the calculation of the behaviour of a micropump. At first we will describe a simple model which gives the quasi-static behaviour. This model includes the valve behaviour in a quasi-static state, where the flow can be calculated for a given valve deflection and pressure difference. It is expected that the flow will be reduced at high frequency because of the finite valve conductances. In this region the pressure will build up inside the chamber during the cycle and will not be released by outflow because of the short cycle time.

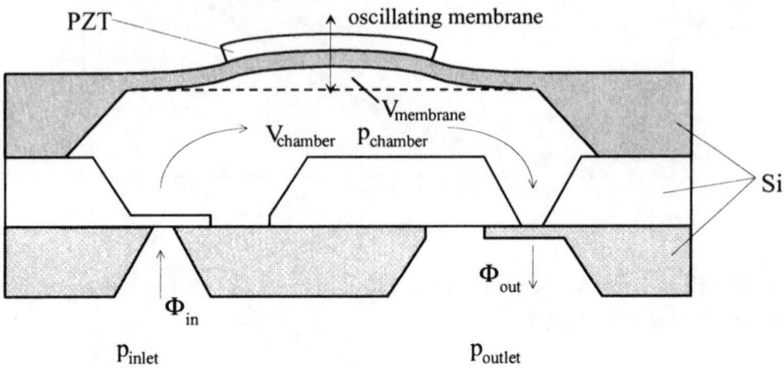

Figure 4.23: Flow within the micropump

The flow inwards and outwards of a micropump (figure 4.23) can be described with the equation of continuity, applied to the full volume V_{chamber} within the pump chamber [43]:

$$-\oint j\,\mathrm{d}A = \frac{\mathrm{d}m}{\mathrm{d}t} \tag{4.45}$$

with j as the net mass flow directed towards the outside of the pump chamber and m as the total mass of the liquid within. This implies that the change in mass within the pump chamber is caused by a mass flow through the valves.

Neglecting dynamic behaviour, the mass flow density j leaving the pump chamber is determined by the static flow $\Phi_{\text{in}}(\Delta p)$ and $\Phi_{\text{out}}(\Delta p)$ of the two passive check valves [43]:

$$-\oint j\,dA = \rho\left(\Phi_{\text{in}}\left(p_{\text{inlet}} - p_{\text{chamber}}\right) - \Phi_{\text{out}}\left(p_{\text{chamber}} - p_{\text{outlet}}\right)\right) \tag{4.46}$$

Additionally, the change in mass of the liquid within the pump chamber can be expressed by the change in volume below the membrane [43]:

$$\frac{\mathrm{d}m}{\mathrm{d}t} = \rho\frac{\mathrm{d}}{\mathrm{d}t}V_{\text{mem}}(t) \tag{4.47}$$

Inlet and outlet flow functions through the cantilever valves are identical, and represented with a cubic spline function $\Phi_{\text{cs}}(\Delta p)$. This function was derived from the previously simulated flow values of the cantilever valve. In the following equations p_{chamber} is described with the variable p and p_{inlet} as well as p_{outlet} are assumed to be constant.

Inserting equations 4.46 and 4.47 into the differential equation (equation 4.45) and rearranging gives [43]:

$$\frac{\mathrm{d}p}{\mathrm{d}t} = \frac{\Phi_{\text{cs}}\left(p_{\text{inlet}} - p\right) - \Phi_{\text{cs}}\left(p - p_{\text{outlet}}\right) - \frac{\partial}{\partial U}\left(V_{\text{mem}}\left(p, U\right)\right) \cdot \frac{\mathrm{d}U}{\mathrm{d}t}}{\frac{\partial}{\partial p}\left(V_{\text{mem}}\left(p, U\right)\right)} \tag{4.48}$$

Because of small deflections of the actuating membrane, a linear behaviour can be assumed. The formula for the displaced volume beneath the membrane is given in the following equation:

$$V_{\text{mem}}(U, p) = \delta_U \cdot U + \gamma_p \cdot p \tag{4.49}$$

where δ_U and γ_p are correlation coefficients calculated in chapter 3. By the use of this equation, further simplification of equation 4.48 is possible:

$$\frac{\mathrm{d}p}{\mathrm{d}t} = \frac{\Phi_{\text{cs}}\left(p_{\text{inlet}} - p\right) - \Phi_{\text{cs}}\left(p - p_{\text{outlet}}\right) - \delta_U \cdot \frac{\mathrm{d}U}{\mathrm{d}t}}{\gamma_p} \tag{4.50}$$

This equation describes the quasi-static behaviour of a micromachined fluid pump. This equation does not include performance-reducing effects like bubbles in the pump chamber or valve deformation and should give an upper bound to the performance of the actual pump.

It is not possible to solve the differential equation analytically. Therefore a numerical solution is used. This can be done using two different approaches. The first one is based on Maple V.4[3] and uses a Runge-Kutta solver of fourth-fifth order. As Maple is an interpretive type of language, the calculations are rather time consuming. As an alternative a Runge-Kutta solver of fourth order was written in the programming language C and the calculation times were reduced dramatically.

The solution of the differential equation is a function of the pressure $p(t)$ with respect to the time. The net transient flow through the pump is then obtained by inserting the pressure function into the flow function of the outlet valve $\Phi_{\text{cs}}\left(p(t) - p_{\text{out}}\right)$. Integration of the flow function for both valves over one actuation cycle at the steady state and multiplication with the frequency yields the pump rate, a figure of merit for the performance evaluation of the micropump.

Results

First we look at the pump rates and the pressures achieved. Several solutions of equation 4.50 were obtained, where frequency, back pressure and actuator geometry have been changed. The two actuator sizes were 7 × 3 ×

[3]Waterloo Maple Inc., Waterloo, Ontario

0.1 mm^3 ($\delta_U = 3.65 \cdot 10^{-13}$ m^3 V^{-1} , $\gamma_p = 2.78 \cdot 10^{-16}$ m^3 Pa^{-1}) and 6.25 × 3.25 × 0.25 mm ($\delta_U = 1.04 \cdot 10^{-13}$ m^3 V^{-1} , $\gamma_p = 0.969 \cdot 10^{-16}$ m^3 Pa^{-1}). The actuation voltage was assumed sinusoidal, and calculations were made for two different voltages.

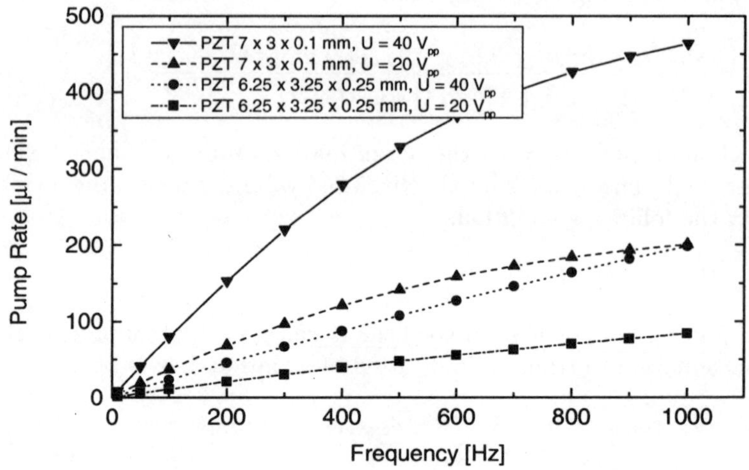

Figure 4.24: Frequency behaviour of the pump as calculated by the quasi-static simulator

Figure 4.24 shows the pump rate as a function of frequency. The pump rate increases linearly for small frequencies. For the 7 × 3 mm^3 actuator the response becomes sublinear and finally saturates at higher frequencies.

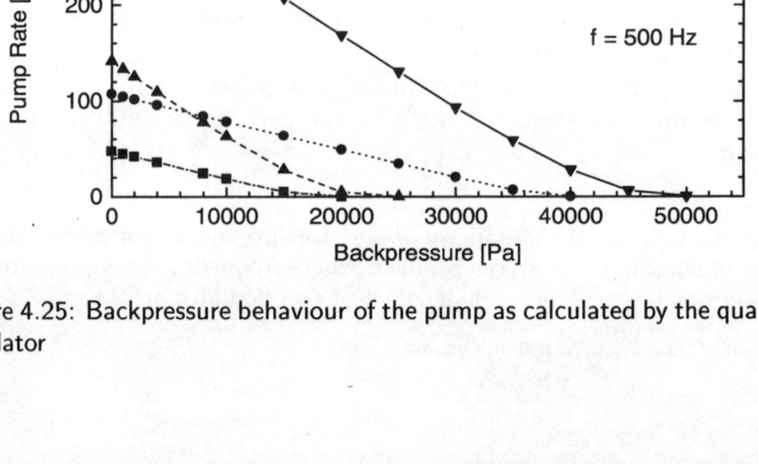

Figure 4.25: Backpressure behaviour of the pump as calculated by the quasi-static simulator

No frequencies above 1 kHz have been modelled, as this simulator is only valid in the quasi-static case. The dynamic behaviour of both actuation membrane (f_{resonant} = 28249 Hz with one side exposed to water) and cantilever (f_{resonant} = 1725 Hz submerged in water) was not taken into account. The inlet and outlet pressure was assumed to be zero for all simulations.

An important result of the model is that the pump rate is higher for the thinner PZT actuator. As the optimisation of the piezoelectric thickness has shown a similar effect on the volume displacement (cf. chapter 3), this result is as expected.

The modelled behaviour of the backpressure dependence for a micromachined pump is depicted in figure 4.25. The actuation frequency for all simulations was set constant at 500 Hz. A nearly linear drop of the pump rate with respect to the applied backpressure at the outlet was observed.

Figure 4.26 shows the transient behaviour of the micropump. The chamber pressure, the flow through the inlet valve and the displaced volume beneath the membrane are given for different frequencies.

The frequencies used were 10 and 1000 Hz. Both inlet and outlet connections were modelled at zero pressure ($p_{in} = p_{out} = 0$). In all cases, the sinusoidal voltage across the piezoelectric layer gives a sine-shaped curve of the chamber pressure with a phase shift of -$\frac{1}{4}\pi$ for 10 Hz increasing to -$\frac{3}{8}\pi$ at 1 kHz. The sine shape is more developed for higher frequencies. The valve was assumed leak free, i.e. with a zero leakage rate in the reverse direction. This can be seen in the graphs of the flow rate through the inlet, as zero flow was obtained for positive chamber pressures.

For small frequencies, the membrane volume changes in phase with the actuation, showing zero displacement when the chamber pressure is maximum. This can be explained by the fact that the membrane develops the largest kinetic energy when passing through zero displacement. At the time of the largest kinetic energy, the largest change in volume below the membrane takes place, which in turn leads to a maximum in the chamber pressure. Again, a shift in the phase angle can be noted for a change to higher frequencies.

The results of figure 4.27 were produced when a backpressure was introduced. The frequency was kept constant at 500 Hz. This shows chamber pressure, flow through the inlet valve and volume displacement at both zero and 20 kPa backpressure. By the use of 20 kPa backpressure, an asymmetry across the centreline for pressure and volume displacement can be noticed and the sine shape of both graphs is rather distorted with backpressure. The maximum flowrate is similar in both cases, but the effect of the backpressure is to shorten the flow period and so produce sharper peaks.

The transient behaviour of the micropump could not be verified with measurements on real devices. Only the time averaged values of the pump rate were measured at different frequencies and backpressures. These results are given in section 4.6.2.4.

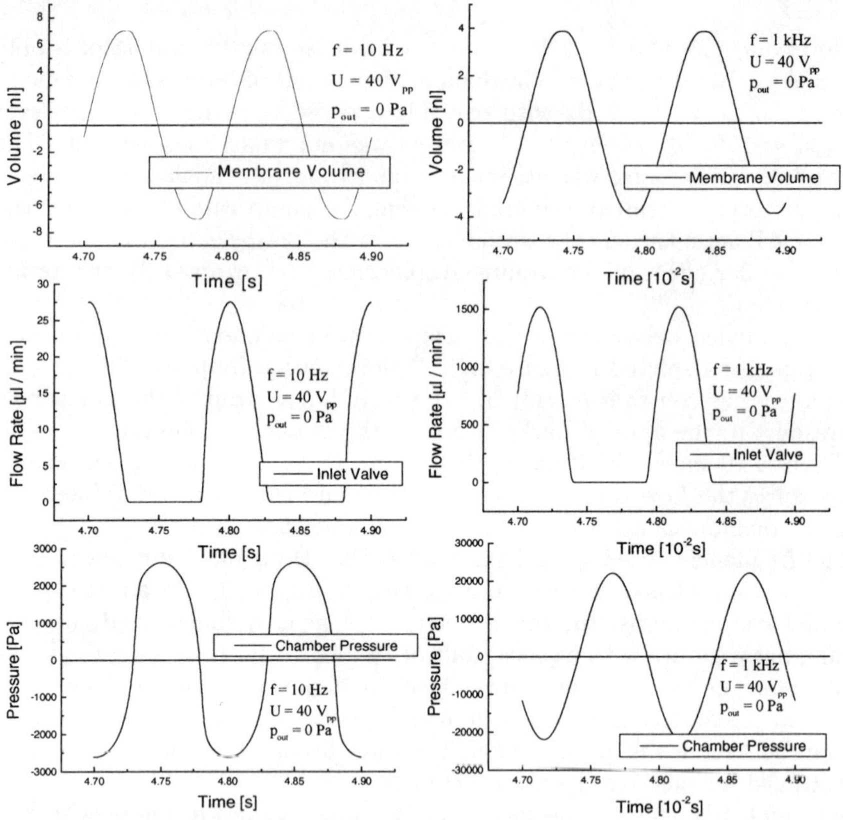

Figure 4.26: Transient simulation results for the micropump actuated at different driving frequencies and zero backpressure

4.6.2.2 Pump Model with Dynamic Valve Approximation

As already mentioned, the models presented above are quasi-static simulations, as they do not include any dynamic effects of membrane and cantilever valves. In this section, dynamic effects are studied and incorporated into the equation describing the flow through the micromachined pump.

At first, the resonance frequencies of both the membrane actuator and the cantilever valve have to be investigated. Apart from FEM studies for both mechanical parts, experimental studies of the cantilever valves resonating in air as well as water have been carried out.

First, the natural frequencies of the cantilever within the valve were determined by using the time independent solution of the wave equation for the flexural wave motion of a uniform cantilever (section 3.1). The density of silicon is $\rho_{Si} = 2.33 \cdot 10^3 \text{ kg m}^{-3}$ and the Young's modulus in <110> direction is $E_S = 168 \cdot 10^9$ Pa. The results were confirmed using Ansys 5.4. The

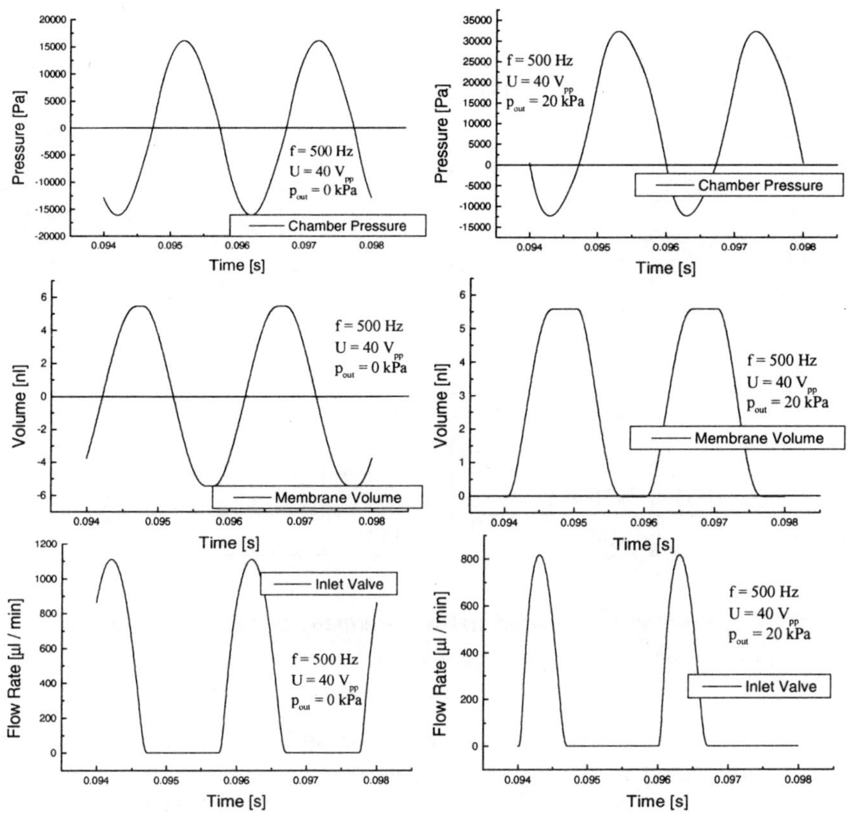

Figure 4.27: Transient simulations of the micropump at 500 Hz for 0 and 20000 kPa backpressure

same matrix components of the Young's modulus were employed as for the simulation of the passive valve (cf. section 3.1). In order to compare theory with practice, the resonance behaviour of the cantilevers has been measured with a laser interferometer. Here, the cantilevers were mounted on top of a piezoelectric actuator and the response of the tip has been measured at different driving frequencies. From the data, the resonance frequency and the damping factor Q could be obtained. All values are summarised in table 4.1 for operation in air.

Note that, according to three-dimensional simulations with Ansys 5.4, additional modes are observable. When the cantilevers operate in liquids with high viscosity, damping occurs which decreases the resonance frequency. This effect can be mathematically described with an 'added' mass m_a (section 3.1). With liquid densities of $\rho_{\mathrm{ethanol}} = 0.789 \cdot 10^3$ kg m^{-3} and $\rho_{\mathrm{water}} = 0.998 \cdot 10^3$ kg m^{-3}, resonance frequency and quality factor Q (resonance frequency divided by bandwidth) in water can be compared with experiments. Here,

mode number	1	11	2	21	22	3
ν_i [Hz], 700 μm (eqn 3.33)	11757	-	73685	-	-	206310
ν_i(Ansys) [Hz]	10644	31148	66050	89513	107972	188332
ν_i(measured) [Hz]	9920	-	$62.6{\cdot}10^3$	-	-	-
Q (measured)	96.3	-	334	-	-	-
ν_i [Hz], 1000 μm (eqn 3.33)	5761	-	36105	-	-	101092
ν_i [Hz], 1000 μm1 (Ansys)	5214	15242	32483	43845	53002	95103
ν_i(measured) [Hz]	4560	-	$29.3{\cdot}10^3$	-	-	-
Q (measured)	84	-	242	-	-	-

Table 4.1: Measured and simulated natural frequencies of micromachined cantilevers operated in air

the measurement setup consisting of cantilever and actuating piezoelectrics needs to be immersed in the liquid, and the displacement at the tip can be measured. The following first natural frequencies ν_1 of submerged cantilevers were obtained as shown in table 4.2.

cantilever length & width	0.7 mm	0.7 mm	1 mm	1 mm
liquid	water	ethanol	water	ethanol
ν_1 (eqn 3.35 – 3.36) [Hz]	2182	2442	899	1008
ν_1 (measured) [Hz]	1725	-	1172	-
Q (measured)	8.5	-	13	-

Table 4.2: Natural frequencies of micromachined cantilevers submerged in a fluid

This study indicates that the quasi-static behaviour of the pump is not valid at frequencies above 500 Hz because of dynamic effects of the cantilever valve. Within the valve, the cantilever will be damped when it moves close to the valve seat. This has not been taken into account in the modelling, and could not be measured due to a rather coarse resolution of the laser interferometer. However, this narrow gap will result in additional damping of the cantilever, reducing the natural frequency and thus further reducing the valid frequency range for the pump simulation.

Similar estimates have to be made for the membrane actuator. Because the PZT does not cover the whole membrane, no exact analytical calculation is possible. Ansys 5.4 can therefore be used in order to determine the natural frequencies of the bimorph, which are shown in table 4.3:

mode number	1	2	3	4	5	6
ν_i [Hz] (L=7 mm)	44860	61616	91551	106343	121837	133923
ν_i [Hz] (L=6.25 mm)	47709	58638	82437	123033	137751	169896

Table 4.3: Ansys 5.4 calculations of the natural frequencies of cantilevers submerged in a fluid

As the membrane behaviour is dominated from the corner region, it is expected that a plate of silicon covered with PZT only at the centre (as is the case for the membrane driving the micropump) can be approximated by the silicon part only. Thus, it is possible to estimate the resonance frequency with equation 3.46. For this the following values are taken: L = 8 mm and h = 0.07 mm are the length and thickness of the membrane and E_{Si} = 168 GPa, ν_{Si} = 0.064, ρ_{Si} = 2.33 $\cdot 10^3$ kg m^{-3} are the material coefficients (Young's modulus and Poisson's ratio in <110> direction [16] and density) of silicon. If the values are inserted into equation 3.46, a natural frequency of 44040 Hz is obtained corroborating the Ansys 5.4 results.

In order to calculate the shift in the natural frequencies for rectangular plates with one side exposed to a fluid, equation 3.35 was employed. The original mass is given by the summation of the mass for the Si layer and the PZT layer, where $\rho_{PZT} = 7.45 \cdot 10^3\,\mathrm{kg\,m^{-3}}$ [17]. The first natural frequency for water and ethanol as the liquid is shown in table 4.4.

fluid	water	ethanol
ν_1 (L_{PZT}=7 mm) [Hz]	28249	30223
ν_1 (L_{PZT}=6.25 mm)[Hz]	35966	37680

Table 4.4: First natural frequency of a diaphragm in contact with liquid

All values are more than a factor of 10 larger than the maximum frequency of the quasi-static simulation for the micromachined pump. Therefore, no interference with the dynamic behaviour of the actuator is expected at low frequencies.

In order to incorporate the dynamic valve behaviour into the simulation of the micropump, a new model is now derived. Here, the differential equation 4.50 is modified as described by Ulrich [40]. The simulator derived here uses a lumped model of the cantilever tip vibrating against the valve seat. An impact model is used to describe the behaviour of the tip of the cantilever. The model used is based on the deflection at the end of the cantilever and is as follows:

- For $y > 0$: The normal differential equation for a damped oscillation is valid. When the term $\hat{y}_{qs}(p)$ is assumed for the deflection of the cantilever at the applied pressure p in the quasi-static case, i.e. when both time derivatives are zero, the ordinary differential equation can be transformed to:

$$\frac{\mathrm{d}^2 y(t)}{\mathrm{d}t^2} + \frac{\omega_0}{Q} \cdot \frac{\mathrm{d}y(t)}{\mathrm{d}t} + {\omega_0}^2 \cdot y(t) = {\omega_0}^2 \cdot \hat{y}_{\mathrm{qs}}(\mathrm{p}) \sin(\omega t) \tag{4.51}$$

- For $y = 0$: The cantilever impacts on the valve seat and bounces off with negative velocity. This behaviour can be described by:

$$\frac{\mathrm{d}y(t+\tau)}{\mathrm{d}t} = -r\frac{\mathrm{d}y(t)}{\mathrm{d}t}; \qquad \tau \to 0. \tag{4.52}$$

  with r as the reflection factor and τ as an infinitely small time delay between impact and reflection. Depending on the value of r, the impact can be described from fully elastic ($r = 1$) to fully inelastic ($r = 0$).

For the simulations of the dynamic behaviour of the valve on its own, the differential equation 4.51 and equation 4.52 are used with a sine-shaped pressure actuation. The numerical solution of this system of equations leads to the deflection $y(t)$ of the valve. From this solution and the use of the pressure function, the volume flow through the cantilever valve and the net flow rate are obtained. The model can be solved with the mathematical program Maple V.4 or a custom built C program. The results obtained are given for three different values of r (0, 0.5 and 0.8). Cubic spline fitting is employed for the simulation results of the valve with cantilevers of 700 μm $\times$ 700 μm size. The simulation was done with a Runge-Kutta solver of fourth order.

Figure 4.28 shows the simulated volume flow through the valve at a pressure of 4000 Pa and actuation frequencies up to 4000 Hz. The model for $r = 0$ shows a weak resonance behaviour for the valve at approximately the resonance frequency of the free cantilever, which was specified as 1745 Hz. A stronger resonance occurs at double the resonance frequency when the reflection factor is increased. After the resonance, the preferred flow direction through the valve changes, and the flow is mainly in a backward direction.

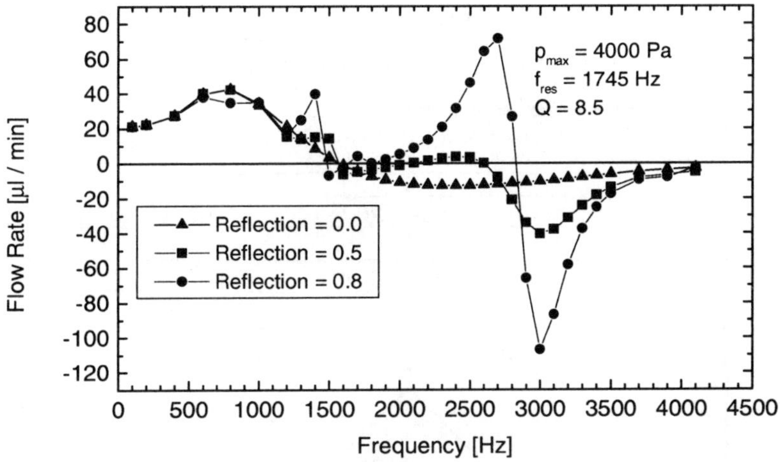

Figure 4.28: Dynamic characteristics of the cantilever valve

This is due to a phase shift of the cantilever deflection compared with the applied pressure.

When the dynamic valve model is included in the micropump simulator, a system of three differential equations has to be solved:

$$\begin{aligned}
\frac{\mathrm{d}p}{\mathrm{d}t} &= \frac{\Phi_{\text{inlet}}\left(p_{\text{inlet}} - p, y_1\right) - \Phi_{\text{outlet}}\left(p - p_{\text{outlet}}, y_2\right) - \delta_U \cdot \frac{\mathrm{d}U}{\mathrm{d}t}}{\gamma_p} \\
\frac{\mathrm{d}^2 y_1(t)}{\mathrm{d}t^2} + \frac{\omega_0}{Q} \cdot \frac{\mathrm{d}y_1(t)}{\mathrm{d}t} + {\omega_0}^2 \cdot y_1(t) &= {\omega_0}^2 \cdot \hat{y}_{\text{qs}}\left(p_{\text{inlet}} - p(t)\right) \\
\frac{\mathrm{d}^2 y_2(t)}{\mathrm{d}t^2} + \frac{\omega_0}{Q} \cdot \frac{\mathrm{d}y_2(t)}{\mathrm{d}t} + {\omega_0}^2 \cdot y_2(t) &= {\omega_0}^2 \cdot \hat{y}_{\text{qs}}\left(p(t) - p_{\text{outlet}}\right)
\end{aligned} \tag{4.53}$$

where y_1 and y_2 are the cantilever deflections of inlet and outlet valve, respectively. Again, a solution is possible with Maple V.4 as well as a custom built C program. Both use the results of the coupled valve simulation.

To include the impact model, the reflection behaviour described above was used, when the cantilever hits the valve seat. A flow diagram for the C program is shown in figure 4.29.

The dynamic simulations of the pump are depicted in figure 4.30. For the model, a 7 × 3 × 0.1 mm-sized actuator ($\delta_U = 3.65 \cdot 10^{-13}$ m^3 V^{-1} , $\gamma_p = 2.78 \cdot 10^{-16}$ m^3 Pa^{-1}) was employed. The resonance frequency of the pump is at about the resonance frequency of the valve for zero reflection. For a reflection coefficient of 0.8 the valve seat seems to nearly double the resonance frequency of the valve. This is due to the fact that the reflection of the valve is in phase with the pressure, allowing larger deflections of the cantilever. At frequencies higher than the resonance, the pump starts to reverse its pumping direction, as was the case for the valve only. This effect

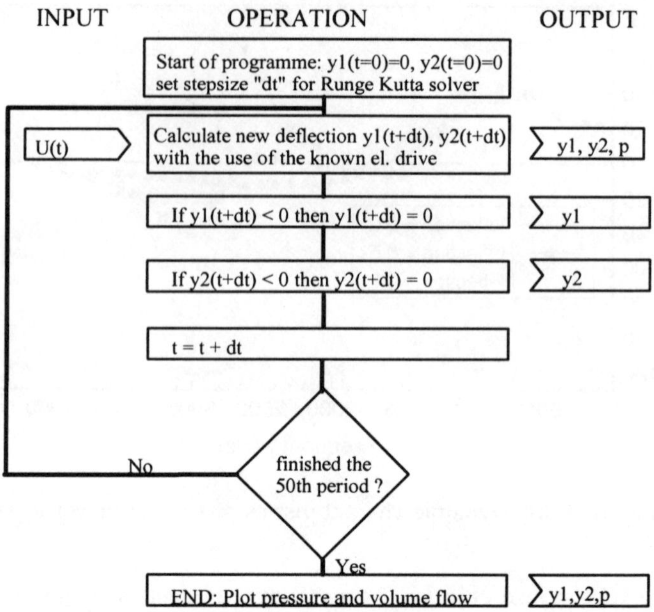

Figure 4.29: Simplified flow diagram of the micropump simulator using a lumped valve model

has been observed by Zengerle [47]. For 5500 Hz and more the pump rates for all reflection coefficients are similar. At this point, the cantilever does not touch the valve seat any more and reflection is irrelevant. A comparison of this model with the measurements of the pump designed by the authors will be given in section 4.6.2.4.

4.6.2.3 Pump Model with Behavioural Model of the Valve

The drawback with the dynamic simulator above is its dependence on the reflection factor, which is difficult to measure. Therefore, a different simulator has been developed for the valve. The approach followed here was to include a behavioural model of the valve.

As the cantilever gets closer to the valve seat, additional squeeze film damping occurs. This is due to the effect that the fluid flow is no longer determined by the flow around the cantilever but by the flow between cantilever and valve seat.

The volume between cantilever and valve seat is simplified to a linear dependence on the deflection y at the tip of the cantilever: $V = 0.5 \cdot A \cdot y$ with A as the area of the cantilever (see figure 4.31). The volume V beneath the cantilever will change by $\mathrm{d}V$ if the pressure difference p between inlet

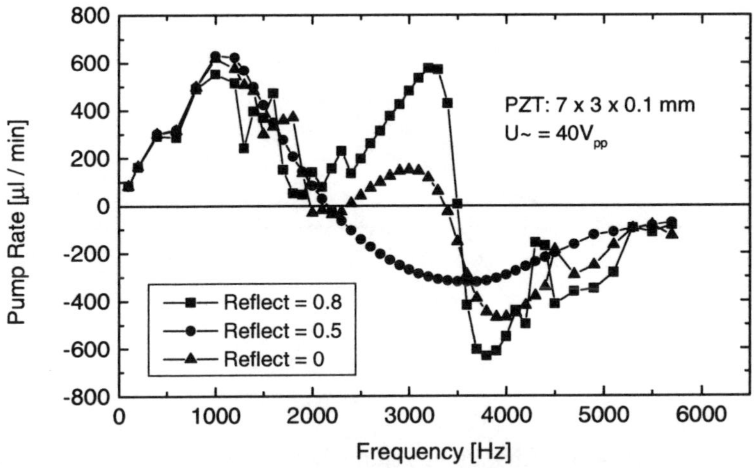

Figure 4.30: Frequency dependent behaviour of the micropump - simulated with the impact model

and outlet of the cantilever is changed, as the static cantilever deflection is a function of the applied pressure. The change in volume can only occur due to a flow of the liquid into this volume. However, this flow Φ will restrict the speed of change in deflection. For simplification, the flow Φ was set to the flow through the cantilever valve at pressure p.
The behavioural model of the valve can be expressed analytically as follows:

- $\Phi = \Phi_{\text{valve}}\left(y(t), p(t)\right)$
- $dy = \min\left((0.5 \cdot A)^{-1} \cdot dt \cdot \Phi\left(y(t), p(t)\right), \quad d\delta(p(t))\right)$

  $\forall\, d\delta\left(p(t)\right) > 0$
- $dy = \max\left(-(0.5 \cdot A)^{-1} \cdot dt \cdot \Phi\left(y(t), p(t)\right), \quad d\delta(p(t))\right)$

  $\forall\, d\delta\left(p(t)\right) < 0$

with dy as the real change in the deflection of the cantilever within the differential equation and $d\delta(p(t))$ as the change in the deflection of the cantilever expected after a time long enough to reach the steady state. The terms $\min(a, b)$ and $\max(a, b)$ are used as operators for minimum or maximum value.
This model can be combined with the differential equation for the pump,

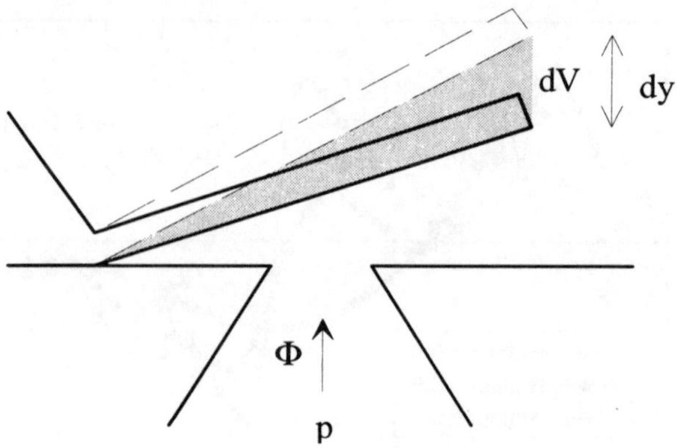

Figure 4.31: Change in volume below the cantilever

giving a system of three differential equations (cf. figure 4.23):

$$\frac{\mathrm{d}p}{\mathrm{d}t} = \frac{\Phi_{\text{inlet}}\left(p_{\text{inlet}} - p, y_1\right) - \Phi_{\text{outlet}}\left(p - p_{\text{outlet}}, y_2\right) - \delta_U \cdot \frac{\mathrm{d}U}{\mathrm{d}t}}{\gamma_p}$$

$$\frac{\mathrm{d}y_1}{\mathrm{d}t} = \begin{cases} \min\left(2A^{-1}\Phi_{\text{inlet}}\left(p_{\text{inlet}} - p, y_1\right), \frac{\mathrm{d}\delta(-p)}{\mathrm{d}p}\frac{\mathrm{d}p}{\mathrm{d}t}\right) \\ \quad \text{if } \mathrm{d}(-p) > 0 \\ \max\left(-2A^{-1}\left|\Phi_{\text{inlet}}\left(p_{\text{inlet}} - p, y_1\right)\right|, \frac{\mathrm{d}\delta(-p)}{\mathrm{d}p}\frac{\mathrm{d}p}{\mathrm{d}t}\right) \\ \quad \text{if } \mathrm{d}(-p) < 0 \end{cases} \tag{4.54}$$

$$\frac{\mathrm{d}y_2}{\mathrm{d}t} = \begin{cases} \min\left(2A^{-1}\Phi_{\text{outlet}}\left(p - p_{\text{outlet}}, y_1\right), \frac{\mathrm{d}\delta(p)}{\mathrm{d}p}\frac{\mathrm{d}p}{\mathrm{d}t}\right) \\ \quad \text{if } \mathrm{d}p > 0 \\ \max\left(-2A^{-1}\left|\Phi_{\text{outlet}}\left(p - p_{\text{outlet}}, y_1\right)\right|, \frac{\mathrm{d}\delta(p)}{\mathrm{d}p}\frac{\mathrm{d}p}{\mathrm{d}t}\right) \\ \quad \text{if } \mathrm{d}p < 0 \end{cases}$$

The system of differential equations is simultaneously solved for the variables p, y_1 being the deflection of the cantilever of the inlet valve and y_2 being the deflection of the cantilever of the outlet valve. Then the flow rates through both inlet and outlet valve are calculated from $\Phi(p, y)$ and the net mass flow at any instant is obtained from $\Phi_{\text{inlet}} - \Phi_{\text{outlet}}$. The pump rate can be obtained by integrating the net mass flow over one cycle and multiplying by the frequency.

Again, the system has been solved with a Runge-Kutta approach of fourth order. The coupled simulation results for the cantilever valve have been interpolated by a cubic spline function and entered in the simulator. The results for a 6.25 × 3.25 × 0.25 mm-sized PZT actuator on a 70 μm-thick silicon layer are depicted in figure 4.32. Three different actuation voltages (200 V_{pp}, 300 V_{pp} and 400 V_{pp}) have been taken for the simulation of the frequency dependence of the micromachined pump.

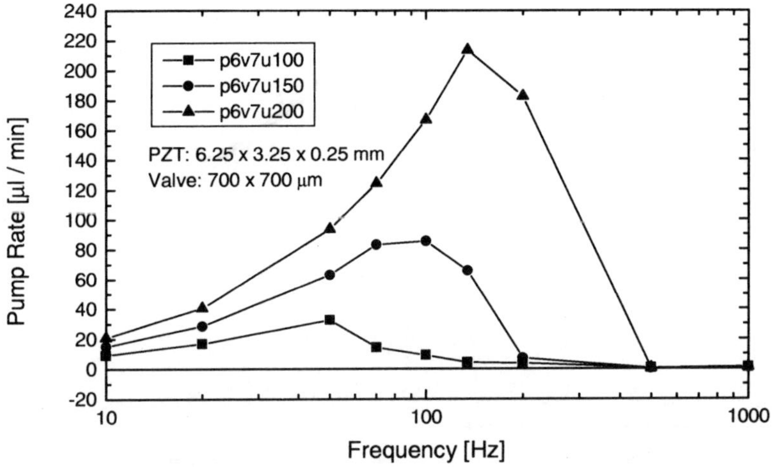

Figure 4.32: Frequency dependent behaviour of the micropump - simulated with the valve behavioural model

At low frequencies, the quasi-static simulations and the new behavioural simulations agree, but at frequencies between 70 and 200 Hz the valve behavioural model predicts a drop in the pump rate which is not seen with the quasi-static model. This drop becomes steeper and shifts to higher frequencies for larger actuation voltages. Beyond 400 Hz, no net pump rate is achieved. The transient behaviour of the pump chamber pressure, the cantilever deflection and the volume flow through the inlet valve are given in figure 4.33 for 20 Hz and 400 Hz actuation. The deflection of the cantilever at 400 Hz is shifted by approximately 0.5π with respect to the pump chamber pressure. This means that there are equal amounts of positive and negative flow through the valve, as can be seen in the volume flow of the inlet valve. Therefore only a small net flow is achieved at this frequency. At low frequencies both chamber pressure and cantilever deflection are in phase and a net flow is realised.

Unlike the quasi-static model, no sine-shape of the pressure is achieved at low frequencies. This is due to the behavioural part of the valve, allowing only "free" cantilever movement at reasonably large deflections from the valve seat.

The sharp decrease of the pump rate at around 200 Hz for a large actuation voltage could be an effect of the simple approximation of the volume below the cantilever. A more accurate but far more complicated model might be to include fluid streaming from both the inlet and the pump chamber to fill the volume below the cantilever. As before, a comparison of all models with the measurements of the pump will be given in section 4.6.2.4. However, only dynamic pump rates have been determined and no transient measurements of chamber pressure or valve flow rates have been made.

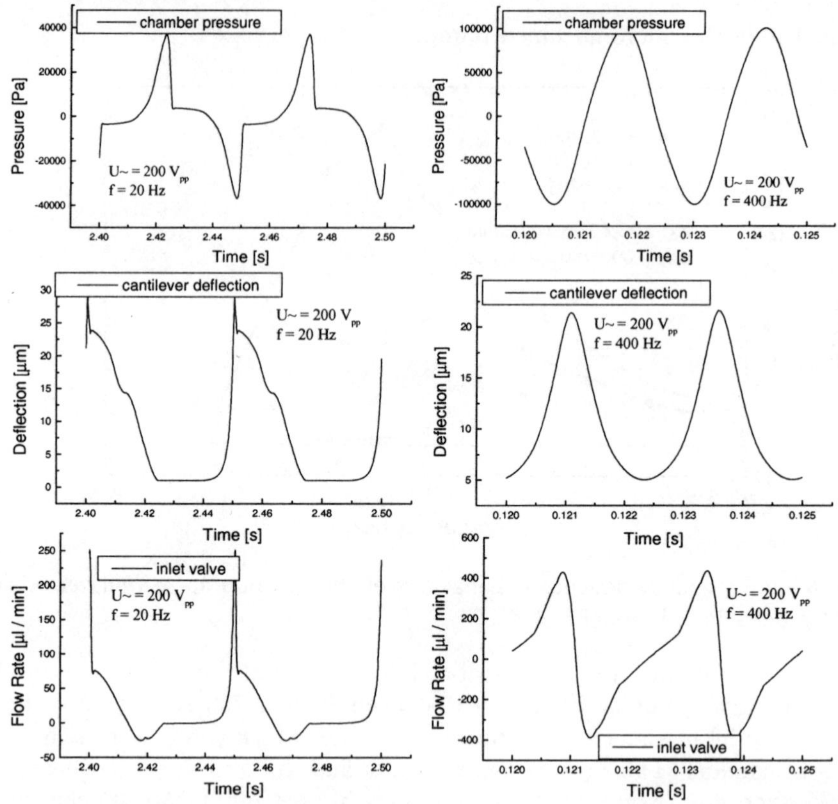

Figure 4.33: Transient simulations of the micropump at 20 Hz and 400 Hz using the behavioural model

4.6.2.4 Comparison of the different Dynamic Solvers

Now the different dynamic extensions for the simulator are compared, with a given actuator of $6.25 \times 3.25 \times 0.25$ mm size and the cantilever of 700 μm side length. The simulation results as well as the measurements on cantilevers of 700 μm side length are plotted in figure 4.34.

All simulations as well as the measurements start at around 20 $\mu\text{l}\,\text{min}^{-1}$ for 10 Hz. This is the quasi-static case as described before which does not depend on the valve behaviour but only on the membrane actuator performance. However, the difference between the models can already be seen at 50 Hz, where both the quasi-static and the impact model show a drastic increase in the pump rate in contrast to the descriptive behavioural model and the real device. Only the behavioural model is able to describe the actual pump rate reasonably accurately up to 135 Hz. At higher frequencies, however, the drop in the pump rate for the behavioural model is larger than shown in the measurements. This shows that the flow behaviour within the valve is more

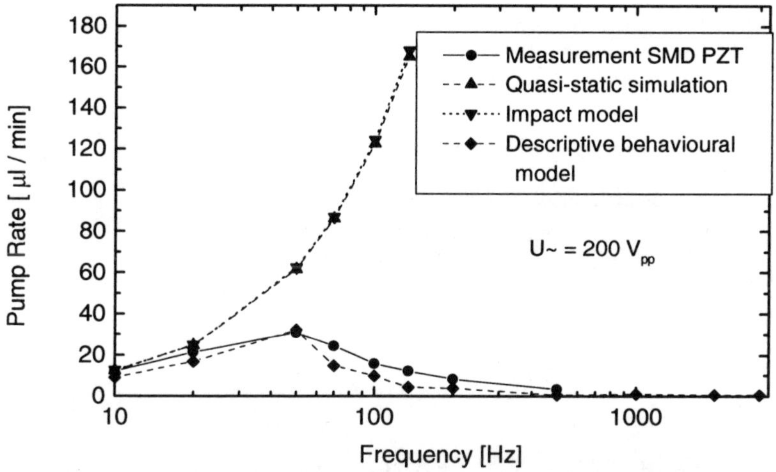

Figure 4.34: Comparison of the pump rate for measurements and simulations

complicated than assumed. Nevertheless, the descriptive model can predict the pump behaviour accurately enough to allow its use for pump development and improvement. As with all models, this can decrease the cost and time for development.

It is interesting to note that the pump with the larger cantilever (1000 μm) follows the quasi-static behaviour and the impact model for 700 μm cantilevers up to 70 Hz. Then the pump rates obtained from both models increase even further whereas the real pump shows a maximum at 135 Hz. The drop at this low frequency emphasises the point that the behavioural model is the best approach for pump simulations.

Zengerle considered additional losses in a micropump system caused by the liquid filled pipes connected to both inlet and outlet [47]. These losses include inertia loss of the accelerated/decelerated liquid within the pipe as well as fluid friction. Both these loss terms have been included in the quasi-static model and the impact simulator but neither shows a drop in the pump rate at frequencies as low as 100 Hz as occurred in the real device.

This has been experimentally verified as well. An increase of the length of the filled tubing at both inlet and outlet from 35 cm to 100 cm did decrease the flowrate at high frequencies by 15 % but this is not sufficient to explain the large reduction of the pump rate with respect to frequency.

4.6.3 Thermal Flowsensors

Another device that can be modelled with analytical equations is a microfluidic flowsensor based on heat flow. The principle of the flowsensor is depicted in figure 4.35. A heater (H) is located between two temperature sensors (T). If the flow is directed to the right, then the temperature sensor on the right hand side will measure a higher temperature than the sensor on the left hand side.

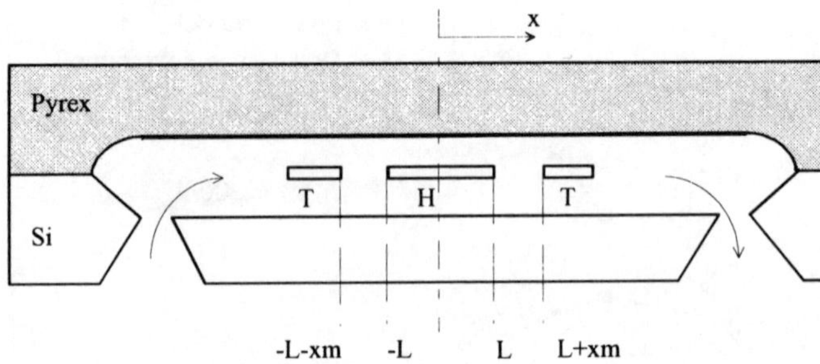

Figure 4.35: Flowsensor employing thermal conductivity measurement

In order to estimate the optimum operating range of this flowsensor, the necessary simulations are now described. The whole analytical model is based on lumped elements and simulates the heater and its surrounding [48]. A cross section of the heater and the equivalent thermal conductance elements is depicted in figure 4.36, where the G's represent the thermal conductance of the heater bridge and the surrounding gas. For simplicity the thermal resistors are assumed to behave linearly with temperature.

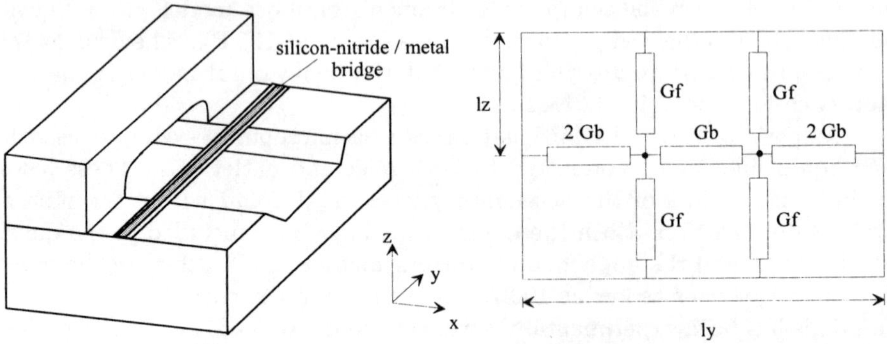

Figure 4.36: Cross section of the heater and equivalent thermal conductance circuit

Several assumptions have been made in order to simplify the analysis [48]:

- The thermal diffusion in the z-direction is much larger than the heat conduction through the bridge
- The thermal diffusion in the z-direction is linear

In order to establish a model for an infinitesimally thin element of the cross section of the channel, the heat current into and out of the element has to be considered. First, the heat power $\mathrm{d}P$ into an infinitesimally thin element $\mathrm{d}x$ with a cross sectional area A (figure 4.37) is dependent on the difference of the heat current density $j(x)$ flowing into the element at location x and out of the element at location $x + \mathrm{d}x$:

$$\mathrm{d}P = A\,(j(x) - j(x + \mathrm{d}x)) = \lambda_f\, A\, \frac{\mathrm{d}^2 T}{\mathrm{d}x^2} \mathrm{d}x \tag{4.55}$$

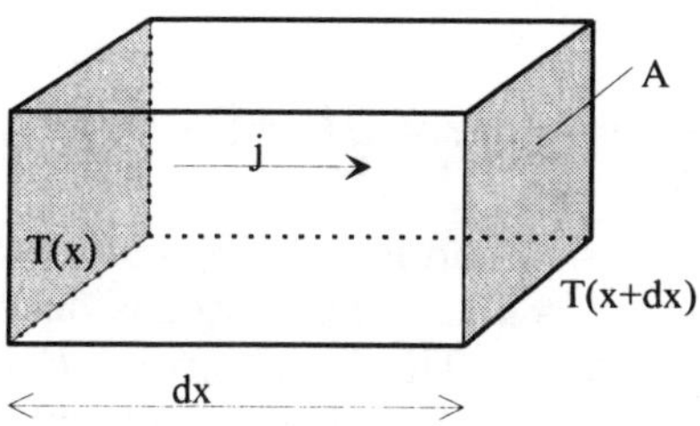

Figure 4.37: Simplified model for a thermal flowsensor

where λ_f is the thermal conductance of the fluid and T is the temperature. The second factor contributing to the heat power $\mathrm{d}P$ is forced convection due to a mass flow of the incompressible fluid:

$$\mathrm{d}P = c_f \cdot \frac{\mathrm{d}m}{\mathrm{d}t}\,(T(x) - T(x + \mathrm{d}x)) = -\rho_f\, c_f\, A\, v\, \frac{\mathrm{d}T}{\mathrm{d}x} \mathrm{d}x \tag{4.56}$$

where c_f is the fluid heat capacity, ρ_f the fluid density and v the velocity of the fluid. Another reason for a change in heat $\mathrm{d}P$ of the infinitesimally thin element is heat diffusion in both z-directions:

$$\mathrm{d}P = 2 \cdot l_y\, \mathrm{d}x \cdot j(z) = -2\lambda_f \frac{l_y}{l_z} T\, \mathrm{d}x \tag{4.57}$$

where l_y and l_z are the channel dimensions shown in figure 4.36.

As mentioned previously, the heat flow through the metal bridge is neglected. The thickness of the metal bridge (100 nm) is small compared with the channel width l_y (300 – 600 μm) and the difference in the thermal conductivity between the liquid (water) and the metal is not more than 50. Thus the heat flow through the fluid in the z-direction is greater than the heat flow through the metal bridge by a factor of 100.

A differential equation for a region away from the heater can now be formulated by combining all these heat flows [48]:

$$\frac{\mathrm{d}P}{\mathrm{d}x} = \lambda_f\, A\, \frac{\mathrm{d}^2 T}{\mathrm{d}x^2} - \rho_f\, c_f\, A\, v\, \frac{\mathrm{d}T}{\mathrm{d}x} - 2\lambda_f \frac{l_y}{l_z} T \tag{4.58}$$

For steady state, the gradient of the heat current is zero, and the differential equation has been solved as follows [48]:

$$\begin{array}{ll} T_1(x) = T_0 \cdot \exp\left(\chi_1(x+L)\right) & (x < -L) \\ T_2(x) = T_0 & (-L < x < L) \\ T_3(x) = T_0 \cdot \exp\left(\chi_2(x-L)\right) & (x > L) \end{array} \tag{4.59}$$

with the assumptions that the heater is located between $-L \le x \le L$ and that the temperature is constant throughout the heater ($T = T_0$ for $-L \le x \le L$). The dimensions of the flowsensor in the x-direction are shown in figure 4.35. The values of χ_1 and χ_2 are given by:

$$\chi_{1/2} = \frac{1}{2\,l_z\,A\,\lambda_f}\left(l_z\,v\,A\,c_f\,\rho_f \pm \sqrt{l_z\,A\left(l_z\,v^2\,c_f^2\,\rho_f^2\,A + 8l_y\,\lambda_f^2\right)}\right)$$

where χ_1 takes the '+'-sign and is positive and χ_2 takes the '−'-sign and is negative. The temperature T_0 at the heater is obtained from the initial conditions for the heated area. The heat current P from the heater is in equilibrium with the difference of heat flow in and out of the heated zone at $x = -L$ and $x = L$ and the flow through heat diffusion in the z-direction:

$$P + j_x(x=-L)\cdot A - j_x(x=L)\cdot A - 2j_y \cdot A_y = 0 \tag{4.60}$$

With the help of equation 4.59 and some rearrangements the temperature T_0 at the heater is obtained:

$$T_0 = \frac{P}{A\,\lambda_f\,(\chi_1 - \chi_2) + 4\lambda_f\,L\frac{l_y}{l_z}} \tag{4.61}$$

The heating power from the metal bridge is given by the following equation:

$$P = U_0^2 \cdot \frac{A_{\mathrm{met}}}{\rho_{\mathrm{met}}\cdot n \cdot l_y} \tag{4.62}$$

where U_0 is the voltage applied to the heater, A_{met} the cross sectional area of the metal wire, n the number of wires passing over the bridge and l_y the width of the bridge.

Equation 4.59 can easily be programmed in Maple V.4 which allows the temperature distribution to be plotted. The size of the example device is as follows: $L = 80$ μm (1/2 length of the heater), $l_y = 600$ μm (width of the heater bridge), $l_z = 40$ μm (distance of heater bridge from top and bottom of the channel), $A = 2l_y\,l_z$ (cross sectional area of the channel), $A_{\mathrm{met}} = 100$ nm × 7 μm(cross sectional area of one metal wire). The properties of water are taken for the fluid and titanium is taken for the metal (bulk values): λ_f = 0.54 W m^{-1}K^{-1} (thermal conductivity), ρ_f = 998 kg m^{-1} (density), c_f = 4182 J kg^{-1}K^{-1} (specific heat capacity); $\rho_{\mathrm{Ti}} = 1.6\cdot 10^{-6}$ Ω m (electrical resistivity).

A plot of the temperature distribution as a function of both velocity and x-location for a voltage of $U_0 = 10$ V and $n = 4$ wires over the heater bridge

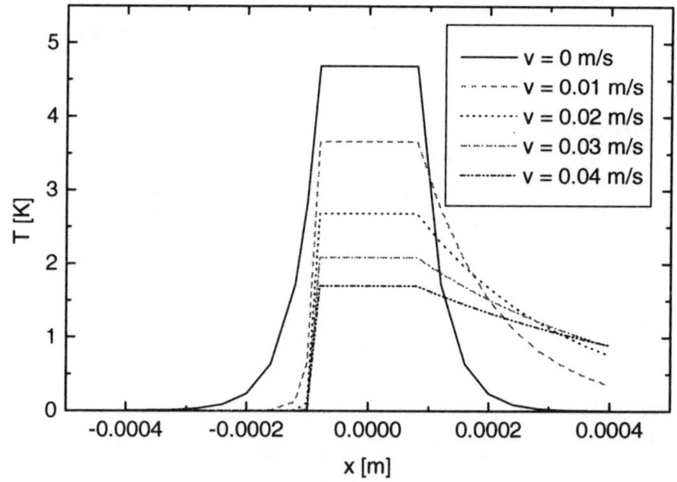

Figure 4.38: Simulation of the temperature distribution along the flowsensor channel

is given in figure 4.38. The velocity is varied from 0 to 0.04 m s^{-1} in steps of 0.01 m s^{-1}. A velocity of 0.01 m s^{-1} is equivalent to a volume flow of 28.8 μl $\min^{-1}$ through the cross section A.

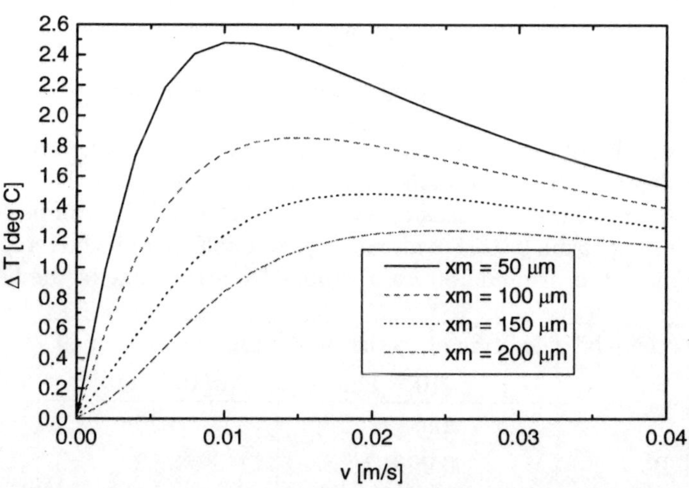

Figure 4.39: Simulated temperature difference of the thermal flowsensor

In order to determine the sensitivity of the sensor, the temperature difference at the location x_m upstream and downstream (distance between heater and temperature sensor, cf. figure 4.35) of the heater is plotted against the velocity of the fluid (figure 4.39). The temperature difference is calculated by $\Delta T = T_3(L + x_m, v) - T_1(-L - x_m, v)$ and is plotted against the distance

x_m. An almost linear region of the temperature difference is obtained for fluid velocities up to 0.02 m s^{-1} for $x_m = 200$ μm (0.01 m s^{-1} for $x_m = 50$ μm). This linear region is most desirable for the flowsensor, as it simplifies the design of the electronics.

A good sensitivity is achieved when the derivative of the temperature difference at a distance x_m upstream and downstream with respect to the velocity v is large. This implies that a small change in fluid velocity leads to a large change in the temperature difference.

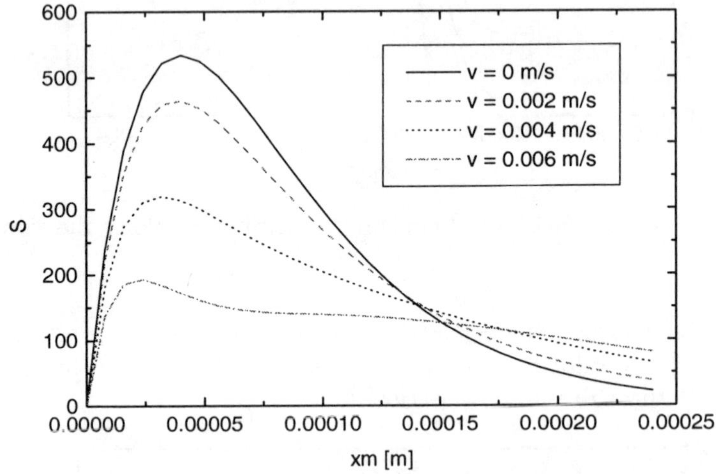

Figure 4.40: Sensitivity of the thermal flowsensor

Several graphs of the sensitivity of the flowsensor for different fluid velocities are depicted in figure 4.40. The sensitivity was calculated by $S = \partial\Delta T/\partial v$. An almost uniform sensitivity between 126 and 141 can be achieved over the flow range from 0 to 17.3 μl min^{-1} (0 to 0.006 m/s) if the temperature sensors are placed at around 150 μm distance from the edge of the heater.

material	specific el. resistance $\rho_{el.0}$ [10^{-6} Ωm]	temperature coeff α [10^{-3} K^{-1}]
Ti	0.475	4.25
Pt	0.0981	3.92
Au	0.02	3.98

Table 4.5: Specific electrical resistance and temperature coefficients for Ti, Pt, Au (bulk values)

The temperature sensors are created in the same way as the heater, except that the current through the wires should be much less than for the heater.

The electrical resistance of a metal is dependent on temperature: $\Delta\rho_{el} = \rho_{el.0} \cdot \alpha \cdot \Delta T$ (table 4.5). Therefore, a measurement of the resistance can be used to measure the temperature. The specific electrical resistance and the temperature coefficient of some common metals in microelectronics is given in table 4.5. Although the values are taken for bulk materials, metals made by thin-film technology behave similarly.

CHAPTER 5

Fabrication Technology

The origins of MEMS technology are found in microelectronics, where electrical devices like transistors, diodes and resistors are integrated to give an electronic system. This electronic system can be a memory device, a microprocessor or any other digital or analog circuit. The milestone events in microelectronics were the invention of the transistor in 1947 and the development of the integrated circuit in 1958.

Micro-electro-mechanical systems have already seen a similar start, but integration is not as far advanced. The first bulk micromachined pressure sensor was for example fabricated in 1962 [49]. Since then, the major markets for MEMS devices have been for pressure and inertial transducers. Owing to the diversification of microtechnology, no standard libraries as in microelectronics have been created so far. This makes micromachining more expensive even for huge mass markets, but allows it to spread into niche markets where mass production at smaller volumes is necessary and cost is not the main issue.

This chapter will give an overview of the different technologies used to produce micromechanical devices in general and microfluidic devices in particular. It starts with a review of the fabrication technologies used in microelectronics and then moves gradually via micromechanical fabrication methods to technologies mainly used for microfluidic devices. For further information on micromechanical processing the books by Heuberger [16], Middelhoek [50], Sze [51] and Gardner [52] are recommended.

5.1 Introduction

After dominating microelectronics due to its useful native oxide, silicon dominates the area of micromachining as well because of its outstanding mechanical properties [53]. A summary of the principal values of the mechanical characteristics of silicon, silicon dioxide, silicon nitride and polysilicon is given in table 5.1. A more extensive overview of the anisotropic properties of silicon can be found in section 3.1.

Material	Si	SiO_2	Si_3N_4	poly-Si
Yield strength [10^9 Pa]	7.0	8.4	14	
Young's Modulus [10^{11} Pa]	$1.9_{<111>}$	0.73	3.85	1.5 – 2
	$1.3_{<100>}$			
Density [g cm^{-3}]	2.3	2.5	3.1	2.3
Thermal expansion [10^{-6} K^{-1}]	2.33	0.55	0.8	2
Thermal conductivity [W cm^{-1} K^{-1}]	1.57	0.014	0.19	0.3

Table 5.1: Mechanical characteristics of silicon (single crystal), silicon dioxide, silicon nitride and polysilicon [53]

MEMS technology covers the complete range of microelectronic fabrication methods, allowing a final integration of micromechanics and microelectronics. Novel techniques have been added, where materials different from silicon need to be used.

Silicon as used in microelectronics is a single crystal, but can also be used in the polycrystalline and the amorphous state. In solid state physics, a direction as well as a surface within the crystal can be expressed by *Miller indices*. Most used in the silicon system are the planes *(100)*, *(110)* and *(111)* where a *1* indicates a intersection of the plane with the x, y or z axis at the unit length and a *0* indicates that the plane and the axis are parallel. A diagram of the three most important planes is given in figure 5.1.

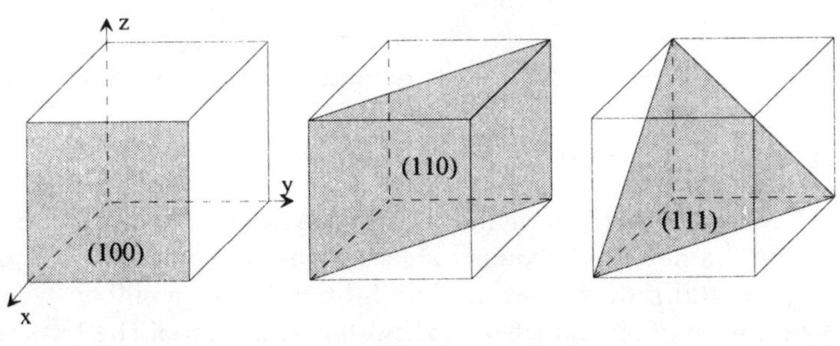

Figure 5.1: Crystal planes within a cubic crystal: (100), (110) and (111)

5.2 Processing Techniques in Microelectronics

A major part of micro-electro-mechanical systems is based on silicon and its microelectronic fabrication techniques. As readers with interests in microfluidics will have different backgrounds it is intended to introduce the microelectronic fabrication techniques in this section before describing micromechanical fabrication methods.

5.2.1 Introduction

Microelectronic fabrication methods are required to provide integrated electronic circuits on a silicon chip. Building integrated circuits (IC's) is done by successive addition (either deposition or growth) of layers which are patterned or protected using photolithography techniques and then etched to remove unwanted material. The above processes are repeated until the required layers are all in place. A simple example of the CMOS transistor fabrication steps is shown in figure 5.2.

The process starts with the definition of the p-tub for the subsequent n-channel transistor. This is done by depositing and structuring a double layer of silicon nitride and silicon dioxide which are used as masking layer for the subsequent implantation of boron into the n-type substrate (figure 5.2.a). With a high temperature step, the boron atoms are diffused and electrically activated to form the p-tub (figure 5.2.b). Then the field oxide and the gate oxide are grown and a layer of polycrystalline silicon is deposited to form the gate. After structuring all these layers (figure 5.2.c), an implantation of As is done to define the source and drain of the n-channel transistor (figure 5.2.d) with the help of a resist mask. The same is then done for the p-channel, where boron is implanted into the silicon (figure 5.2.e). Note that this process is self-aligned as the polysilicon gate acts as a mask around the implantation and subsequent diffusion of the dopants. The final stages in the CMOS process are metal deposition and structuring (figure 5.2.f) to form the interconnects of the transistors and deposition of an oxide and a nitride to protect the device.

5.2.2 Crystal Growth

Integrated Circuits for logic or memory chips are fabricated on silicon wafers of 150 to 200 mm diameter. Micromachined devices are generally still processed on wafers up to 100 mm diameter. The fabrication of a silicon wafer starts with the generation of silicon (98% purity) by reduction of the basic material quartz with carbon in furnaces at about 1500°C:

$$SiO_2 + 2\ C \longrightarrow Si + 2\ CO \tag{5.1}$$

The next step is a reaction of the silicon with hydrochloric acid at a temperature of around 300°C to yield trichlorosilane, which is then cleaned with several distillations:

$$Si + 3\ HCl \longrightarrow SiHCl_3 + H_2 \tag{5.2}$$

A reduction of trichlorosilane with hydrogen at temperatures as high as 1000°C produces finally an electronic grade polycrystalline material of silicon atom

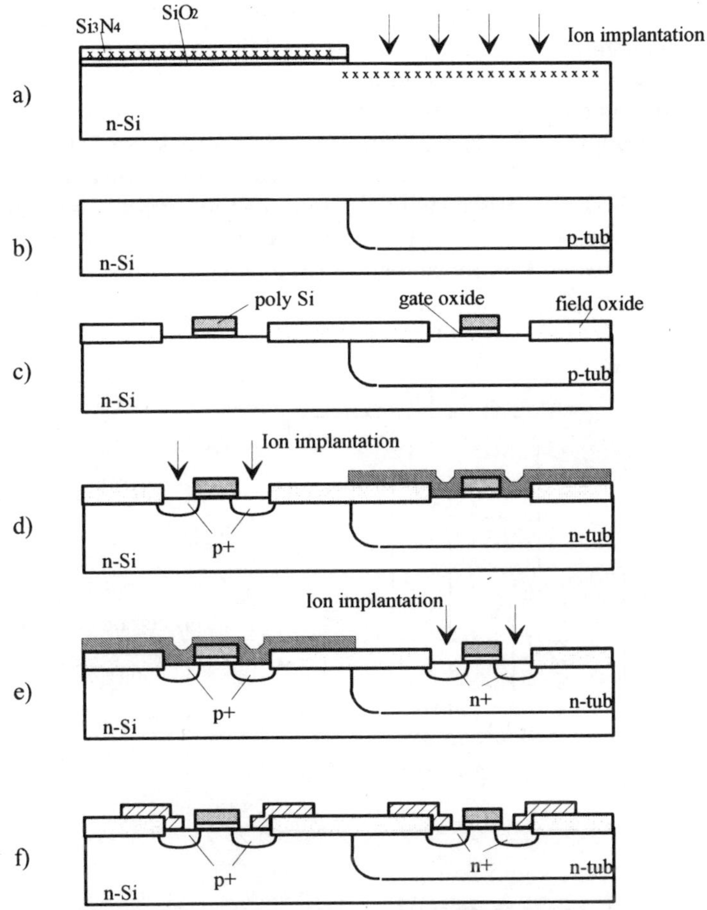

Figure 5.2: CMOS fabrication process

concentrations of approximately 10^{22} cm^{-3} with doping impurity concentrations of parts per billion and carbon contents in the range of parts per million:

$$SiHCl_3 + H_2 \longrightarrow Si + 3\ HCl \tag{5.3}$$

In order to produce silicon wafers, a rod of single crystal silicon is grown from molten silicon in a crucible (Czochralsky method, abbreviation: CZ) by pulling a seed crystal slowly from the melt (see figure 5.3). The growth has to be done in an inert gas or vacuum as silicon oxidises rapidly in air. Also, impurity levels have to be very low as they degrade the semiconducting behaviour of the single crystal silicon. This is achieved by using a quartz inner crucible and a carbon outer crucible. The carbon crucible however leads to a carbon content of up to 10^{17} cm^{-3}. Although this is a rather large concentration compared with typical low level doping levels (10^{14} – 10^{15} cm^{-3}) it does not affect the semiconducting behaviour of the silicon as carbon is electrically neutral within the silicon crystal. Also, oxygen is inserted into the silicon through the quartz crucible. The concentrations of this contaminant is around $5{\cdot}10^{17}$ cm^{-3}. Oxygen usually acts as a getter centre for metal contaminants, which is advantageous in the bulk of the crystal, but can lead to bad electrical behaviour at the surface of the wafer. The latter can be avoided by annealing the wafer before the actual processing in order to create a denuded zone of oxygen at the surface.

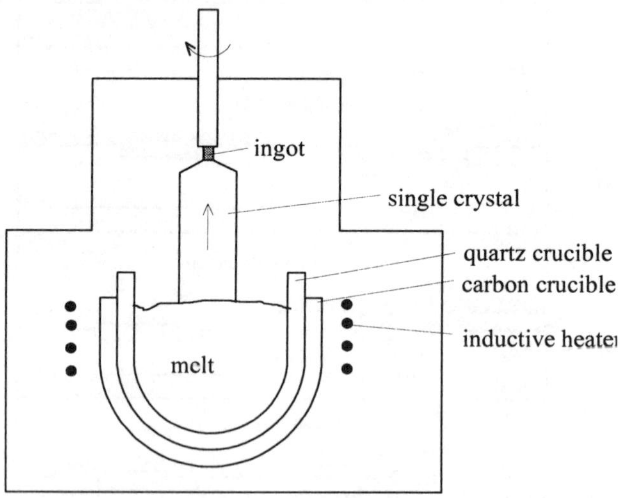

Figure 5.3: Fabrication of single crystal silicon with the Czochralski method

A different growth method uses the float zone technique (abbreviation: FZ) where a polycrystalline rod is moved slowly through an inductively heated zone. Here, the material is first melted and then recrystallised as the molten zone passes through the rod. The polycrystalline rod needs to be of the size of the final wafer, which makes it more complex for larger diameters. The advantage of this technique is however that contaminants can be reduced, as most dopants have a higher solubility in the melt than in the single crystal silicon. Thus, the dopants can be collected in the molten zone and swept to the end of the crystal.

After the growth, the silicon rod is sawn into thin wafers of 350 to 700 μm thickness. Finally, the surfaces are polished to achieve a smooth and optically flat surface suitable for photolithography. In microelectronics, this is done only for one side, as the electronic circuits are made on the surface of one side. For micromechanics, the wafers are often polished on both sides, as the fabrication is quite often from both sides of the wafer.

5.2.3 Lithography

The most important technology in micromachining is lithography, because all patterns are generally defined in photoresist and then transferred onto the appropriate layer. For the lithography, an organic polymer resist is locally exposed to light. Depending on its light affinity the polymer chains are then broken (positive resist) or linked together further (negative resist). This changes the solubility of the resist in an organic solvent acting as a developer. The exposed areas become more soluble for a positive resist whereas the exposed areas become less soluble for a negative resist. Once the resist is developed, i.e. some parts of it locally removed, the desired structure has been transferred into the resist as shown in figure 5.4.a.

A very important characteristic of the resist is its resistance to dry or wet etching methods. This characteristic is improved by hard baking the resist at 95°C for 30 minutes before exposure to these chemicals. A similar treatment is necessary before implantation, i.e. the firing of dopant atoms into the silicon substrate. For some harsh dry etches like 20 – 30 minutes in CHF_3 the resist can also be exposed to deep UV light before a hardbake at 190°C in order to harden the surface of the resist even further.

Three different exposure techniques are available in microelectronics. These are: contact/proximity exposure, projection printing with a wafer stepper and electron beam writing. For the first two techniques, a mask with the intended structure is necessary. The highest achievable resolution for contact/proximity printing is around 2.5 μm for positive resist and therefore no longer enough for today's microelectronics industry. However, micromechanical and microfluidic structures are often larger than this and can therefore easily be printed with this technique. The minimum feature sizes for negative resist are around 2 – 3 times the resist thickness due to swelling of the exposed areas during the development procedure. This is greatly reduced for the unexposed areas of the positive photoresist, which therefore allows a greater resolution.

Photoresists for optical exposure are often PMMA (poly-methyl methacrylate) based. This allows an exposure at shorter wavelength where a higher resolution is possible. The wavelengths used with these deep UV based resists are less than 250 nm. For proper adhesion to the silicon substrate, the wafers are primed in HMDS (hexamethyldisilazan) vapour, which produces a hydrophobic surface.

The basic principle of contact/proximity printing is depicted in figure 5.4.a. A mask consisting of quartz with a structured chrome layer is pressed against the wafer which is covered with resist. The smaller the distance between mask and wafer, the higher the achievable resolution. Standard distances are 10 – 40 μm from the silicon wafer. If the mask is always in contact with the resist, the chrome structure can be damaged which is not particularly good for high volume printing. This can be resolved by the use of projection printing, where the mask does not come into contact with the wafer. The minimum resolution of this technique is around 0.2 μm and therefore an optimum technique for today's microelectronics industry. It is based on a reticle which contains the mask of one chip with a magnification of 5 to 10 (see figure 5.4.b). With high precision optics, parts of the wafer are then exposed. After one exposure, the target area of the wafer is moved to allow the next exposure (wafer stepping).

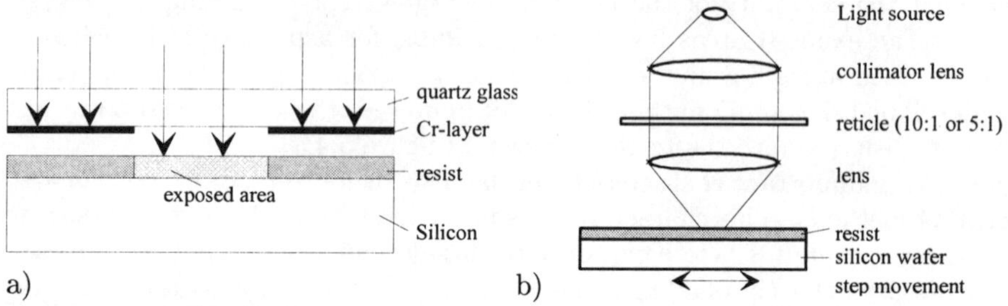

Figure 5.4: a) Contact/Proximity Printing, b) Wafer stepping with projection printing

Unlike deep UV light for contact/proximity and projection printing, an electron beam is used for the third technique of exposing resist. Electrons are ejected from an electron gun, accelerated and focused onto a specific point of the wafer. No mask is necessary as the beam can be steered with electron optics. However, the disadvantage is the time that it takes to expose all the features of one whole wafer. The small wavelength of 10 – 50 keV electrons does not limit the resolution which is about 0.1 μm. The resolution is greatly dependent on electron scattering, producing secondary electrons and X-rays. Thus, a smaller feature size is possible with thinner resists. This technique is used for microoptics [54] as well as for writing the masks necessary for contact/proximity and projection lithography. The same types of resist (positive and negative) are available for electron beam lithography.

Apart from deep UV and electron beam techniques, X-ray and synchrotron radiation can also be used for lithography. Synchrotron radiation has opened the possibility of achieving structures with very large height/width ratios. This technique is however mainly used in micromachining and will therefore be covered later.

A technique for removing unwanted material is the *lift-off technique* (see figure 5.5). Here a reverse image is patterned in the resist layers. The developed windows have re-entrant edge profiles produced by a subsequent hardbake and flood exposure. Evaporated metal films do not form a continous layer over the resist and the metal sitting on the resist is lifted off when the resist is dissolved in acetone.

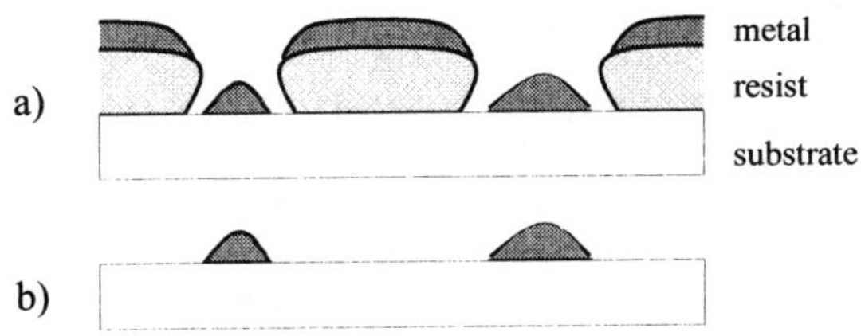

Figure 5.5: Lift-off technique: a) Image-reversed resist with evaporated metal on top, b) Metal structure after lift-off

5.2.4 Oxidation of Silicon

Growth of silicon dioxide in a furnace at elevated temperatures is a process used in both microelectronics and micromechanics. Grown silicon dioxide can be used for many purposes. Some examples are: masking against implantation or diffusion of dopants, the gate oxide of MOS transistors, isolation of individual devices on a chip, stress compensation layer for a successively deposited silicon nitride, etch mask for wet etches of silicon and many more. Silicon dioxide is mainly grown in a wet or dry oxygen atmosphere at elevated temperatures as well as via rapid thermal oxidation (high temperature, 1150°C, for a very short period of time). However, there are several other techniques available, such as chemical vapour deposition, plasma anodisation or wet anodisation. The latter techniques produce a high-charge density level and are therefore not used when the silicon dioxide has to perform as a gate oxide. They do, however, allow inter-metal isolation, passivation and smoothing of the surface.

When silicon is exposed to air, an oxide layer rapidly develops because of the high affinity of silicon to oxygen. This layer is called "native oxide" and is usually several nm thick. For thicker layers, the wafers are oxidised at higher temperatures. Depending on the atmosphere, silicon reacts with O_2 or H_2O to form silicon dioxide and hydrogen in the case of wet oxidation. During the course of the oxidation, the SiO_2 - Si interface moves into the silicon surface. The thickness of silicon consumed is 44% of the final oxide thickness.

The oxidation of silicon can be described by a model developed by Deal and Grove, which is valid for temperatures between 700 and 1300°C and oxide thicknesses between 30 and 2000 nm. From the assumption of steady state of the gas flows at the gas/silicon dioxide interface, diffusion through the SiO_2 and the reaction rate at the SiO_2/Si interface (see figure 5.6), a model can be built linking the thickness of the silicon dioxide d_0 with the oxidation time t:

$$d_0^2 + A d_0 = B\,(t + \tau) \qquad (5.4)$$

where A and B are constants depending on the oxidation temperature

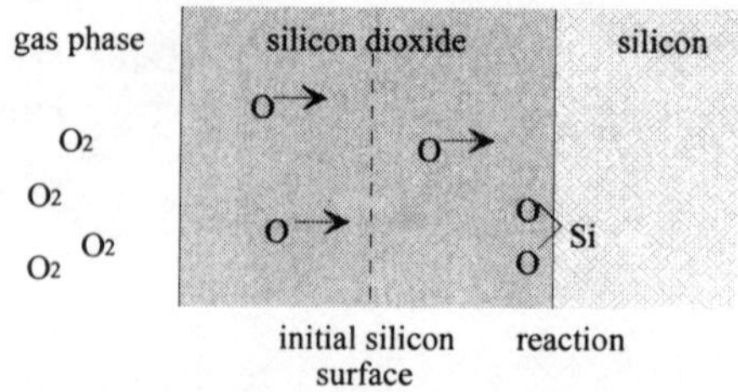

Figure 5.6: Silicon oxidation model after Deal and Grove

and τ represents a time shift for the initial oxide layer. Some values for the constants are given in table 5.2.

Temperature [°C]	atmosphere	A [μm]	B [$\mu m^2\ h^{-1}$]	τ [h]
920	dry	0.235	0.0049	1.40
1000	dry	0.165	0.0117	0.37
1100	dry	0.090	0.027	0.076
1200	dry	0.040	0.045	0.027
920	wet	0.50	0.203	0
1000	wet	0.226	0.287	0
1100	wet	0.11	0.510	0
1200	wet	0.05	0.720	0

Table 5.2: Constants for wet and dry oxidation of silicon [51]

This table is valid only for undoped silicon substrates. If the silicon wafer is doped with group III and IV elements at high concentration levels, the oxidation rates can be accelerated . Also, a dry O_2-HCl mixture shows an increased oxidation rate. Typical HCl additions range from 1 to 5% and are used to passivate ionic sodium, improve the breakdown voltage and getter impurities and defects in the silicon.

Before the oxidation, a cleaning step is important to eliminate both organic and inorganic contaminants from the surface of the silicon wafer. A common cleaning technique is based on H_2O-H_2O_2-NH_4OH to remove organic contaminants and group I and II metals. For the elimination of heavy metals, a solution of H_2O-H_2O_2-HCl is used as a second cleaning step. Both are performed at temperatures between 75 and 85°C for 10 minutes. This cleaning procedure is called “RCA” clean. After cleaning and drying, the wafers are inserted into the furnace at 750°C and the temperature is then ramped in a nitrogen atmosphere to the desired value. Cooling down to 750°C in nitrogen is the final step before the wafers can be removed from the furnace.

Because the diffusion of the usual silicon dopants in SiO_2 is slow it acts as a good masking material. As an example, table 5.3 lists the diffusion constants of some dopants in silicon. In micromachining, silicon dioxide can be used as a mechanical layer. Due to the oxidation, however, the film shows a compressive stress of up to 300 MPa which can lead to buckling of beams and membranes [51].

Dopants	D SiO_2 [$cm^2\ s^{-1}$]	D Si [$cm^2\ s^{-1}$]	D poly-Si [$cm^2\ s^{-1}$]
B	$3.4 \cdot 10^{-17} - 2.0 \cdot 10^{-14}$	$1.5 \cdot 10^{-13}$	$6.6 \cdot 10^{-8}$
P	$2.9 \cdot 10^{-16} - 2.0 \cdot 10^{-13}$	$1.43 \cdot 10^{-13}$	$1.4 \cdot 10^{-10}$
As	$1.2 \cdot 10^{-16} - 3.5 \cdot 10^{-15}$	$2.06 \cdot 10^{-14}$	$0.85 - 4 \cdot 10^{-12}$

Table 5.3: Diffusion constants in SiO_2, Si and effective diffusion constant in poly-Si at 1100°C [51, 55]

5.2.5 Physical and Chemical Vapour Deposition

Deposition of materials (Si, SiO_2, Si_3N_4, and metals) is a technique for realising conducting or dielectric layers [16, 51]. A distinction is made between chemical vapour deposition (CVD) for Si_3N_4, SiO_2 and Si and physical vapour deposition (sputtering, evaporation) of metals and dielectrics. The thickness of these layers ranges from several tens of nanometers up to a few microns.

Regarding the source of information, the book published by Sze [51] is a comprehensive work. A very interesting source of information regarding thin film properties is also the ISI webpage http://mems.isi.edu

5.2.5.1 Chemical Vapour Deposition

The materials which are most commonly deposited by chemical vapour deposition (CVD) are silicon dioxide, silicon nitride, silicon carbide and polycrystalline silicon. Of these, the first two are dielectrics, but polycrystalline silicon may be made to be either conducting or nonconducting, depending on the deposition conditions. In chemical vapour deposition, a reaction of gaseous components at a surface leads to the deposition of a solid film. Depending on the conditions, the methods can be classified into low pressure CVD (LPCVD), plasma enhanced CVD (PECVD) and atmospheric pressure CVD (APCVD). These process conditions affect the uniformity of the deposition (see figure 5.7).

Conformal deposition is very important for protection layers during wet chemical etches as thin layers are more easily penetrated by chemicals like HF or KOH. A uniform coverage is obtained if the reactants or reactive intermediates adsorb on the surface and rapidly migrate along before reacting. Generally, conformal deposition, i.e. step coverage, is best for LPCVD layers,

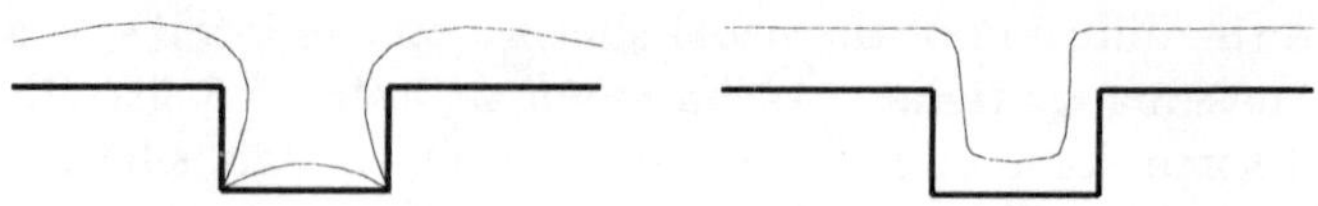

Figure 5.7: Uniformity of the deposition: a) non-conformal deposition, b) conformal deposition

reasonable for PECVD layers and not very good for APCVD deposition. Low pressure CVD is usually performed in hot-wall furnaces where the whole tube is heated and the gases stream through the gaps between the wafers (figure 5.8). Plasma enhanced deposition is most often done in a parallel-plate reactor, where the plasma is generated between the plates, and the lower sample plate can be heated. APCVD uses open reactors, where the wafers pass under the reactant gas nozzles. In order to avoid any gases in the environment, the reaction gas is surrounded by N_2 and immediately pumped through an exhaust.

Polycrystalline silicon (polysilicon, poly-Si) is usually deposited at 575 – 650°C under low pressures at around 7 Pa with silane as the reaction gas. At temperatures of around 560°C amorphous silicon is formed which can be annealed at 900°C to give polysilicon. Polysilicon is used in microelectronics as the gate of MOS structures, whereas in micromachined devices it is more likely to be used as a mechanical layer, an electrical heater for flowsensors and actuators or for other devices. The compressive stress in poly-Si material is usually in the range between 0.9 and 1.5 GPa but can be reduced by annealing to less than 10 MPa. In-situ doped poly-Si shows a minimum resistivity of 0.005 Ω cm (B-doped) and 0.01 Ω cm (P-doped) for doping temperatures above 535°C (B-doped) and 625°C (P-doped). Other properties of poly-Si are its density of around 2.3 g cm^{-3}, its Young's modulus of approximately 150 – 200 GPa, its coefficient of thermal expansion of $2 \cdot 10^{-6}$ K^{-1} and its temperature coefficient of resistance of $1 \cdot 10^{-3}$ K^{-1}.

The deposition of single crystal silicon of the same orientation as the wafer is called epitaxy. It is usually done in a similar way to poly-silicon but requires temperatures in the range between 950 and 1250°C depending on the reaction gases. The layer is mainly used for bipolar devices but can also be employed for MOS structures. In the fabrication of SOI materials the thin single crystalline silicon layer on top of an oxide is sometimes thickened by epitaxial growth.

Silicon nitride is the next material, and it can be deposited by both LPCVD and PECVD methods. The process for silicon nitride depends either on silane or on dichlorosilane reacting with ammonia to form silicon nitride. The density of LPCVD material is usually higher and less hydrogen is in-

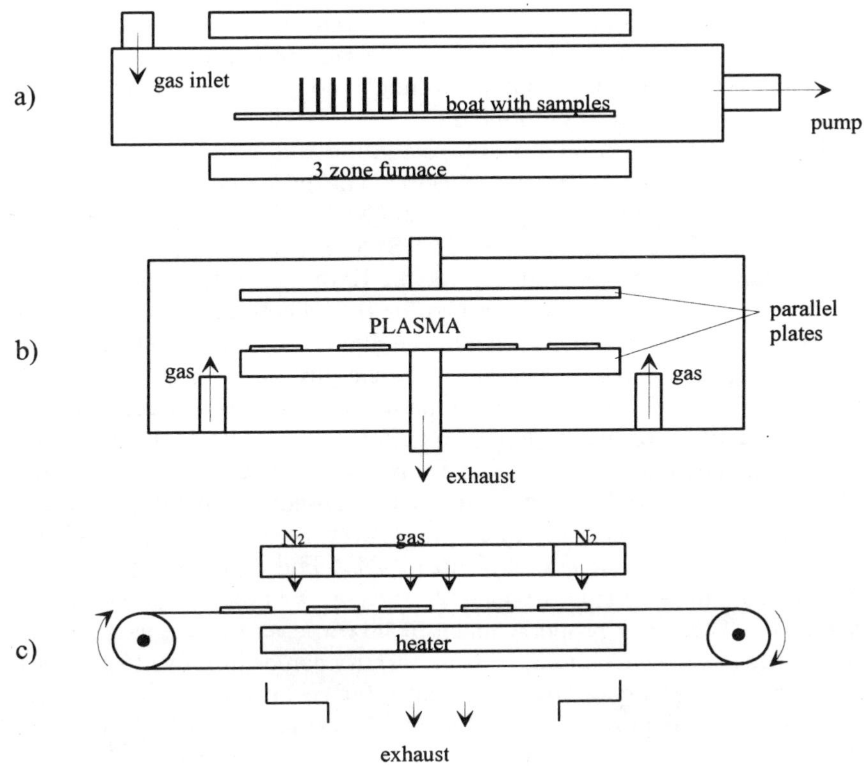

Figure 5.8: Equipment for chemical vapour deposition: a) hot-wall reactor, b) parallel-plate reactor, c) APCVD reactor

corporated in the material which is advantageous for protecting silicon from wet etchants like KOH, HF etc. Deposition temperatures between 700 and 900°C make it unsuitable as protection for aluminium films or other metals of low melting temperatures. PECVD silicon nitride requires only temperatures up to 350°C but contains a rather large hydrogen proportion of around 20 – 25 atom%. The stress of a LPCVD silicon nitride layer is generally tensile and of the order of 1 GPa whereas PECVD silicon nitride shows stresses between 200 MPa (compressive) and 500 MPa (tensile). This variation is process dependent and can be controlled by temperature, deposition rate and ionic bombardment from the plasma. Other properties are the density which is around 2.9 – 3.1 g cm^{-3} for LPCVD films and 2.4 – 2.8 g cm^{-3} for PECVD films, and the Young's modulus of 290 GPa (LPCVD) and 210 GPa (PECVD).

Silicon carbide is a material which shows similar protection properties against wet etchants to silicon nitride. It may be deposited as a single crystal, polycrystalline or amorphous film. For this, LPCVD and PECVD reactors are applicable, although a large hydrogen content is obtained in the PECVD deposited layer. The internal stress for SiC varies from 0 to 250 GPa (tensile) depending on the deposition and the Young's modulus of the film is approximately 190 GPa for polycrystalline SiC.

Silicon dioxide can be grown by oxidation and also deposited by chemical vapour deposition as described here. The processes for the layer fabrication are either PECVD deposition at 200°C, silane oxidation at 400°C, decomposition of tetraethoxysilane (TEOS) at 700°C as well as a reaction of dichlorosilane and nitrous oxide at 900°C. Hydrogen can be found in both low temperature processes as described for the silicon nitride process above. Additionally, a conformal deposition of the film is only guaranteed at elevated temperatures. The stress of the layer is generally between compressive 300 MPa and tensile 300 MPa for low temperature processes, whereas the stress is only compressive for higher deposition temperatures and of the order of 100 to 300 MPa. The density of the film is around 2.2 g cm^{-3} for all deposition methods. In order to planarise a surface for subsequent metal deposition or as a sacrificial oxide beneath a mechanical layer, phosphorus-doped silicon is employed with P-concentrations of up to 4%. This layer is deposited with APCVD and shows usually a bad uniformity as well as a large etch rate in buffered HF. The latter property makes it however very useful as a sacrificial layer in surface micromachining. The uniformity can be improved by reflowing the glass at elevated temperatures around 1100°C. This planarises the surface and allows easier deposition of subsequent metal layers. Additional doping with boron at a concentration of up to 4% leads to boro-phospho-silicate glass (BPSG) which shows flow behaviour at temperatures as low as 850°C.

A summary of the properties of all the different materials described can be found in table 5.4.

material	poly Si	Si_3N_4	Si_3N_4	SiO_2	SiO_2
deposition	LPCVD	LPCVD	PECVD	PECVD	TEOS
T_{dep} [°C]	575 – 650	700 – 900	350	200	700
stress [GPa]	0.9 – 1.5 (comp.)	1 (tensile)	0.2 (c) – 0.5 (t)	0.3 (c) – 0.3 (t)	0.1 (comp.)
density [g cm^{-3}]	2.3	2.9 – 3.1	2.4 – 2.8	2.3	2.2
E-mod [GPa]	150 – 200	290	210	75	74

Table 5.4: Properties of dielectric and conducting layers deposited with chemical vapour deposition

5.2.5.2 Physical Vapour Deposition

Physical vapour deposition (PVD) is the process of depositing atoms on a target surface by evaporation or sputtering. This technique is mainly used for metals in microelectronics, but can also be used to deposit insulators like silicon dioxide and silicon nitride as well as piezoelectric materials like zinc oxide and lead zirconate titanate (PZT).

For evaporation, an electron beam hits and evaporates the source metal, which is placed in a carbon crucible. Due to a low ambient pressure of around 10^{-4} Pa, the atoms have a very long free path length before colliding with air molecules. Thus the atoms will condense on the target, which is located vertically above the crucible, without hitting any air-molecules on the way. The stream of atoms from the crucible to the target is therefore directed, and a conformal deposition is possible on surfaces perpendicular to the beam. Reasonable side-wall coverage may be achieved by arranging the target wafers and evaporating source on the surface of a sphere, and having a continuous planetary motion of the target holders. Evaporation is possible for a variety of metals like aluminium, gold, chromium, platinum, titanium and many more.

The second technique is to sputter deposit the material. An argon plasma is generated between two parallel plates, which form a capacitor. Argon ions are accelerated towards the negative plate, and remove or sputter neutral atoms from the surface of the plate by virtue of their kinetic energy. In the case of metals, a d.c. voltage may be applied directly to the plates (d.c. sputtering), but for dielectrics it is necessary to use an a.c. voltage in order to create the plasma (a.c. sputtering). In this case one of the plates is made smaller than the other and as a result acquires the negative surface bias necessary to effect the sputtering. The sputtered atoms from the source move towards the other plate of the capacitor by virtue of their kinetic energy, which they acquired during the initial bombardment. As the ambient pressure is between 0.1 and 1 Pa for the sputtering process, the atoms from the target collide with particles from the plasma. This means that they will deposit on the other plate from random directions, guaranteeing a more conformal deposition than obtained with evaporation. However, at vertical walls of trenches the deposition remains problematic but coverage is good for some metals. A picture of a 5 × 5 μm trench with sputtered titanium is shown in figure 5.9. It can be clearly seen that the sputter technique for titanium achieves good side-wall coverage. Compared to this, Al proves to give less-conformal deposited layers for the same trench geometry. This can be important for certain applications in silicon micromachining.

Sputter deposition may be used for the deposition of dielectric, piezoelectric and other layers, which are of interest for actuators in micromechanics/microfluidics. A very interesting piezoelectric material is PZT (lead zirconate titanate). PZT has been successfully sputter deposited at thicknesses up to 1 μm on a Pt/Ti/SiO_2/Si substrate [56]. In order to achieve a dense film and a Perovskite phase, the wafers have to be post annealed at tempera-

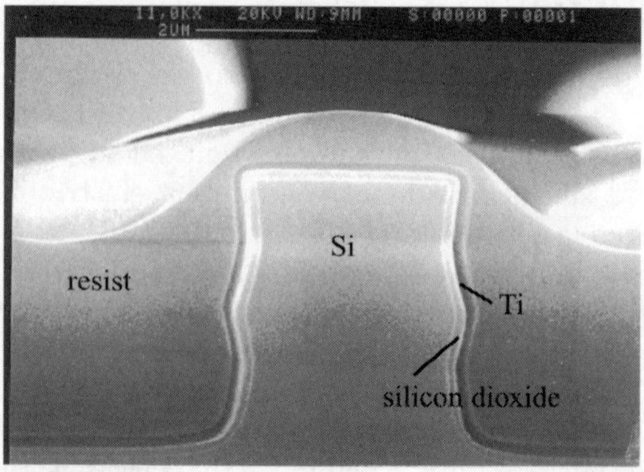

Figure 5.9: Ti sputter deposited between resist and oxide layer. The titanium has been selectively removed and the silicon dioxide layer as well as the resist are still visible in order to improve the contrast between titanium and oxide.

tures between 450 and 650°C. Some characteristics of sputter deposited PZT are its piezoelectric coefficient d_{31} of up to -100 pC N^{-1} [57], its pyroelectric coefficient of 210 C m^{-2} K^{-1} [58], its Young's modulus of around 75 GPa [57], its stress which can be between compressive 60 MPa and tensile 380 MPa, its dielectric coefficient between 300 and 1207 as well as its remanent polarisation which is in the range of 12 – 25 μC cm^2.

Another material which is important in micromechanics is zinc oxide (ZnO). It is used in both actuators and chemical gas sensors. Its piezoelectric coefficient is around 10.5 – 11.5 pC N^{-1} [59], the resistivity is 2.5 – 10 MΩ m, the internal stress is between 170 and 300 MPa and its Young's modulus is 14 – 23 GPa [59]. Also, materials like silicon dioxide and silicon nitride have been sputter deposited [60] but lack the density of thermally grown or LPCVD films.
Finally, the properties of PZT and zinc oxide are summarised in table 5.5.

material	PZT	ZnO
stress [GPa]	0.06 (c) – 0.4 (t)	0.17 – 0.3
d_{31} [pC N^{-1}]	-100	10.5 – 11.5
ϵ_r	300 – 1207	10.8 – 11
density [g cm^{-3}]	2.2	5.5
E-mod [GPa]	75	14 – 23

Table 5.5: Properties of piezoelectric layers deposited with physical vapour deposition

5.2.6 Diffusion and Implantation of Dopants

In order to change the electrical and chemical properties of silicon crystals, atoms from the third and fifth group of the periodic table (B, P, As, Sb) are introduced. This is realised either by diffusion, where at elevated temperatures atoms diffuse into the silicon wafer from a gas phase, or by implantation, where atoms are fired and embedded into the Si surface.

Diffusion of dopants is used in MOS devices for the generation of source and drain areas, adjustment of the transistor threshold voltage and for doping poly-Si lines. In bipolar devices, diffusion is necessary to generate base and emitter areas as well as resistors. In micromechanics, diffused dopants are employed for piezoresistive transducers, which can measure stress in the silicon layer. A high dopant concentration is useful as an etch stop layer during anisotropic silicon etching.

The one dimensional partial differential equation for diffusion, valid for small to medium concentrations c, is formulated as follows [51]:

$$\frac{\partial c(x,t)}{\partial t} = D \cdot \frac{\partial^2 c(x,t)}{\partial x^2} \tag{5.5}$$

where x denotes the direction perpendicular to the wafer surface, t is the time and D is the diffusion constant of the dopant in silicon. The diffusion constant itself is again dependent on the activation energy E_A and the temperature T and is given by $D = D_0 \cdot exp\left(\frac{-E_A}{kT}\right)$ with D_0 as the temperature independent diffusion constant and k as the Boltzmann constant. Both D_0 and E_A are given in table 5.6 for the most important dopants.

Dopants	B	P	As	Sb
D_0 [cm^2 s^{-2}]	0.76	3.85	22.9	0.21
E_A [eV]	3.46	3.66	4.1	3.65

Table 5.6: Temperature independent diffusion constants and activation energy for dopants in Si

The above differential equation can be solved for two boundary conditions. For constant source diffusion, the doping concentration at the surface of the wafer is constant and equal to c_0 leading to equation 5.6, whereas if the total number of atoms per unit area N_{D0} stays constant for an implantation with subsequent diffusion, equation 5.7 applies:

$$c(x,t) = c_0 \cdot \mathrm{erfc}\left(\frac{x}{2\sqrt{D\,t}}\right) \tag{5.6}$$

$$c(x,t) = \frac{N_{D0}}{\sqrt{\pi\,D\,t}} \exp\left(-\frac{x^2}{4\,D\,t}\right) \tag{5.7}$$

Doping the silicon wafers with boron or phosphorus wafers (often also called tiles) in a furnace at elevated temperatures is mainly used for deep diffusion where high concentrations are required. This is for example the case in boron doped etch-stop layers which require a concentration of $3{\cdot}10^{19}$ cm^{-3} in order to show a significant drop in the etch rate in KOH. Apart from solid source diffusion, implantation is a widely used technology to introduce dopant atoms. During the implantation, dopant atoms are ionised, accelerated and fired into a silicon substrate. When entering the silicon crystal, the dopant atoms lose their kinetic energy by colliding with silicon atoms and displace them. Then they come to rest at some depth within the lattice. In order to repair the damage caused to the crystal by the incoming dopant atoms, a heat treatment is necessary. The vertical distribution of the concentration c from an implantation of an area far away from masked edges can be described analytically:

$$c(x) = \frac{N_{D0}}{\sqrt{2\,\pi}\sigma_P} \exp\left(\frac{-(x - R_P)^2}{2\,\sigma_P^2}\right) \tag{5.8}$$

where N_{D0} is the total dose of the implant, R_P the depth at which the maximum concentration occurs and σ_P the standard deviation from the depth R_P at which the concentration drops to e^{-1} times the maximum concentration. Both R_p and σ_P can be obtained from process models such as SUPREM or implantation tables (see Sze [51] for more details).

A subsequent heat treatment is necessary not only to reduce the damage in the silicon crystal, but also to electrically activate the dopant atoms, i.e. to place them on substitutional sites in the silicon crystal. This can be done with conventional furnace processes at elevated temperatures (eg. at 900°C), or, to avoid large diffusion and guarantee shallow doped regions, by rapid thermal annealing (RTA). RTA is normally used in solid phase regrowth mode for the regrowth of amorphous layers. Temperatures range from 750 °C to 1100 °C depending on the application.

An important application of dopants in micromachining is the *piezoresistive effect* for pressure sensors, accelerometers and others. Mechanical strain in silicon leads to a change in the overall electrical resistance because of both the deformation of the structure as well as producing a change in the specific electrical resistance. Because the latter effect is caused by a change in the occupation of conduction and valence bands, the piezoresistive effect is sensitive to the doping level as well as the temperature. The fractional change in the resistance $\Delta R/R$, neglecting the effect of dimensional changes, can be denoted by [16]:

$$\frac{\Delta R}{R} = \pi_l \sigma_l + \pi_t \sigma_t \tag{5.9}$$

where σ_l and σ_t denote the longitudinal and transversal stress, R and ΔR are resistance and its change under the strain, and π_l and π_t are the longitudinal and transverse piezoresistance coefficients. They can be obtained from tensor calculations as the piezoresistive effect in silicon is direction dependent.

5.2.7 Dry and Wet Etching Techniques

To remove material, wet or dry etching is used. Etchants can be classified as isotropic or anisotropic depending on the directional dependence of the etch. Isotropic etchants have the same etch rate in all directions, whereas anisotropic etchants show orientation dependent etch rates (figure 5.10). Traditional electronic processes used isotropic wet etches, but modern processes use dry etching which can be either isotropic or anisotropic, depending on whether the etch process is mainly chemical or physical.

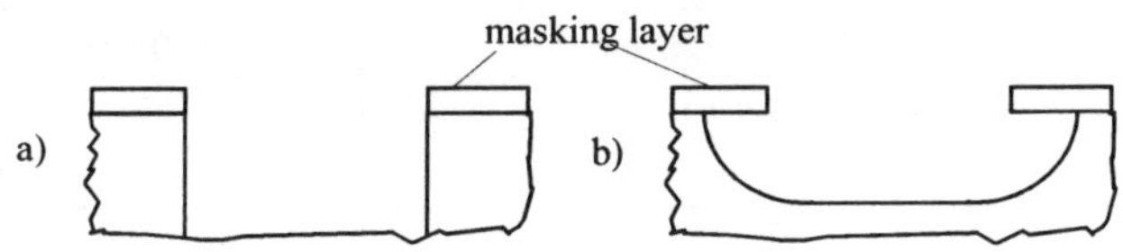

Figure 5.10: Etch profile: a) anisotropic, b) isotropic

There are several important factors for both dry and wet etching techniques. These are the etch rate of the layer, and the selectivity to any overlying or underlying layers as well as the anisotropy of the etch. The etch rate is simply defined by etch depth over time. The selectivity is the ratio between the etch rates of mask layer and operational layer and the anisotropy factor is given by the difference between horizontal and vertical etch rate divided by the vertical etch rate.

Wet etching is not used very much in submicron microelectronics fabrication facilities as it tends to be isotropic and is not suitable for defining small features. In micromachining, where structures are often larger than $10\mu m$, this limitation does not apply and the cheaper wet etch process is often favoured. A short list of important wet etchants is given for various layers in table 5.7. A big advantage of wet etchants is their high selectivity to other materials and the absence of radiation and other damage which is produced by plasma etching. This is advantageous for solar cell production for example, where recombination centres created by radiation damage decrease the efficiency.

Dry etching can be classified into several different techniques. The first technique of plasma etching is fully dependent on the chemical nature of the radicals from the plasma. Here the intention is that the plasma ions which arrive at the substrate are of low energy so that a chemical reaction rather than sputtering is favoured. The conditions necessary for this technique are a long life-time of the radicals, enabling them to travel from the plasma to the

Material	Etchant	Process Information
Pt	570 ml HCl (37%) 30 ml HNO_3 (70%)	boiling + stirred, 50 nm min^{-1} at 70°C masklayer: resist, only structures $> 10\ \mu m$
Al	H_3PO_4, HNO_3 (4:1) CH_3COOH, H_2O (4:1)	35 nm min^{-1} masklayer: resist
Au	4 g KI, 1 g I_2, 40 ml H_2O or: HNO_3, HCl (1:3)	0.5 – 1 $\mu m\ min^{-1}$ masklayer: resist 25 – 50 $\mu m\ min^{-1}$ at 70°C
Cr	1 ml of 1g NaOH in 2 ml H_2O 3 ml of 1g $K_3Fe(CN)_6$ in 3 ml H_2O	25 – 100 nm min^{-1} masklayer: resist
Ti	H_2O, H_2O_2, HF (20:1:1)	880 nm min^{-1}
Si_3N_4	H_3PO_4	10 nm min^{-1} at 180°C resist mask not possible
SiO_2	HF, H_2O, NH_4F (28:170:113)	100 – 250 nm min^{-1} masklayer: resist

Table 5.7: Wet etch solutions for different deposited layers

wafer, and a gaseous reaction product of the radical and the layer material. The etch is therefore fully isotropic and generally shows a high selectivity.

The second dry etching technique relies on a chemical/physical effect and is called reactive ion etching (RIE). The etch reaction is initiated by the kinetic energy of the ions in the plasma when they hit the wafer. As an example, a parallel-plate reactor can be used where the pressure is rather high, enabling a reasonable anisotropic etch. For better anisotropy a reactive ion etcher is used where the gas pressure is smaller and the kinetic energy of the ions is larger than 100 keV. The process is called reactive ion beam etching if the generation of the plasma and the acceleration of the ions towards the substrate are locally decoupled. The advantage of RIBE is that the kinetic energy and the current density of the ions can be adjusted independently and that the pressure is quite small. This guarantees that the products of the reactions are very likely in a gaseous form. A brief summary of the etchants is given in table 5.8.

The etch rates for other materials and the selectivity against masking materials for both wet and dry etching methods can be found in the review paper on etching by Williams [61].

Material	Etchant	Typical Process Parameter
Si	SF_6	150 mTorr, 100 W, 1 μm min^{-1} isotropic, mask: resist
	$Cl_2 + SiCl_4$	100 mTorr, 80 W, 0.2 μm min^{-1} highly anisotropic, mask: SiO_2
Si_3N_4	Ar + CHF_3	25 mTorr, 200 W, 33 nm min^{-1} partially anisotropic, mask: resist etch rate SiO_2: 14 nm min^{-1}
SiO_2	Ar + CHF_3	40 mTorr, 250 W, 35 nm min^{-1} partially anisotropic, mask: resist

Table 5.8: RIE dry etchants for different deposited layers

5.3 Micromechanical Fabrication Methods

As already stated, silicon micromechanics has been developed from microelectronic processing techniques. As microelectronics only uses the surface of the silicon, other processes had to be developed to allow the exploitation of the third dimension. Most micromechanical devices are produced from silicon because of its good mechanical properties. This, however, limits the variety of materials and arbitrary shapes to be used. Recent work in electroforming and ablation have however allowed plating and deposition of metals and other materials into a preformed mould thereby increasing the shapes and aspect ratios available for micromechanics.

5.3.1 Wet Isotropic and Anisotropic Etching of Silicon

In addition to microelectronic etching technology, micromachining uses bulk silicon etching with anisotropic or isotropic wet etchants. Isotropic etching produces equal vertical and lateral etch rates as shown in figure 5.11.a. The result is an undercut of the masking material. Other etches are orientation dependant. For potassium hydroxide (KOH) the (111) planes etch slowest and pyramidal pits bordered by (111) planes can be produced as shown in figure 5.11.b. The (111) plane etch rate is several orders of magnitude less than that of other orientations [62, 63]. Due to the orientation of the (111) planes in a (100) Si wafer, every arbitrary shape of window in the masking layer eventually leads to a rectangular structure due to underetching effects. All known anisotropic etchants are bases, including KOH [62, 63], TMAH (Tetra methyl ammonium hydroxide) [64, 65] and ethylene diamine pyrocatechol (EDP) [62, 63]. As resist dissolves in bases, SiO_2, Si_3N_4 or SiC may be used as a masking layer. We will now discuss etch rates, masking layers, roughness and other characteristics for both KOH and TMAH.

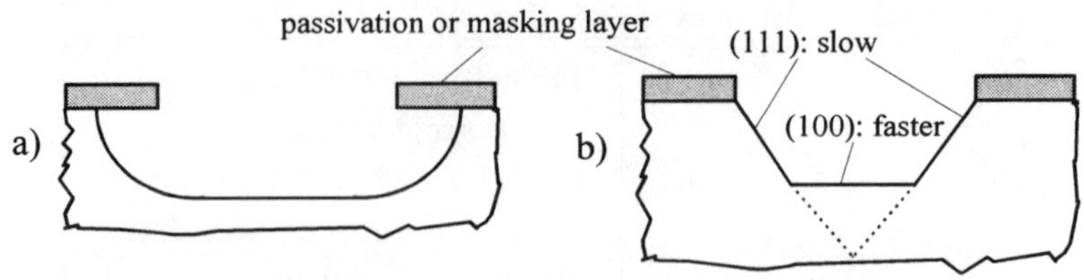

Figure 5.11: Etch profile: a) isotropic, b) anisotropic

Figure 5.12 shows the etch rates for both KOH and TMAH at various concentrations and temperatures. It can be seen that an etch rate maximum is obtained at 2 and 20 % wt for TMAH and KOH respectively. The etch reaction depends on both hydroxide and water molecules. According to Seidel [62], the etch rate is proportional to the product $[OH^-]^{0.25} \cdot [H_2O]^4$, which explains the existence of a maximum in the etch rate. Also, as the etch reaction follows the Arrhenius law, the etch rate increases exponentially with temperature.

A look at the passivation layers shows etch rates between 50 and 100 nm h^{-1} for KOH at 60°C and between 250 and 400 nm h^{-1} for KOH at 80°C. This can be drastically reduced by a change to TMAH. Here, etch rates between 5 and 30 nm h^{-1} have been measured for temperatures up to 80°C. This favours TMAH for wafers with an oxide etch mask. For silicon nitride, however, KOH is almost as good as TMAH. The etch rate of LPCVD nitride is around 0.5 nm h^{-1} for KOH at 60°C and PECVD nitride etches between 2 and 5 nm h^{-1}. The same layers are etched between 1 and 5 nm h^{-1} (LPCVD) and between 4 and 9 nm h^{-1} (PECVD) in KOH at 80°C. With TMAH, LPCVD nitride etches with 0.2 – 3 nm h^{-1} for temperatures between 40 and 80 °C. PECVD shows etch rates up to 12 nm h^{-1} for temperatures as high as 80°C.

Another interesting aspect is the roughness at the bottom of the etched (100) plane. For KOH at 60°C, the surface is smoothest for a concentration of 30% (average mean roughness of 2.5 nm for an etch depth of 145 μm). For both higher and lower concentrations, the mean roughness increases to values up to 20 nm for an identical etch depth. The same behaviour can be found for KOH at 80°C. The smoothest surface for TMAH is obtained at 10 % wt and degrades drastically for lower concentrations, but only slightly for higher concentrations.

Due to different etch rates at the centre of the etched plane and near the walls, where (100) and (111) planes meet, a notching effect takes place. For KOH and TMAH, the height difference between centre and corner (notch depth in figure 5.13) is almost linearly dependent on the average etch depth (0.5 – 3.5 μm for varying KOH concentrations and structural sizes at 200 μm

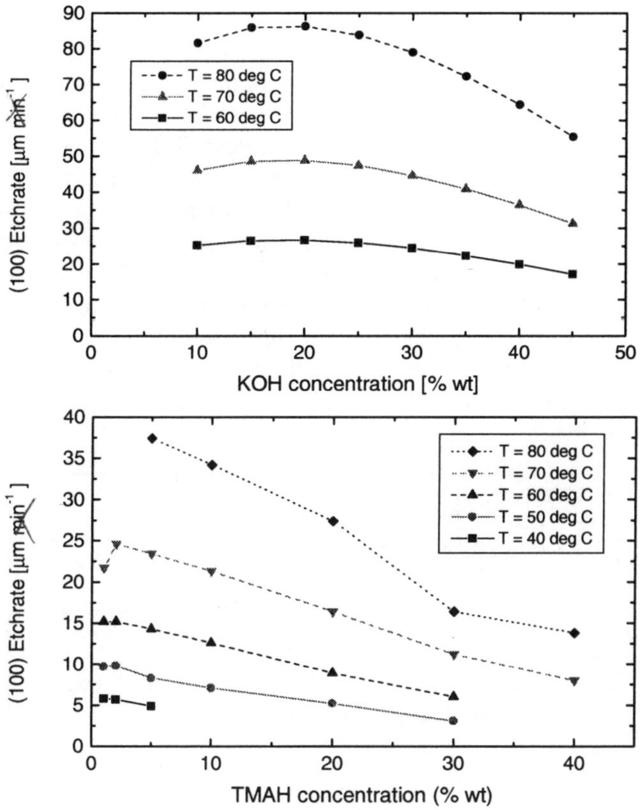

Figure 5.12: (100) Si etch rates for KOH and TMAH

etch depth). Also, the notch depth approaches a maximum value (ca. 6 μm for 25% KOH solution down to 1 μm for 45% KOH) asymptotically for larger windows in the protection layer [65].

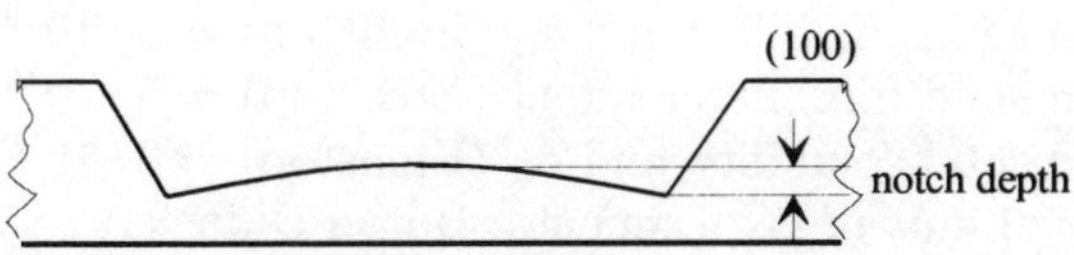

Figure 5.13: Notch depth for anisotropic etching

A very important aspect of anisotropic etchants is the dependence of the etch rate on the boron concentration, allowing the use of etch stop layers. For boron concentrations above $3 \cdot 10^{19}$ cm^{-3} [62,63] the (100) Si etch rate in KOH decreases by several orders of magnitude. It has been shown by Seidel [62], that this effect is even sharper, i.e. the decrease in the etch rate with respect to the boron concentration is steeper, for higher KOH concentrations. This makes it possible to realise a variety of boron doped structures, eg. beams of several mm length, bridges over almost 0.5 mm-wide trenches, diaphragms of several mm size, and other supported elements which can be used for microfluidic valves, pumps and flowmeter construction. The etch stop layer thickness has been experimentally determined for a diffusion time of 3, 4, 5 and 6 hours, using a 10%, 70°C KOH solution. The thickness of the etch stop layer for different diffusion times was compared with Suprem simulations, and a boron etch stop concentration between 5 and $8 \cdot 10^{19}$ cm^{-3} could be determined (figure 5.14).

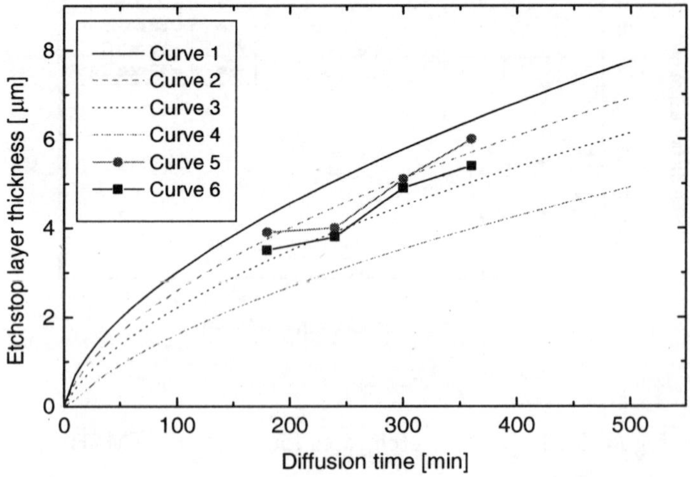

Figure 5.14: Simulation and experimentally determined etchstop layer thickness for different diffusion times at 1150°C

Curve 1: Boron solid solubility and c_B (Etchstop) = $1 \cdot 10^{19}$ cm^{-3}
Curve 2: Boron solid solubility and c_B (Etchstop) = $3 \cdot 10^{19}$ cm^{-3}
Curve 3: Boron solid solubility and c_B (Etchstop) = $5 \cdot 10^{19}$ cm^{-3}
Curve 4: Boron solid solubility and c_B (Etchstop) = $8 \cdot 10^{19}$ cm^{-3}
Curve 5: Higher Limit of Experimental Boron Diffusion
Curve 6: Lower Limit of Experimental Boron Diffusion

The same effect can be achieved electrochemically, where a p- or n-doped layer is positively biased relative to the solution [66,67]. Etching silicon is an electrochemical process and therefore the etch rate can be controlled by a voltage applied to the silicon and the use of a platinum counter electrode and a Calomel reference electrode. The etch process can be effectively stopped with voltages in excess of 0.6 or 1 V for p or n types respectively. If this voltage is applied to a thin n-doped layer, a back etch of the p-doped substrate will allow a simple mechanism of stopping the etch at the n-doped layer and thus creating membranes or other structures of a well defined thickness (figure 5.15). A three electrode solution is preferred in order to overcome the problem of an ill defined solution potential with a two electrode configuration. Thus, a potentiostat in conjunction with the counter electrode and the reference electrode is employed to maintain a constant solution potential with respect to the reference electrode.

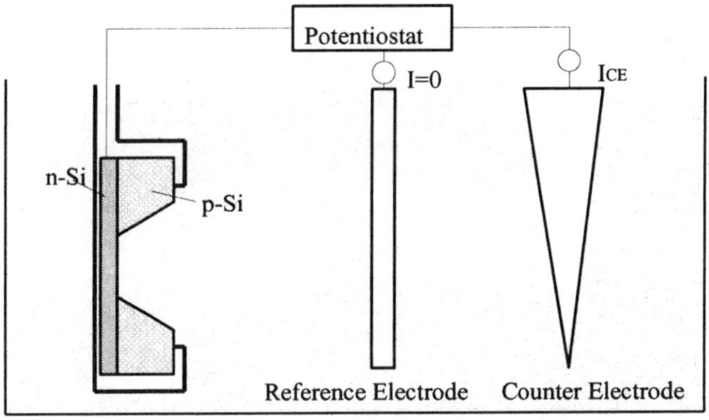

Figure 5.15: Three electrode configuration for the pn etchstop

A very interesting characteristic of TMAH is its low etch rate of aluminium if the solution is highly doped with silicon. This has particular advantages in the integration of microelectronics and micromechanics when aluminium tracks are used for interconnection of the circuits. At 80°C, the silicon content required to stop aluminium from being etched is around 14 g l^{-1} for a TMAH concentration of 5 % wt. and 120 g l^{-1} for 25 % wt [64] and shows almost linear behaviour with respect to the concentration between these values.

The second group of bulk Si etchants shows isotropic etch behaviour where the etch rate is independent of the crystal orientation. This behaviour allows for example the use of round windows in the mask and does not lead to rectangular structures. The disadvantage of this process is the underetching behaviour at the mask edge due to the isotropic nature of the etchant. Isotropic etchants contain an oxidant and an oxide etching species. The most used is the system HNO_3 (70%) and HF (49%) with water or acetic acid as solvent [68–71]. Because HF etches silicon dioxide, layers like silicon nitride, silicon carbide or metals like gold are employed as a masking layer. The surface roughness of the etched area is very dependent on the ratio of the chemicals. A very rough surface with numerous pits and trenches is created by high HF concentrations as well as by high concentrations of the solvent.

Smooth surfaces are only achievable at HNO_3 (70%) concentrations between 35 and 85 % as well as water concentrations up to 15%. This is slightly different if acetic acid is used as the solvent. Here, the HNO_3 concentrations have to be between 40% and 90% and the acetic acid concentration can be up to 30% to achieve smooth surfaces and round corners. The etch rates of the system with water as a solvent are depicted in figure 5.16. It is clearly visible that the highest etch rates are obtained for HNO_3 concentrations between 30 and 40 % .

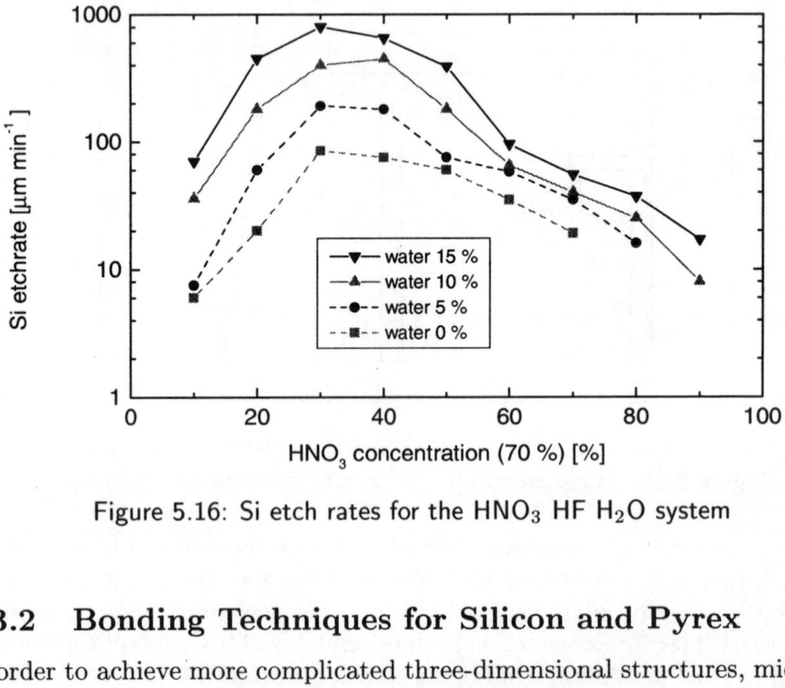

Figure 5.16: Si etch rates for the HNO_3 HF H_2O system

5.3.2 Bonding Techniques for Silicon and Pyrex

In order to achieve more complicated three-dimensional structures, micromachining offers the possibility of stacking two or more wafers on top of each other and achieving bonding without adhesives. To date, several ways have been used for chemically bonding wafers together:

Silicon Fusion Bonding [72]: Two silicon wafers, with pre-cleaned hydrophilic surfaces, are pressed together at room temperature. This leads to a pre-bonding of the two wafers by attractive interactions between hydrophilic surfaces [73]. An additional annealing step at temperatures above 800°C is needed to achieve a permanent chemical bond between surface atoms of the two wafers.

The dangling bonds at the surface of a silicon wafer are usually filled with OH groups. These groups attract further H_2O molecules from the atmosphere. If two wafers are brought into contact, the Van-der-Waals force of the water molecules is responsible for a pre-bonding of the wafers. When the temperature is raised to around 200 °C, the water molecules form a water cluster, which shows four times the bonding strength of the simple Van-der-Waals attraction. At temperatures of 700°C and above, the water molecules break and diffuse into the silicon crystal. This results finally in Si-O-Si bonds at the interface [74]. Particles, surface non-uniformities and trapped gas are problematic during the bonding process, as these are responsible for large unbonded areas on the wafer. Some of these problems can be reduced with a thorough cleaning stage immediately before the room temperature contact.

If alignment errors of around 20 μm can be tolerated, the process for the wafer bonding is rather simple. For good adhesion, one wafer needs to be coated with a 500 nm-thick wet oxide with the other wafer being covered with a native oxide only. However, it has been shown that silicon wafers covered with LPCVD nitride or only native oxide result in a uniform bond, too [73,75]. Bonding these surfaces requires temperatures of 800 °C and above, as the same surfaces have been reported not to bond at temperatures of around 400°C [72]. In order to keep the misalignment small, both wafers need two additional perpendicular flats sawn along the outer scribelines of the design. This guarantees that the flats are exactly oriented to the structures of the devices. The hydrophilic cleaning procedure is best done with an RCA clean. Immediately after the clean, room temperature bonding has to be carried out on a jig, where the perpendicular flats are aligned. Now the permanent bond is achieved for example by annealing the silicon-stack in oxygen atmosphere at 1100°C for 4 hours. Ramping up and down from 400°C has to be done very carefully to avoid any stresses in the wafers, which could lead to a failure of the bond. Good results are achieved with ramping times of 6 hours in either direction. By using this process up to three wafers can be bonded together in one step, with permanent adhesion.

For 1 μm accuracy of alignment of bonded wafers special equipment is necessary that allows x and y adjustments of the two wafers. Such a system is available from both Electronic Visons Inc., USA, and Karl Suss GmbH, Germany. These bonders allow direct alignment of the two wafers with a bottom splitfield microscope. The two wafers are then fixed in hard-, soft-, or vacuum contact in the bond chuck for a pre-defined time to guarantee room-temperature bonding. After this, the pre-bonded wafers are fusion bonded in a chamber, where the process conditions can be set (pressure, high temperatures etc.).

Low temperature bonding can be achieved using an intermediate layer. If a reactive sputtered oxide [76], an APCVD boron oxide [77] or spin-on sodium silicate, aluminium phosphate or ammonia-silica glass [78] are deposited on one wafer, the annealing temperature for obtaining a suitable bond can be decreased drastically to approximately 300°C.

For APCVD deposited boron oxide, grown with a gas mixture of N_2 : O_2 : B_2H_6 (2000 ppm) of 6500 : 70 : 405 sccm at a temperature of 425°C, good results are possible with either SiO_2 covered wafers, silicon nitride covered wafers or wafers with a top layer of aluminium or Cr/Au of 0.5 and 1.0

μm thickness respectively [77]. The annealing temperature after the room temperature bonding has been as low as 450 to 475°C for 10 min.

A second possible low temperature bonding technique uses spin-on of either a diluted solution of sodium silicate or aluminium phosphate on one of the two wafers to be bonded. The solutions are prepared with up to 2% wt. of a concentrated and filtered silicate/phosphate solution in water and should be used soon after the preparation to avoid ageing [78]. Good results are achievable for 2% sodium silicate solution at temperatures of 200°C and higher as well as for 1% aluminium phosphate solution at temperatures of 400°C and higher [78].

Anodic bonding: If a sodium silicate glass (Pyrex #7740) is sputtered on one wafer, the bonding step can be performed by applying a voltage difference of around 50 V between the two wafers and heating to about 350 – 500°C [79]. The second wafer can be coated with silicon dioxide, silicon nitride, aluminium, polysilicon and native oxide. No significant difference in the bond strength is noted for any of these materials [80]. The conventional fusion or low temperature bonding is different to electrostatic bonding in that the wafers are pulled together through the electrostatic field at the interface between the two wafers. This field can be quite high over a very short distance, as the sodium ions move away from the interface towards the cathode leaving negative charges very close to the interface. Once the wafers are in contact, the extremely high fields are responsible for a transport of oxygen atoms out of the glass to bond with the silicon. An improvement of the low deposition rates of sodium silicate glass with sputtering can be obtained by evaporating the glass with an electron beam gun onto one of the silicon wafers [81].

The same technique can also be used for bonding a Pyrex glass wafer to a silicon wafer [79]. Again, the negative potential needs to be connected to the Pyrex wafer. The silicon wafer can be covered with either native oxide, thermal silicon dioxide or silicon nitride (PECVD as well as LPCVD) to allow reasonable bonding, although the bonding voltage needs to be increased for silicon nitride layers. The surface of both glass and silicon wafer needs to be quite smooth to guarantee leak-free bonding. In case of metal tracks, around 100 – 200 nm-thick layers still give reasonable results but often lead to tenting (figure 5.17), i.e. a void between the silicon surface and the Pyrex wafer either side of the metal track. This void can extend up to 30 μm away from the metal track.

Alignment of wafers for both electrostatic bonding techniques can be aided by additional flats in both wafers or with special bonding equipment as described previously.

5.3.3 Surface Micromachining

A whole class of technologies is grouped under the name "surface micromachining" [82]. Instead of using the whole wafer thickness, structures are created only on the surface. In all cases a sacrificial layer is used below a me-

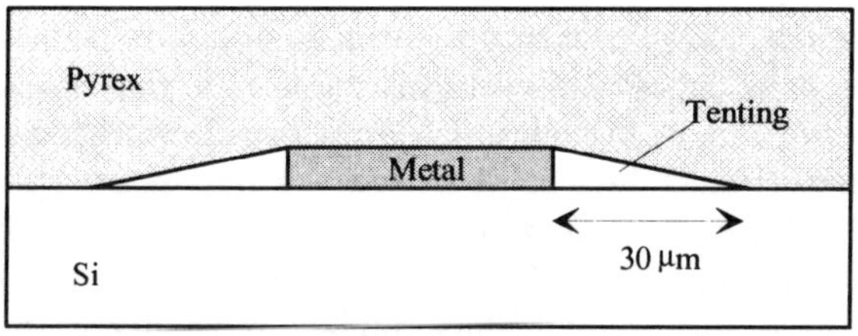

Figure 5.17: Tenting during the anodic bonding of silicon and Pyrex with intermediate metal tracks

chanical layer (metals, poly-silicon, silicon nitride and polyimide thin films). This sacrificial layer is removed after the deposition of the mechanical layer to create suspended structures. Suitable sacrificial layers are APCVD oxide with poly-Si [83,84] or silicon nitride as the mechanical layer, as well as resist with aluminium [85].

The basic principle of surface micromachining is shown in figure 5.18, where poly-Silicon is deposited over a structured oxide layer (figure 5.18.a). Using buffered HF or dry etching, the oxide beneath the poly-Si is removed (figure 5.18.b), and thus only anchored at the poly-Si/Silicon interfaces.

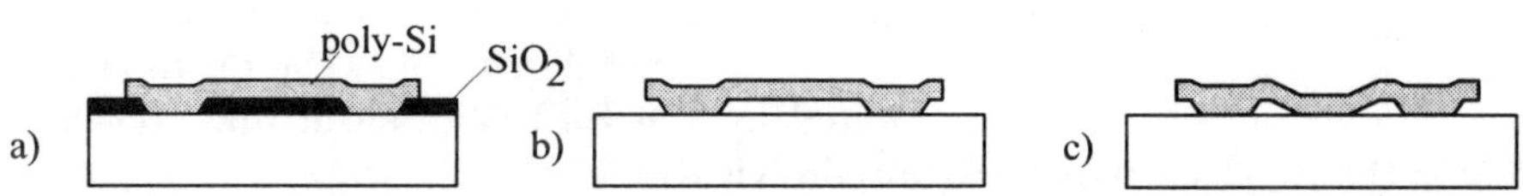

Figure 5.18: Surface micromachining: a) Poly-Si on sacrificial oxide, b) After removal of the sacrificial oxide, c) Sticking of a surface micromachined structure

With this technology a huge number of devices has been fabricated, ranging from devices that are also produced by bulk micromachining to devices only made with surface micromachining. These include accelerometers [86], pressure sensors [83], micromotors [87], microgrippers [84], a windmill [88], microvalves and micropumps [89].

If wet etching is used for the stripping of the sacrificial layer, sticking of the mechanical layer to the substrate (figure 5.18.c) due to surface tension forces is a major problem for the creation of bridges or cantilevers. This problem can be reduced by freeze-drying [90], the use of non-polar solvents [91], resist coverage and ashing or, with limited success, by heating the water in which the wafers are rinsed [92]. If possible, the whole problem can be avoided by removing the sacrificial layer with an isotropic dry etch.

The surface tension of the withdrawing liquid film is large enough to pull any structure and its opposite substrate surface together while drying a surface micromachined device. This can be prevented by immersing the wafer in a liquid that is then frozen to a solid which sublimates, i.e. changes thermodynamically from the solid state directly to the gas phase without going through the liquid phase. Thus, surface tension no longer occurs. As an example the liquid can be t-butyl which is in the solid phase at room temperature [90]. By rinsing the wafer with t-butyl at elevated temperatures and subsequent cooling in a refrigerator, the solvent is frozen. If the wafer with the frozen t-butyl solvent is now placed into a vacuum generated with a rotary pump, the solvent sublimates and does not create any adhesion problem.

Another method to prevent sticking is the use of Toluene (C_7H_8) or n-hexane (C_6H_{14}). Both organic chemicals are hydrophobic liquids. Because of the similar electronegativity of the carbon and hydrogen atom, no dipole as in the water molecule is generated. After the rinse in water, the silicon surface has already developed a native oxide surface and therefore repels any Toluene or n-hexane molecule. This means that the surface micromachined structure is not dragged towards the silicon surface due to surface tension and thus does not stick. However, as Toluene as well as n-hexane are not miscible with water, an intermediate rinse in alcohol is necessary because water mixes with alcohol and alcohol mixes with Toluene.

An example of a surface micromachined structure is depicted in figure 5.19. With the use of hinges and several layers of sacrificial / mechanical layers, movement can be achieved in the third dimension. The example is a micromirror, which has been created by the Berkeley Sensor and Actuator Center (http://bsac.eecs.berkeley.edu/visitors.html).

5.3.4 Deep Dry Etching

Deep reactive ion beam etching is similar to microelectronic dry etching but achieves aspect ratios of at least 15:1 with almost vertical walls up to 400 μm depth. It is based on RIE but through special sidewall protection during the etch, high aspect ratios are achievable.

One example of such a process has been developed by STS Inc. and Robert Bosch GmbH to allow deep trench etching. Using an environmentally friendly fluorine based chemistry, the process has been developed on a multiplex inductively coupled plasma system which produces high plasma densities at low pressure with good uniformity. It allows silicon etch rates of approximately 3 μm min^{-1}, has a selectivity to resist of greater than 75:1 and provides aspect ratios of greater than 40:1. A picture of a 80 μm-deep trench etched using an STS equipment is given in figure 5.20.

This process can be used for controlled deep anisotropic etching to buried oxide layers. The absence of undercutting at the oxide interface is one of the interesting features of this process. Together with wet or dry undercutting of

Figure 5.19: Micrograph of a micromachined movable reflector (Berkeley Sensor and Actuator Center)

the buried oxide material this technique allows the creation of free standing structures for capacitive inertial sensors [93]. Here, SF_6 has been applied as the etch gas to create structures of 45 μm depth and almost vertical walls.

Deep RIE is a versatile technique for silicon micromachining. As another example, a vertical mirror was fabricated with an etch depth of 75 μm and a maximum width of 2.3 μm [94] (figure 5.21.a). In order to create force-balanced accelerometers, the use of silicon fusion bonding and deep RIE [95] (figure 5.21.b) is possible. Here, two wafers are bonded together after a cavity has been etched into the handling wafer. The wafer on top of the handling wafer is then polished down to 20 – 200 μm. Now, circuitry can be integrated using standard IC technology and the top wafer structured with DRIE to form trenches between capacitive plates. The advantage of this technique is that no removal of an underlying oxide is necessary as in the other cases, thus avoiding sticking problems.

5.3.5 Ion Beam Milling

Ion beam milling is a way of etching a material in a purely physical way by the use of a collimated, uniform beam of inert ions which sputters the surface of a substrate. The inert ions are usually Ar and are generated from a plasma discharge. The momentum of the incoming ions is large enough to break the bonds of the atoms on the substrate surface. If the area is protected by a mask layer, the substrate will not be etched. However, the etch rate ratio between substrate and mask layer is often around 1:1 due to the physical nature of the etch. Thus, for structure depths of several microns, a resist of several microns

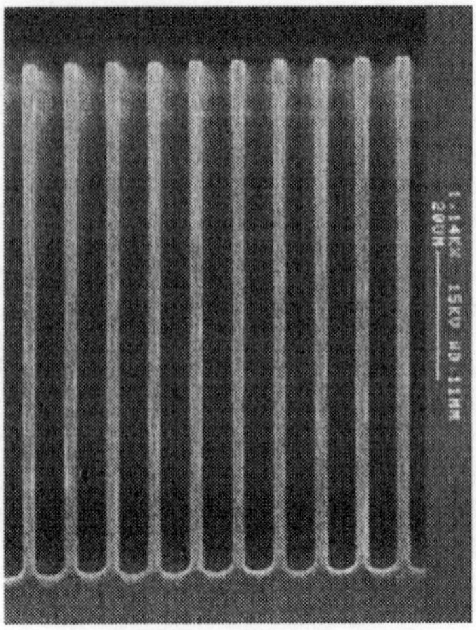

Figure 5.20: Micrograph of silicon trenches 80 microns deep, 4.5 micron trench widths, 2 micron line widths, aspect ratio = 18:1, SiO_2 mask (STS Inc.)

has to be employed.

Due to its simplicity, a Kaufmann source is widely used in the generation of the ions. It shows good performance for moderate beam diameter and non-reactive gases. In this type of source, a plasma is generated in an electric field between cathode and anode. The ions from this plasma are then accelerated through a grid towards the target. A neutraliser is placed before the target to reduce the possibility of removed atoms re-depositing on the substrate. Due to sputtering of the cathode by ion bombardment and chemical erosion a Kaufmann source has a limited lifetime, and there is also a problem of contamination of the beam with tungsten from the cathode. In contrast to this, capacitively coupled RF ion sources are simple, do not have any electrodes and therefore have longer lifetimes.

In order to allow long milling times the substrates have to be cooled. This can be done by passing either liquid helium or liquid nitrogen through the substrate chuck. It is well known that the optimum milling of an ion beam system requires that the substrate be rotated in the beam and that the impact angle needs to be greater than 0°. The general etch rate of resist, Ni, Fe, Si and SiO_2 is around 32 nm min^{-1} for a substrate normal to the beam and an argon ion beam acceleration voltage of 600 eV. Aluminium and gold show a value of 34 nm min^{-1} and 92 nm min^{-1} respectively for the same angle. The optimum milling rate for Al, Ni, Fe, Si, resist and SiO_2 is obtained for angles in the range between 40 and 60 °. Gold shows a continuous superlinear decline of the etch rate for increasing incidence angles. A brief summary of the etch rates of various materials at an acceleration voltage of 600 eV and a beam current density of 0.6 mA cm^{-2} is given in table 5.9.

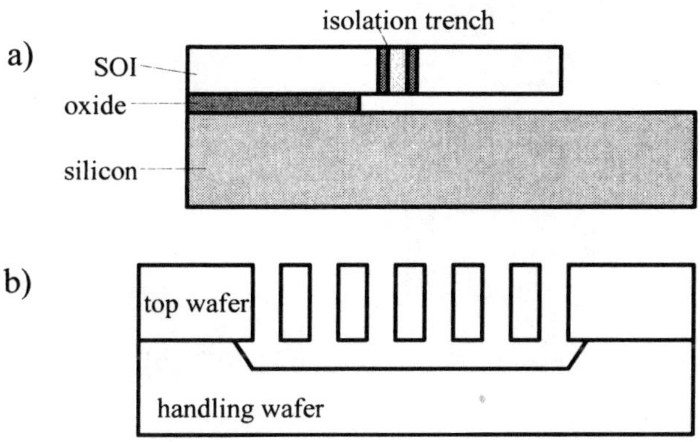

Figure 5.21: Devices with DRIE: a) DRIE and underetching of the buried oxide for a vertical deflectable mirror [94], b) silicon fusion bonding and DRIE for a force-balanced accelerometer [95]

Material	Al	Au	Pt	Si	SiC	SiO_2
etch rate [nm min^{-1}]	34.3	92.4	51.4	24.4	21.1	26.4
Material	Si_3N_4	Ti	resist	glass	GaAs	Ge
etch rate [nm min^{-1}]	25	21.1	19.8	13.2	99	59.4

Table 5.9: Etch rates of various materials (normal beam incidence)

Systems for ion beam milling are available from Oxford Instruments, Nordiko, Veeco and other suppliers. Some of these systems allow the combination of IBE, Reactive IBE and Chemical Assisted IBE in one reactor.

Ion beam milling is often used in structuring metal layers on semiconductor devices. More interesting however are the applications in micromechanics where IBE allows the creation of three-dimensional structures. As an example, an optical lens has been fabricated (figure 5.22) by reflowing patterned photoresist islands and subsequently transferring these microlens shapes into a quartz substrate by ion beam milling [96]. A 30 μm lens height is produced from a 19 μm-high resist structure, after milling for 10 h. The difference in height and hence in the focal length of the resist and quartz lens is due to the different milling rates of quartz and resist.

5.3.6 Sol-Gel Deposition

The capital costs for sol-gel deposition are low, as no complicated reaction chambers are necessary. This is especially interesting for multi-component systems like PZT. The production of films by sol-gel deposition is done via

a) 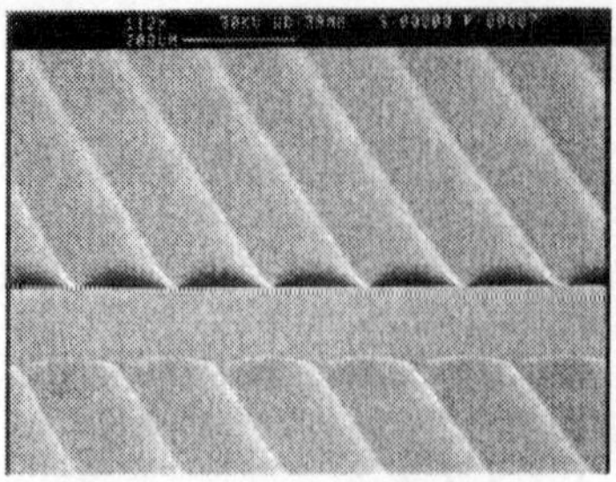 b) 

Figure 5.22: a) Resist lenses in 19 μm-thick STR1075, b) lenses milled into quartz

spin-coating a substrate with an organic solution of the components and firing at temperatures between 500 and 700°C to convert the spin-coated layer into a crystalline oxide layer. In addition to low cost, sol-gel offers the advantage of retaining the purity and chemical homogeneity of the starting film. A disadvantage is the maximum layer thickness which is usually not greater than 10 μm.

PZT is one of the most interesting materials for both microelectronics due to its high dielectric constant and micromechanics due to its actuation and sensing abilities. The method of sol-gel deposition is described here using the creation of a PZT layer as an example. To date several ways of preparing PZT layers have been described. The first method is the methoxy-ethanol method which yields single layer films of up to 0.1 μm thickness [97]. A second method is based on acetic acid and gives thicknesses of up to 2.5 μm that are rather porous [98]. The third method uses diol based sols to form PZT layers which are up to 1 μm thick [99]. To achieve thicker layers it is possible to repeat the sol-gel deposition up to 10 times without any major loss in performance. Recently, another sol-gel route has been described to give thicknesses of up to 60 μm by spin and dip coating multiple layers [100]. The technique involves mixing PZT powder with conventional PZT sol gel to form a paint which is then coated onto a flat substrate [101].

The generation of PZT via the diol sol-gel route is now described in more detail as it is very common. Gel precursor solutions for PZT (Zr 53 / Ti 47) are formed from lead acetate trihydrate [$Pb(CH_3COO)_2 \cdot 3H_2O$], titanium di-isopropoxide bisacetylacetonate [$Ti(OC_3H_7)_2(CH_3COCHCOCH_3)_2$] (TIAA) and zirconium n-propoxide [$Zr(O\text{-}n\text{Pr})_4$]. To arrive at the stock solution, the zirconium n-propoxide is refluxed with acetylacetone for 20 min and then TIAA and 1,3 propanediol are added and the whole is refluxed for 2 hours. At the same time lead acetate trihydrate is refluxed with 1,3 propanediol for 2 hours. The two solutions are then mixed together and refluxed for 5 h. During this time, the solution is distilled after 1 h and 4 h at 80 – 90 °C to remove any unwanted by-products. Now the sol is spin-coated on platinised silicon substrates and pre-fired at 300 – 450 °C before a final firing stage in air at 600 – 700°C. This sol-gel route can lead to high-quality single-layer films of up to 1 μm thickness. In order to increase the thickness further, 10 single layers on top of each other are possible yielding a thickness of 10 μm. Owing to the growth from a Pt-Ti-SiO_2-Si substrate and homogeneous nucleation, the PZT layer shows the preferred (111) orientation. With this technique, dielectric constants of 250 – 400 are possible [99]. The piezoelectric behaviour of this PZT layer has not been determined.

PZT films based on the 2-methoxyethanol route need three spin - dry - firing cycles to reach 1 μm thickness. The piezoelectric coefficient d_{33} for these films has been determined at 130 – 200 pC N^{-1} after poling [102]. PZT films derived from the mixture of powder and sol-gel show even higher values of 250 pC N^{-1} for the d_{33} coefficient and 900 for the dielectric constant [100] for a layer of 18 μm thickness.

5.4 Special Microfluidic Processes

Although the following processes are technologies that can be used in either micromechanics or microfluidics, they have been separated out here as many of the publications and presentations on these topics have been on microfluidic applications. The thick-film-on-silicon technique has been extensively used for micropump actuation [103, 104]. Isotropic etching of Pyrex wafers is a common process in microfluidics and the micro total analysis systems [105] for example make use of etched channels in Pyrex or quartz wafers. LIGA is an abbreviation of lithography, electroplating and moulding and is a very general technology. Because of the extensive use of LIGA in microfluidics (amplifiers [106] and micropump [107] for example) it is also included in this section.

5.4.1 Thick-Film-On-Silicon Technique

One way of fabricating piezoelectric actuators on silicon membranes is to bond a bulk piezoelectric plate on the membrane with the use of adhesive. A better way of mass producing thick piezoelectric layers on silicon wafers is to use thick-film technology and print the piezoelectric layer. With this technique, thicknesses of up to 150 μm can be achieved.

5.4.1.1 Introduction to Thick-Film Technology

Thick-film printing is a way of using screen printing for depositing patterned thin conducting, resistive or dielectric layers onto a substrate. The substrate is most often alumina, although some applications, like anti-reflection coatings on solar cells, are fabricated on silicon. The pastes or inks for screen printing are generally composed of the active element, a binder (glass frit) and a vehicle (organic solvent), to allow printing. A layer is realised by printing through a mesh, i.e. the screen, which is usually filled with a UV-sensitive resist that has been exposed with the intended structure and developed. The wire-mesh fulfils two tasks, which are holding the screen together, and determining the thickness of the printed layer, as the thickness of the screen is usually the final thickness of the layer. The screen is placed at a certain distance from the substrate (~ 1 mm, figure 5.23.a), and a squeegee (similar to a windscreen wiper) is swept over the screen (figure 5.23.b) to bring it into contact with the substrate and press some material through the very fine mesh (figure 5.23.c). Once the squeegee has passed, the screen jumps back to its original position, and the deposited material stays in place due to adhesive forces (figure 5.23.d).

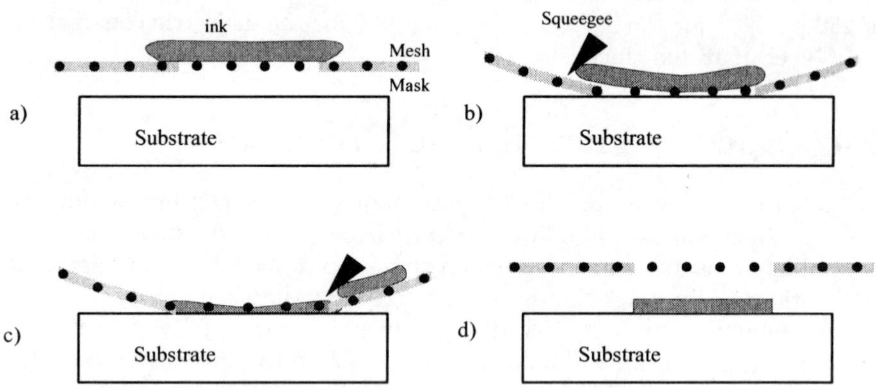

Figure 5.23: Screen printing technology

After printing, the substrates are dried in air for a short time to allow the ink to settle, before a subsequent drying process at 100 – 150°C removes most of the organic vehicle. Finally, a high temperature firing cycle follows where the rest of the organics is removed and the active material is sintered to give good mechanical contact within the grains of the material and between the material and the substrate. This can be done at temperatures as high as 1000°C.

If thicker layers are required, it is possible to deposit another layer of the ink on top of the first one. However, the second deposited layer is most often thinner than the first.

Generally, there are two different types of binder available. Inks with the first binder are usually called *cermet* inks, as this binder consists of both a ceramic component and a metallic component. Both components are mixed with the active element and the organic vehicle to form the paste as described above. The binding mechanism of these inks is more mechanical than chemical, and the difference in printed, dried and fired thicknesses of the layer is only small. The second ink types are called *resinate*, or *metallo-organic* inks, as they contain a metal and an organic compound dissolved in a large quantity of solvent. The binding mechanism of these inks is mainly chemical. The wet thickness of the printed layer should generally not exceed 13 μm in order to achieve a uniform bubble-free layer. After firing, this decreases to less than 1 μm due to the large amount of solvent evaporating during the firing process.

Suitable inks for producing conducting or dielectric films are readily available, but there is no commercially available piezoelectric paste. The most promising material for piezoelectric behaviour is PZT (lead-zirconate-titanate), due to its large piezoelectric coefficients and good mechanical stiffness. Researchers have, however, made a printable ink from PZT powder [108] and used it in thick-film sensors on alumina [109]. A characterisation of screen printable piezoelectrics has been published by Dargie [110], examining the piezoelectric properties of the materials after firing pastes with different binders. The best composition on alumina substrates was found to be with 97% PZT powder and 3% lead(II)oxide.

It is not the intention of this work to give a more detailed review of thick-film technology and the possible applications in the electronic industry. For a more extensive description of thick-film technology or a summary of sensing devices that can be fabricated with this technology see the review papers by Brignell and Middelhoek [50, 111].

5.4.1.2 Process Sequence for a Piezoelectric Actuator

In order to print the piezoelectric actuator (figure 5.24) on silicon a sandwich of electrode - piezoelectric layer - electrode needs to be printed. As a first layer, the bottom electrode is deposited. This can be done by screen printing or physical vapour deposition. Best results for the piezoelectric material have been achieved with a 10 μm-thick gold cermet ink and a 1 μm-thick evaporated platinum layer. The cermet ink is printed twice, dried at 150°C for 10 min and fired in a conveyor belt furnace for 1 h with a peak temperature of 950°C.

The bottom electrode needs to be made larger than the diaphragm. The reason for this is to reduce stress in the membrane due to temperature coefficient mismatch. A list of common thermal coefficients of expansion is given in table 5.10. Most commercially available thick-film inks are designed to match the TCE of alumina, which is different from silicon by a factor of 3.

Material	Si	Al_2O_3	Ag	Pd	Au	bulk PZT
TCE [10^{-6} K^{-1}]	2.33	6.8	19.3	11.6	14.1	2.5

Table 5.10: Thermal coefficient of expansion TCE at T=300K [17, 25]

The alignment of the first thick-film screen to the silicon membranes can be done by the use of alignment structures on both silicon wafer and screen. Unlike the alignment in contact printing on silicon (1 μm alignment error) and the sawing (10 μm alignment error), the horizontal alignment resolution for thick-film processing is about 100 μm.

The next step is to print the piezoelectric layer. The ink for this layer consists of 19 g PZT-5H powder (supplier: Morgan Matroc) with a grain size of 6 μm mixed with 1 g lead zinc borosilicate powder (Corning 7575) and 5 ml of organic vehicle (Electro-Science Laboratory ESL 400). This gives a concentration of 5% lead zinc borosilicate in the PZT paste. Alignment of the PZT layer to the underlying electrode layer is possible with alignment structures on both the bottom electrode and the PZT screen. Again, alignment errors of 100 μm and less are achievable.

In order to realise 100 μm-thick layers of PZT, the ink has to go through two cycles where it is printed twice, dried and fired. Drying is done for 10 min at 135°C in an infrared dryer. The ink has to be fired in a conveyor belt furnace for 1 h with a peak temperature of 950°C to give optimum piezoelectric coefficients. Adhesion of the PZT ink after the second firing step is good as no parts come off during scratch tests.

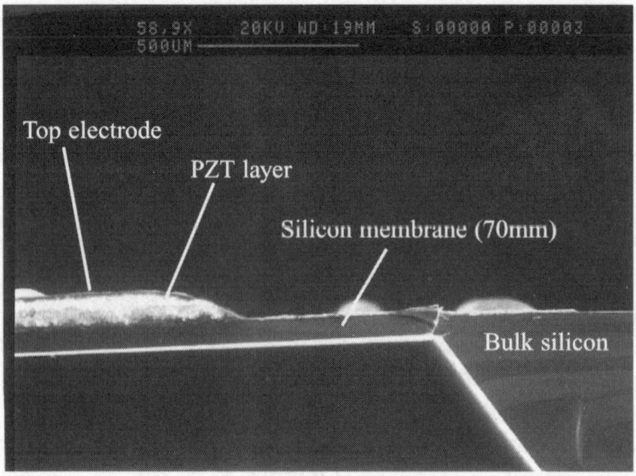

Figure 5.24: SEM of a cross section of the hybrid actuator

The structure is completed with the printing of the top electrode onto the PZT layer (figure 5.24). Usually, the same electrode material is printed as for the bottom electrode. However, in order to keep the processing temperature low, a conducting polymer layer can be used.

The dielectric constant of PZT with a cermet gold bottom electrode on 500 nm wet silicon dioxide is 461±26. This figure of merit is a factor of 7 lower than the bulk values (3400 for PZT-5H [17]). The cermet gold electrode for this process was fired at 950°C in a dry atmosphere as described before. Displacement measurements of a 8 × 4 × 0.07 mm silicon membrane with a 7 × 3 × 0.09 mm PZT layer gave 450 nm_{pp} at 100 V_{pp} for low frequencies around 500 Hz and 650 nm_{pp} at frequencies around 3 kHz. The piezoelectric charge constant for PZT on cermet gold electrodes was 103 pC N^{-1} or 1/6 of the bulk value [17].

For lower firing temperatures of the PZT layer on a cermet gold electrode, the piezoelectric activity drops. At a firing temperature of 850°C for 1 hour the dielectric constant is around 370±10 and measurements of the piezoelectric charge constant give values around 92 pC N^{-1}. Both factors indicate that the higher firing temperature is necessary to properly densify the printed ink.

Further reduction of the temperature is only possible by combining the PZT powder with 5% of a low-melting-temperature glass (FEG EG2760). This reduces the firing temperature to 570°C. The dielectric constant is reduced further to 83±12. The membrane actuator, as described before, is functional in the range of 8 kHz to 12 kHz but central displacements are less than 150 nm for 100 V.

With the use of a polymer binder (Lea Ronal OPSR 5100G) even lower firing temperatures are achievable. Although the reduction in firing temperature decreases the dielectric constant and thus the output power of a membrane actuator, this could be useful for other sensing or actuating devices that include for example microelectronics on chip. The ink needs to be printed twice, dried at 90°C for 30 min, exposed with UV light for 10 min and then cured at 150°C for 2 hours. After curing, adhesion of the PZT/polymer ink to a smooth evaporated Pt surface is not satisfactory, as the piezoelectric layers peel off in scratch tests. This seems to be due to a difference in the adhesion mechanism. Whereas fired inks achieve a chemical bond to the bottom electrode, a UV curable ink relies purely on a mechanical bond. Other tests of the PZT/polymer ink on alumina substrates however achieve good adhesion to a cermet gold electrode.

Apart from gold cermet electrodes, evaporated platinum electrodes give good piezoelectric coefficients. The dielectric constant of a PZT layer on 500 nm platinum and 500 nm of wet oxide is around 485±67, which is slightly larger than the values for a gold cermet ink. For 85 μm-thick PZT on a silicon membrane and low actuation frequencies the peak-to-peak displacement is 350 nm_{pp} for 100 V_{pp}, rising to 600 nm_{pp} at 3 kHz. This is in the range of the gold cermet electrodes. A 135 μm-thick PZT layer on Pt achieves only 350 nm peak-to-peak displacement at 3 kHz, being half the value of the thinner layer.

5.4.1.3 Diffusion of Lead into Silicon

A problem of the thick-film-on-silicon technique for PZT is that lead diffuses out of the piezoelectric layer during the firing step at elevated temperatures [112] and into the underlying dielectric layer (SiO_2), converting the insulator into an electrical conductor.

Figure 5.25 shows a SEM cross section of the interface between silicon and PZT with an intermediate SiO_2 layer. As backscattered electron imaging is used, materials with a high atomic number result in bright points and materials with low atomic number result in darker points.

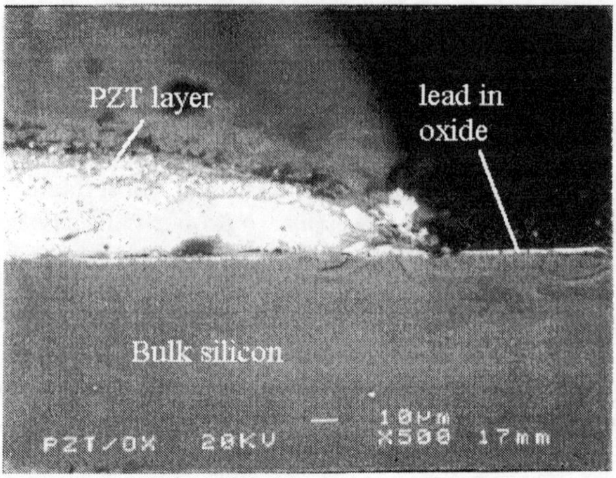

Figure 5.25: SEM micrograph of the PZT/SiO_2/silicon interface

Both the picture and EDX measurements confirm that the silicon dioxide layer is lead enriched. The lead can only originate from the PZT. No atom concentration greater than 1% of lead could be found 10 μm below the silicon surface, showing that the effect is only on the surface [112].

In microelectronics, Si_3N_4 is often used as a diffusion barrier. When used instead of SiO_2 in the above structure however, this was found to result in some sort of bubbles beneath the electrode. These can have a diameter of around 70 μm and a height of 15 μm above the silicon surface. The large density of such bubbles in a small area disqualifies Si_3N_4 as a suitable insulator.

In order to screen print conducting layers onto metal structures [113], a commercially available insulating ink (IP211 by Heraeus) needs to be printed as the first layer. This paste contains a devitrifying glass component that crystallises on initial sintering. Owing to this effect, the film does not soften in subsequent firing steps. This leads to a stable foundation layer for further printing. It has been shown that the ink can be successfully printed, dried and fired on top of a 500 nm thick SiO_2 layer [112]. The ink prevents the surface of the wafer becoming conductive, but introduces major stresses into the silicon. This does not affect any devices on 500 μm-thick silicon but can lead to the destruction of membranes of 100 μm thickness and less. This is due to the thermal coefficient of expansion for IP211. As it was formulated by Heraeus to be printed onto steel, the TCE matches that of steel (10 to $16 \cdot 10^{-6}$ K^{-1}), whereas the TCE for silicon is $2.33 \cdot 10^{-6}$ K^{-1}. This difference leads to a considerable mismatch in thermal expansion during the firing stages.

A change in the layout, however, can help to overcome the problem of lead diffusion and its effect on the conductivity of the silicon dioxide. As long as the top electrode is isolated from the lower electrode by the PZT layer, the conducting SiO_2 does not affect the operation. This means that the PZT has to be extended beneath the top electrode up to the bonding pads. A picture of such a layout is given in figure 5.26.

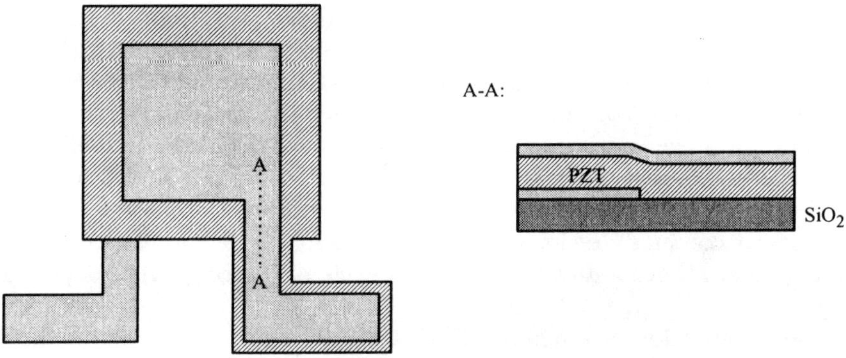

Figure 5.26: Top view and cross section of the improved design for the actuator

Apart from the insulation, this structure has another advantage in that the top electrode always runs on the PZT. For printing metal tracks, which run over PZT and silicon, a large printing pressure has to be applied. This can lead to the destruction of the membranes during printing. In the case of the top electrode always being on the PZT, the printing pressure and thus the danger of membrane destruction can be significantly reduced. The purpose of the silicon dioxide layer is thus reduced to guarantee a good bond of the inks to the substrate and to prevent the formation of a eutectic of silicon with gold.

5.4.1.4 Characterisation of the Inks

For the electrode materials other conducting layers are possible as well: silver palladium, gold resinate, platinum cermet, platinum resinate. Additionally, the bottom electrode can be made of evaporated gold. As silicon nitride is unsuitable due to the bubble generation, the insulation layer can only be silicon dioxide in different thicknesses. However, all these combinations result in smaller piezoelectric and dielectric coefficients than gold cermet and evaporated platinum.

Silver palladium is printable on 500 nm wet SiO_2 and native oxide covered silicon wafers. The dielectric constant on both substrate insulators is between 50 and 70, independent of the substrate insulator. This has to be compared with the bulk value for PZT-5H, which is 3400 at room temperature [17], being nearly sixty times larger. An actuator with a 100 μm-thick PZT layer on a 70 μm-thick membrane is functional in the frequency range between 8 and 12 kHz at a voltage of 100 V_{pp} but shows central deflections of less than

150 nm_{pp}. The same inks, when printed on alumina substrates, achieve a dielectric constant of 373±11.

There are several possible ways to deposit gold layers: apart from thick-film technology with resinate gold inks, thin-film deposition is an attractive method of layer formation. Any deposition of gold onto pure silicon wafers, only protected by the native oxide, fails because gold reacts with silicon to form a eutectic. Thus the only substrate insulator which can be used is a 500 nm-thick oxide. All resinate gold (ESL D8085) bottom electrodes printed onto the silicon wafer with an intermediate 500 nm thick silicon dioxide layer show local failures in conductivity after two prints, drying and firing. Even additional prints of the resinate gold cannot improve the layer. Unlike the local loss of conductivity for the bottom electrode, the top electrode is not conductive at all even after repeating the cycle of printing twice, drying and firing.

The evaporation of a 500 nm-thick gold bottom electrode is, unlike the resinate gold ink, a fully compatible thin-film process that can be carried out in any cleanroom where the silicon membrane is etched. After firing the PZT, the exposed gold of the bottom electrode changes its colour from metallic to dull. A similar effect occurs when printing the piezoelectric material directly onto silicon dioxide, where the oxide around the PZT changes its colour. As the impact on the gold is similar, this can very probably be taken as the result of the lead diffusion. The dielectric constant of the PZT layer on thin-film gold reached a value of 126±7 which is already an improvement compared to the AgPd screen printed electrodes. Yet this is still a factor of 24 less than the bulk value.

It is not possible to print a continuously conducting resinate platinum layer (Heraeus RP 1001-145B). Even after four prints on 500 nm wet SiO_2 the layer is not conducting.

With a cermet platinum ink (ESL 5545) the problem of conduction can be solved. After this, the PZT is printed by the standard technique as described before. Compared with gold cermet layers, the dielectric constant of the PZT is rather low at 58±8. As in the case of the AgPd electrodes, a PZT-on-silicon membrane actuator is functional in the frequency range between 8 and 12 kHz but displacements are below 150 nm_{pp} for 100 V_{pp}.

Table 5.11 gives a summary of all the prints with different binders and bottom electrodes, and shows that gold cermet inks and evaporated platinum achieve the best performance.

5.4.2 Isotropic Etching of Pyrex

An important technique for making structures in glass is the isotropic etching of Pyrex wafers. Typical structures might be through holes, surface channels or flat recessed areas. In particular, capillaries to be used for electrophoretic/osmotic pumping might be made this way because of the superior voltage insulation of glass compared with alternative structures made on sil-

Bottom Electrode	AgPd cermet	Au cmt.	Au evap.	Pt cmt.	Pt evap.	Au cmt.	Au cmt.
PZT binder	*BS*	*BS*	*BS*	*BS*	*BS*	*BS*	*LTG*
Diel. Constant	60	461	126	58	485	370	83
d_{33} [pC N^{-1}]	x_1	103	x_1	x_1	95	x_1	x_1

Table 5.11: Summary of some of the combinations for a hybrid actuator.
x_1: d_{33} not measured but membranes functional between 8 and 12 kHz.
BS: 5% Lead zinc borosilicate glass, *LTG*: 5% Low temperature firing glass

icon [114]. Other examples of the use of this technique, in conjunction with anodic bonding, are flowsensors [48] and a capacitive accelerometer [115].

The pre-treatment of the glass wafers has a significant effect on the surface quality of the etching. Depending on the supplier of the Pyrex material, the surface is often mechanically stressed after polishing. This stress can only be relieved by annealing the glass for 1 – 2 hours at 570 °C. A very ragged edge is obtained in the trenches if the Pyrex wafers are etched without prior removal of the surface damage. The only way to avoid the ragged edges is to remove the mechanical stress in the wafer surface. Before the annealing, the glass is normally cleaned with fuming nitric to avoid any stains on the surface.

Where masking is required against etching, the Pyrex wafer is coated with a 100 nm-thick Au layer using a 20 nm-thick Cr layer as adhesion layer. Both layers can be patterned through a photoresist that has been exposed. After hardbaking the resist at 115°C for 10 min, the wafers are first immersed in potassium iodide for 2 min to etch the gold layer and then in a chrome etch for another 2 min. Without removing the resist, the whole wafer is placed into 48% HF for 5 min to give 40 μm-deep trenches. Thus the etch rate for this etch is around 8 μm min^{-1}. Now the layers of resist, gold and chrome can be removed successively and the trenches in the glass are finished. A picture of the etched surface of pre-annealed Pyrex is shown in figure 5.27.

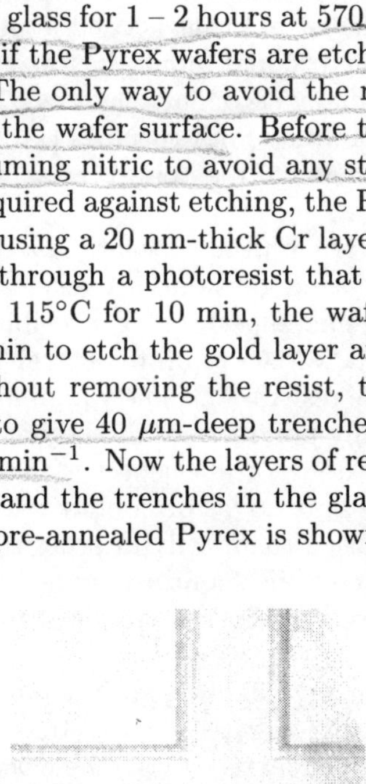

Figure 5.27: Pyrex wafer after annealing and wet isotropic etching in 48% HF. Etchdepth = 40 μm, structure width = 500 μm

For double-sided deep etching of the Pyrex wafers, thicker gold layers are necessary. In order to protect the smooth surface beneath the chrome/gold mask layer for etches up to 250 μm or half way through the wafer, a gold layer thickness of up to 700 nm is necessary. Even then it is recommended that the resist is left in place during the etching. Keeping the resist on both sides is not possible for double sided etching as the lithography and etch step of the chrome/gold layer needs to be performed for each side individually. Thus it is recommended to spin-strip the resist from the less important side with acetone in order to leave the resist on the more important side.

Another possible Pyrex etch is a mixture of 30 % HNO_3 and 70 % HF at 48 °C [115]. Again, a chrome-gold mask is necessary when smooth consistent results with minimum undercutting are to be obtained.

5.4.3 LIGA Technique

Another technology for fabricating microfluidic and other devices is LIGA. This is the German abbreviation for lithography, electroplating and moulding. Unlike most other technologies, LIGA does not rely on silicon as a substrate material but can use any substrate that is conducting or provides a conducting surface layer. Once the original mould has been formed, it can be used for injection moulding of many secondary devices. The technology has been successfully used for the production of turbines [116], pumps [117], microfluidic amplifiers [106] and chemical reactors [118] as well as other devices like accelerometers.

The LIGA process starts with the exposure of a resist with radiation from a synchrotron. The synchrotron guarantees a very parallel beam of radiation, making structures of extreme aspect ratios possible. The resist is usually polymethyl-methacrylate (PMMA) of up to several 100 μm thickness (figure 5.28.a). As the stress in the resist is extremely high for these thicknesses when spun on, the resist is usually polymerised directly at the substrate surface to avoid mechanical stress. An important aspect is the adhesion of the resist to the substrate. Best results have so far been obtained by the use of a titanium surface layer which is oxidised chemically in order to achieve a micro porous surface to allow the resist to mechanically adhere to the substrate. After exposure, the resist is developed in the standard tank development process (figure 5.28.b).

It is important to have a conducting surface on the substrate for the following stage of electroplating (figure 5.28.c). Here, several problems occur and have to be solved independently: First, the electroplated material needs to have good adhesion to the substrate. Again, the microroughness of the chemically produced TiO_x layer is advantageous for the adhesion and, provided $x < 2$, the layer shows the electrical conduction which is necessary for electroplating. Second, the intrinsic stress needs to be low in order to avoid bending of structures with large aspect ratios. The intrinsic stress of the plated film can be controlled by the deposition temperature and the current

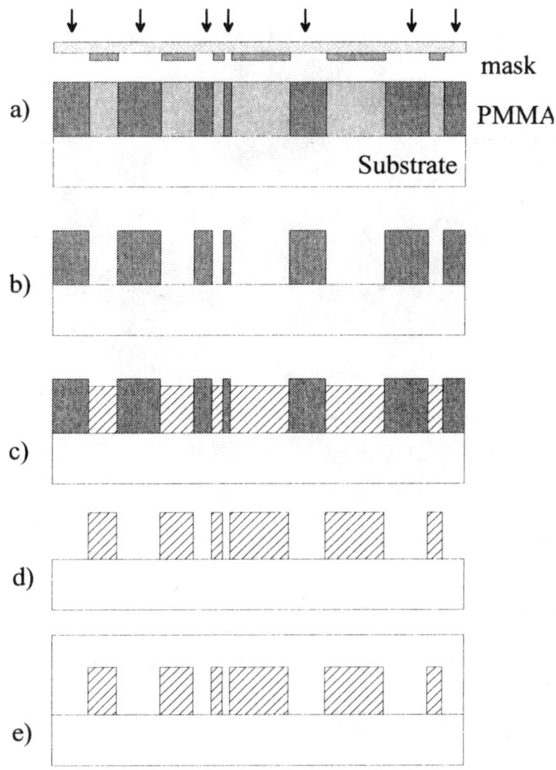

Figure 5.28: Overview of the process steps for LIGA technology

through the electrochemical cell. For example, the intrinsic stress is almost zero for structure heights greater than 50 μm and current densities of around 500 A m^{-2} [119]. Bubble generation during the electroplating is a problem too, as is the deposition in narrow trenches, which can be improved by wetting agents.

The third step in LIGA is the removal of the resist (figure 5.28.d). This is a relatively simple process and similar to normal lithography, where the resist is removed by acetone, fuming nitric acid or in an oxygen plasma. Now the metal form is finished and can be used without damage for up to 1000 injection mouldings [119]. As in conventional large scale moulding, microstructures can be formed by reactive injection moulding, injection moulding or pressing.

As an example, an integrated planetary gear system which has been fabricated with the LIGA technology is depicted in figure 5.29.

In order to mass produce microstructures with the LIGA technology, the final moulding step is usually the formation of a plastic structure from the previously generated metallic mould (figure 5.28.e). Reaction injection moulding

Figure 5.29: SEM micrograph of a microgear fabricated in LIGA technology (IMM Mainz, Germany)

is a process where different components are injected simultaneously and polymerise within the mould. This has the advantage that the viscosity is very low and therefore injection temperature (40 °C) as well as pressure (100 kPa) can be kept low, thereby reducing the possibility of damaging the original mould. The second technique is injection moulding, where molten polymers are injected into the mould. This requires temperatures of around 200 °C and pressures of 1 MPa. In order to achieve the high aspect ratios of microsystems, the mould has to be kept above the glass transition temperature to prevent hardening before the plastic has filled all structures. This, however, increases the moulding time and therefore the production costs. The successful removal of the plastic from the mould for both injection moulding techniques is only possible if the walls are very smooth. This can be guaranteed by the use of the synchrotron radiation for exposure.

The third method of moulding is pressing. In this case the mould is pressed into the softened plastic which retains the shape of the mould when it is cooled. This technique has been used for the fabrication of metal structures on silicon. In this case PMMA is deposited on the silicon surface and the mould is pressed into the PMMA layer at 160°C and allowed to cool to 80°C. The mould does not press all the PMMA away from the surface and therefore a thin layer of PMMA is still present. This layer can be removed by RIE to expose the underlying conducting layer. The conducting layer can now be used for selectively electroplating a metal onto the silicon wafer. Finally, the PMMA structure is stripped and the metal layer remains.

CHAPTER 6

Microfluidic Devices I: Flow Control

6.1 Introduction

The first research in microfluidics was carried out in the year 1977. At Stanford University [120] a gas chromatograph was developed, and at IBM [121] research was on an inkjet printer nozzle. Since then research in this area has undergone an enormous growth and many different microfluidic devices have been developed. These include valves [122, 123], pumps [10, 28, 29, 124–128], flowsensors [48, 129, 130], fluidic mixers [131–134] as well as chemical and biochemical sensors [135]. Recently, work has been carried out on integrating several devices on a single substrate. The eventual aim is to produce an on-chip chemical analysis system [127, 136, 137].

In order to cope with the rapidly growing area of microfluidics, the discussion of microfluidic devices is split into two chapters. This chapter deals with inkjet nozzles, amplifiers, active and passive valves as well as all types of micropump and the techniques used to measure flow rates and pressure. The next chapter will concentrate on particle handling techniques and (bio)chemical devices.

6.2 Inkjet Nozzles and Flow Amplification

The first topics we consider are flow directing and amplification. This includes nozzles for inkjets and fuel injection systems as well as fluidic amplifiers working like transistors in the fluidic domain. The latter can be employed as a fluid power switch to drive an actuator or tools like a micro-endoscope.

6.2.1 Nozzles for Inkjet and Fuel Injection

Microfluidics started with inkjet nozzles in 1977 and continues today because of the 21% microfluidic market share of these devices [2]. The first nozzles for inkjets were through-holes etched in silicon by KOH and passivated with silicon dioxide (figure 6.1.a [121]). This has been improved by highly boron doping the front of the wafer at all places apart from a smaller, desired orifice (figure 6.1.b [138]). As described in chapter 5, EDP shows the characteristic of a highly reduced etch rate for deep boron doped silicon. Thus, with the use of EDP for the through etch, a p^+ doped silicon membrane remains with an orifice in the centre.

At a characteristic distance from the orifice, the jet breaks up into numerous small droplets. Thus the next development stage for inkjet printing was to charge the ink droplets in order to allow electrostatic deflection (figure 6.1.c [139]). Therefore the walls of the through-hole are highly doped and metal pads are used to contact to the diffused area, which charges the ink at the break up point.

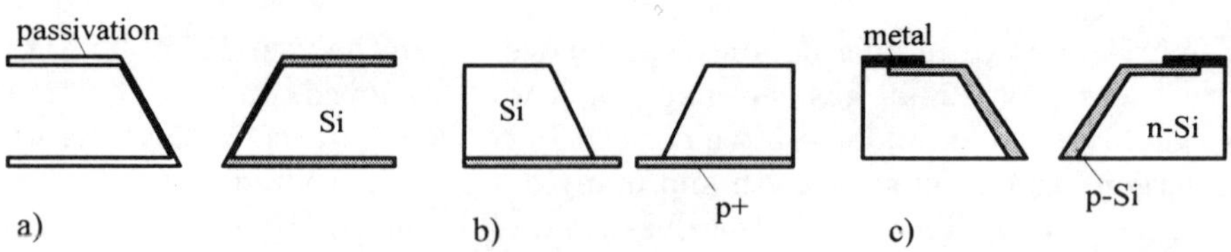

Figure 6.1: Inkjet nozzles: a) simple nozzle [121], b) highly doped orifice [138], c) charging of ink droplets [139]

The next step in inkjet development was the realisation of an integrated inkjet assembly with nozzle, ink cavity and piezoelectric actuation [140] as shown in figure 6.2. Isotropic wet etching with HF, HNO_3 and acetic acid is employed to generate the orifice of the nozzle. The second step is to create the ink cavity as well as vertical-walled trenches in the (110) silicon wafer which form the face of the exit of the nozzle. Now a thin Pyrex plate is anodically bonded to seal the cavity and a piezoelectric plate is mounted on top of the plate. The piezoelectric plate is driven with an a.c. voltage to dispense the ink in regular time steps through the nozzle.

A somewhat more modern approach to inkjet printing is described by Chen [141] who uses the second actuation technique commonly applied in inkjet printers: thermal actuation. Thermal actuation is the dominant actuation technique used in the inkjet heads on the market today. Heaters, which are close to the nozzle, are used to locally evaporate the ink and thus drive some ink through the nozzle. The whole device is totally integrated with the actuator, the nozzle and ink reservoir on the same chip and can achieve a

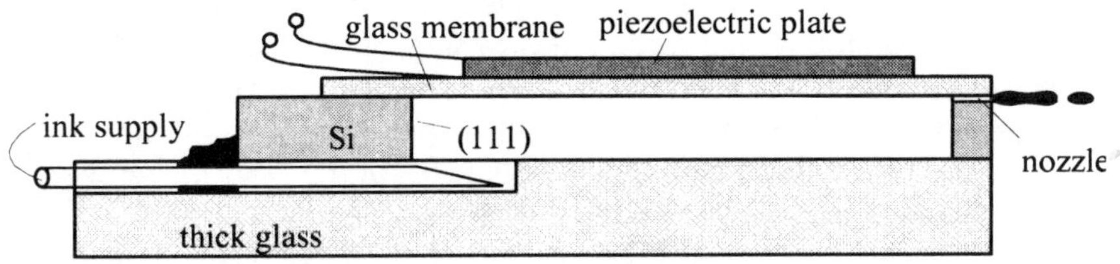

Figure 6.2: Inkjet head with piezoelectric actuation [140]

resolution of 1016 dots per inch. The technology is a mixture of surface and bulk micromachining and can be described as follows. Initially, the whole surface is highly boron doped except for small chevron ribs (see figure 6.3) in the area of the nozzle and the fluidic interconnects. This allows undercutting of these areas with EDP to form bridge-like structures and the channel. After etching, the channel is closed on top with a combination of wet oxidation (0.5 μm thickness) and LPCVD deposition of an oxide (0.6 μm-thick). Both layers completely join the chevron ribs and close the top of the channel. In order to reduce the thermal resistance the layer is thinned with a planarisation and etch back process. Now the poly-Si heaters are deposited followed by the deposition of LTO (silicon dioxide deposited at low temperatures around 400°C) and silicon nitride. Areas are then opened for contact and Cr/Au is inlaid to reduce the contact resistance. The next step is to create an ink inlet orifice from the back of the wafer through to the channels on the front. This is done by another wet etch with EDP where the structures on the front (silicon nitride and Au) are only marginally etched. The chips are finally diced and the horizontal nozzles are opened. The inkjet print head is depicted in figure 6.3.

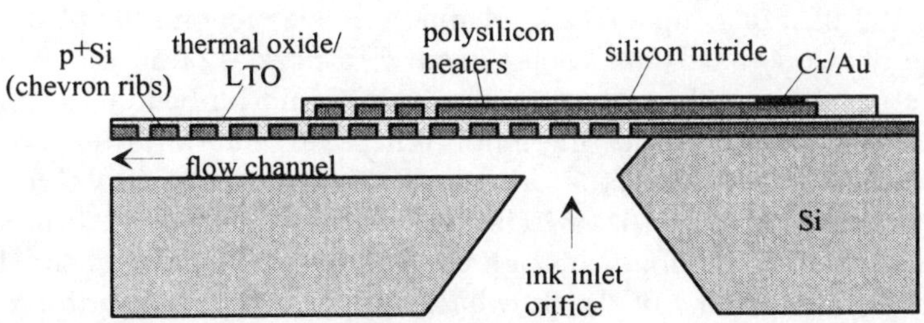

Figure 6.3: Integrated inkjet head with thermal actuation [141]

A similar device has been proposed by Westberg, who developed a CMOS compatible fabrication process for the generation of inkjet heads [142] (see figure 6.4). Thermal actuation is again used and the resolution of the inkjet is around 1000 dots per inch. The process is based on a 0.8 μm CMOS process requiring only two additional mask steps to the conventional CMOS process. The deposited metal (Al), used in CMOS for wiring, acts as a sacrificial layer for the generation of the nozzle and the flow channels. The standard process allows the use of a metal layer of either 0.5 μm or 1.5 μm thickness, which yields the channel height. The doped gate polysilicon layer in CMOS is used as a resistive heater underneath the channels. A thin layer of TiNi on top of the gate, used as diffusion barrier in the CMOS process, is employed as a protection layer for the corrosive ink. The first post-processing stage is to define the end of the nozzle structure laterally. Here, a Cr mask is used to etch through nitride, oxide and metal layers of the continuing tube to generate a horizontal opening of the tube. With this, any layers are removed from the silicon next to the nozzle opening. The exposed silicon is then etched with EDP to undercut the nozzle. Finally, the metal from the channel is removed with HCl, H_2O and H_2O_2. The final device is depicted in figure 6.4.

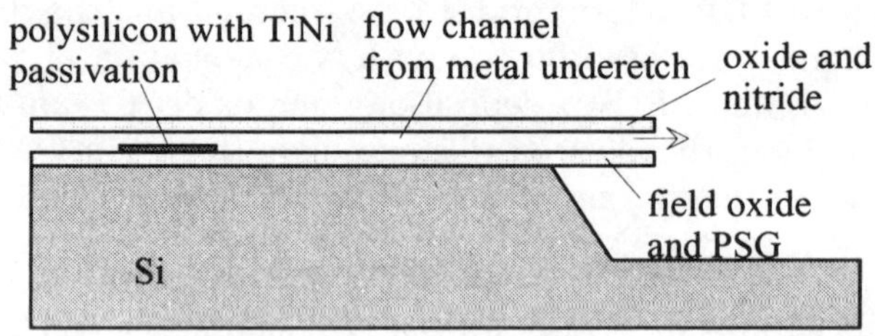

Figure 6.4: Integrated inkjet head with thermal actuation [142]

A slightly different device has been presented by Ederer [143] who has achieved a droplet generator which has its main application in fuel injection. With this device it is possible to generate a flow of microscopic fluid drops in air from 100 pl min^{-1} up to 11.9 ml min^{-1} at ejection speeds of up to 7 m s^{-1}. The device is partly micromachined with nozzles of 50 μm in a silicon plate. Actuation is achieved by a piezoelectric bimorph bending from and to the nozzles and thus transporting fluid. When the bimorph bends away, fluid flows into the gap between actuator and nozzle through capillary forces. Then the piezoelectric actuation is switched off and the bimorph returns towards the nozzles pressing the fluid through the nozzles. A diagram of the device is shown in figure 6.5. One of the drawbacks of the droplet generator is that it does not self-prime when it is filled with gas but has to be rinsed with liquid before operation.

6.2.2 Microfluidic Amplifiers

Transistors are simple amplifiers in electronics. Here, a small current is used to control a large current. The same principle can be transferred to microfluidics.

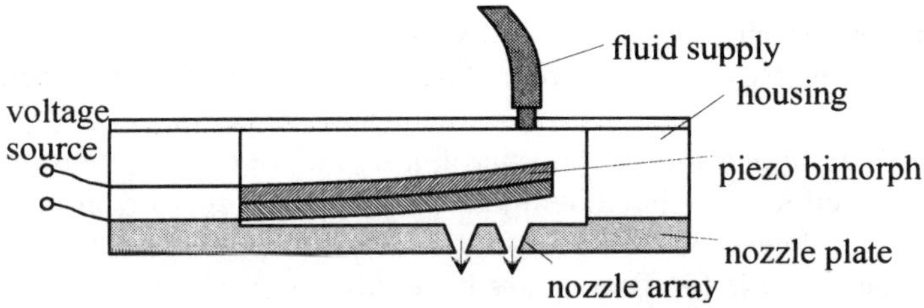

Figure 6.5: Droplet generator for fuel injection applications [143]

Here, the first amplifier was presented by Vollmer [106]. The basic principle of the microfluidic amplifier is depicted in figure 6.6. A fluid streams out of a supply nozzle and attaches to one or the other wall some distance away due to the Coanda effect. Thus most of the fluid passes through the outlet closest to the attached wall. This condition is stable up to the point when a small amount of fluid from the control port pushes the main stream from one wall to the other, so initiating a switch from one output to the other. The vent chamber is necessary for the Coanda effect to be achieved, especially when the outlets are closed.

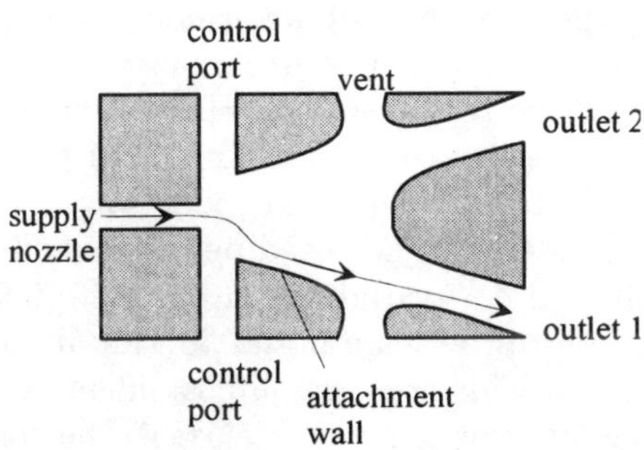

Figure 6.6: Schematic of a wall-attachment microfluidic amplifier

The device has been successfully fabricated in LIGA technology. The LIGA process is very suitable for this as it allows a large structural height so that the influences from top and bottom surfaces are greatly reduced and the flow is mainly determined by the side walls. The side walls for the fluidic amplifier can be up to 500 μm in height.

For the design to work, large enough fluid velocities and pressure differences between supply nozzle and outlet (> 30 kPa) are required to achieve a bistable element. With nitrogen as a test fluid and a supply pressure of 33 kPa, pressures of around 1.2 kPa are required at the control ports to switch the output pressure between 0 and 4 kPa at each outlet. Higher supply pressures require higher switch pressures. The switch pressure increases drastically for all gases (air, nitrogen and carbon dioxide) above 500 to 600 kPa supply pressure.

A possible application of the microfluidic amplifier is a microfluidic flip-flop. It has been shown that one amplifier can drive up to 8 similar devices. This number of devices is called fan-out in electronics where it is quite important. The microfluidic flip-flop makes it possible to use logic devices at places of high voltage and in other environments unsuitable for electronic circuits.

Two other microfluidic amplifiers have been presented by Furlan [144] based on ordinary silicon micromachining. The general principle of the first device does not differ from the wall-attachment amplifier of Vollmer. However, a second device has been presented which does not utilise the Coanda effect. Instead, the fluid streams into both exits in equal proportions unless a pressure is applied at one of the two control ports. This deflects the main flow towards one or the other outlet. Unlike the digital behaviour of the wall-attachment amplifier, this so-called jet-deflection amplifier is an analogue device which changes the flow between the two outlets gradually.

Identical fabrication technology is used for both silicon amplifiers. The channels are made by isotropic silicon dry etching (depth around 31 μm) and connection through-holes are etched from the back with KOH. Finally, the channels are sealed with Pyrex using anodic bonding. Compared to the perfect switching behaviour of the wall-attachment amplifier with the LIGA technique, the same device with silicon technology shows only analogue behaviour. This is mainly due to the small surface to which the flow attaches. The achievable channel depth in LIGA is almost 10 times larger and therefore the surface is much larger. This behaviour was not very good as even the silicon jet-deflection amplifier achieved better switching results. The jet-deflection amplifier allowed flow variations of up to 50 % for supply pressures of 210 kPa, control pressures between 0 and 30 kPa and nitrogen as the test fluid. In construction both devices are almost identical, although the jet-deflection amplifier has its vent ports very close to the control ports to avoid an initial bias of the fluid flow towards one side.

Another silicon micromachined fluidic amplifier has been presented by Kim [145]. By the use of dry etching with Cl_2 and BCl_3, a structure depth of 39 μm has been realised. After the dry etching step, through-holes are etched with KOH from the back and the structures on the front are sealed with a Pyrex cover plate. Again nitrogen was used for experimental verification of the device operation. Here, an input pressure difference between the two control ports of -6 to 6 kPa can control an output pressure difference of -20 to 20 kPa. The pressure at the supply nozzle was 100 kPa and the pressure difference between the control ports was 10 kPa. An almost linear behaviour was observed for a variation of the pressures at the control port, confirming the behaviour described above.

A further step on from a bistable switch is a fluidic micro oscillator where

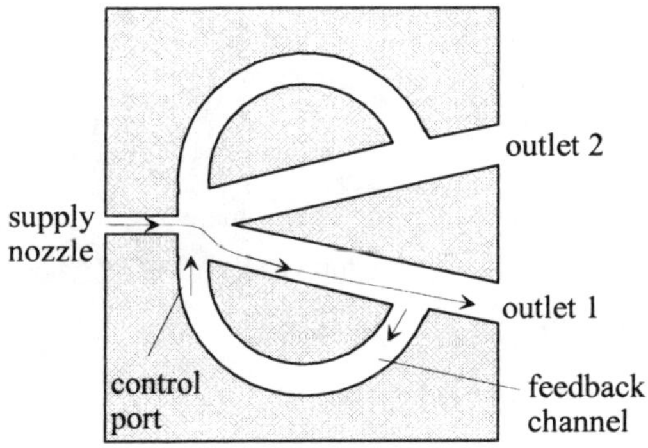

Figure 6.7: Schematic of a fluidic micro oscillator [146]

the flow switches from one exit to the other with a characteristic frequency [146]. A diagram of the working principle of such a device is depicted in figure 6.7. The fluid enters through a supply nozzle and attaches to one wall due to the Coanda effect. When it travels to the corresponding outlet, the flow is also fed back through a channel to the control port. The feed-back fluid then pushes the jet from one side to the other with a characteristic frequency. The device has been fabricated with the use of the LIGA process and successfully tested. For the geometry used, a switching frequency in the range between 250 and 390 Hz has been measured by varying the supply pressure between 60 and 250 kPa.

Together with a piston actuator having an inlet on either side of the piston, a micro actuator system can be fabricated [146]. This system could be employed for the actuation of a cutting tool in minimal invasive cardiac therapy in medicine.

6.2.3 Micromachined Turbines

Microturbines are also interesting devices in microfluidics. The first turbine has been presented by Mohr [116] and was fabricated with the LIGA process. The diameter of the axle is 80 μm, that of the rotor 250 μm and the height is approximately 100 μm. The fabrication process makes use of a sacrificial layer technique, where the rotor is processed on the sacrificial layer and the axle is fixed to the substrate. The material used for the turbine is nickel. The devices have been successfully tested with a lifetime of 10^7 rotations with external air flow. The lifetime could be increased further by improving the shape of the axle to generate an air cushion between rotor and axle. The basic principle of the device is shown in figure 6.8.

Apart from LIGA, bulk micromachining has also been employed to fabricate microturbines [147]. The fabrication is based on SOI material with a

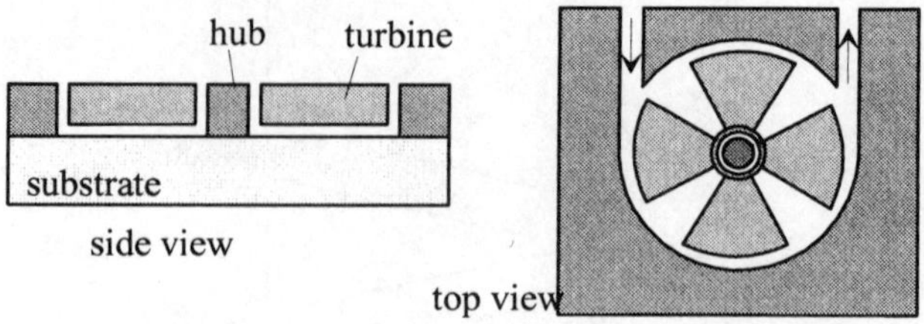

Figure 6.8: General layout of a microturbine [116]

20 μm-thick silicon layer and a 2 μm-thick intermediate oxide. The turbine is etched with RIE and released with HF. The device can be enclosed by electrostatically bonding a Pyrex wafer on top. A microturbine with a radius of 275 μm and a gap between rotor and chamber walls of 9 μm has been fabricated with this process. Rotating the microturbine has been possible by directly blowing air into the devices with a perforated needle. The general device layout is quite similar to figure 6.8.

A newer device has been described recently by Tsai [148]. It is based on surface micromachining but also uses bulk micromachining and stress engineering of microstructures. In order to achieve a three dimensional formed microturbine in surface micromachining, stress engineering is a suitable technique. With this technology it is possible to use a near-zero-stress polysilicon layer together with a high-stress silicon nitride film. If structures which are made out of these two layers are released from the substrate, curl-up effects occur and three-dimensional bent structures are fabricated. A diagram of the microturbine after all fabrication steps is depicted in figure 6.9. The process is started with deep boron diffusion on one side to form the final support areas. Next, a sacrificial layer of LTO with a thickness of 2 μm is deposited and structured with buffered HF to form dimples and anchors. Now a poly silicon film of thicknesses between 1.6 and 3.6 μm is deposited and defined with lithography and dry etching. Then a 150 nm-thick LTO is deposited and structured to create anchors to the polysilicon for the subsequent silicon nitride deposition. Again, the polysilicon is now structured to define the turbine followed by another 1 μm-thick LTO sacrificial layer and the hub formation with a 1 μm-thick polysilicon. Finally, the front of the wafer is protected with wax and the back is etched in KOH and HF to remove the silicon and eventually the LTO. Due to the bending force of the stress-engineered structures stiction is no great problem. The device has been tested with a continuous flow of nitrogen gas and operated successfully.

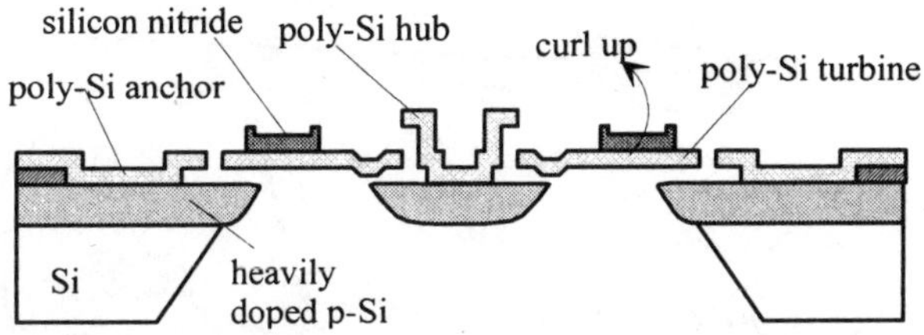

Figure 6.9: Microturbine fabricated with surface micromachining and stress-engineering methods [148]

6.3 Active and Passive Valves

Many different designs for micromachined valves can be found in the literature. These range from simple passive check valves to electrostatically driven active diaphragm valves. In this section a summary of all the different valve types is given as well as an overview of the fabrication techniques for the different devices. Some measurements of flow rates through these valves are described, too.

6.3.1 Passive One-Way Valves in Bulk and Surface Micromachining

The first class of valves is passive and does not include any actuation. These valves are usually designed to give a high flow rate in one direction and a rather small flow rate in the opposite direction. The main use is in micropumps with mechanical actuation, as these pumps need such valves to direct the flow of the fluid. Such valves could also allow the filling and closure of chemical reaction chambers to achieve isochoric reactions.

6.3.1.1 Cantilever Valves

Cantilever valves (see figure 6.10) are passive one-way valves allowing the flow of a liquid or a gas in one direction and prohibiting it in the other. This type of valve is either fabricated macroscopically with cutting tools or micromachined. Both techniques give a similar device with an identical working principle. However, micromachining allows the generation of small-scaled devices in the sub-millimetre and even micrometre range as well as low-cost mass production.

A cantilever valve generally consists of a thin flap sitting on top of a duct [28,39,122]. The flap controls the flow through the valve-system. In the case of a low pressure on the "free" side ($p_2 < p_1$), the cantilever bends into an open state and allows the fluid to pass through. For a high pressure on the

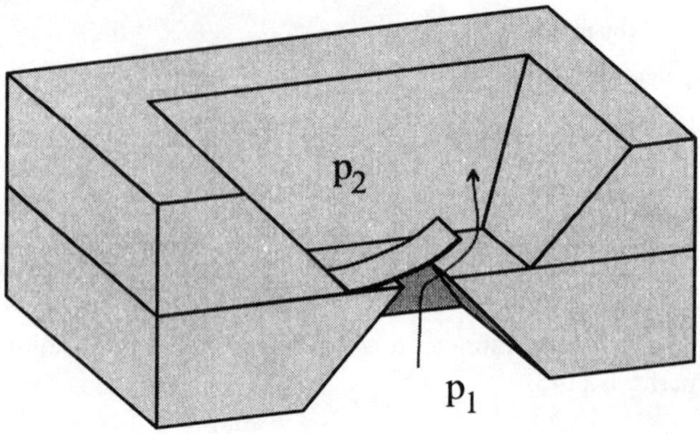

Figure 6.10: Principle of the passive cantilever valve

free side ($p_2 > p_1$), the flap is pressed against the valve seat and the passage is sealed.

Several fabrication principles for micromachined cantilever valves have been described in the literature. The fabrication of the cantilever can be either with a time controlled etch stop [28], with a high boron diffused p^+ etchstop, or electrochemically controlled with a pn-etchstop [44]. Bonding a second wafer with a nozzle anisotropically etched through the whole wafer completes the valve. Emmer [122] used fusion bonding of a wafer containing 30 μm-thick Si cantilevers to a second wafer incorporating the duct. Zengerle [28] stuck cantilever-chip and duct-chip together using glue at the border of the chips.

Sticking, i.e. bonding of two valve surfaces through drying of an intermediate liquid, is a major problem with this type of valve, and therefore a recess can be added to reduce the problem [44] as shown in figure 6.11.b. An additional rim (figure 6.11.c) on the valve seat can be achieved with a somewhat more complicated process [43]. The advantage of this rim is a simpler handling procedure as stiction is greatly reduced and leakage is negligible at small backpressures compared to valves with a recess. The disadvantage is that another photolithographic step is added, making the process more complicated. The three types of cantilever valve are shown in figure 6.11.

The fabrication of a simple cantilever valve is now described in more detail: Through-hole and cantilever are generated on the same wafer. Using symmetrical placement, the cantilever on one wafer joins the duct of the second wafer. Because processing is required from the front and back of the wafer surface, an IR-aligner is used which can see through the wafer. The process sequence is shown in figure 6.12: Deposition and patterning of SiO_2 and Si_3N_4 is followed by deep B^+ diffusion at 1150°C (Design 1, figure 6.12.a). Boron concentrations above $3 \cdot 10^{19}$ cm^{-3} provide a good etch-stop in KOH bulk etching, and give a controlled thickness for the cantilevers. The diffusion time

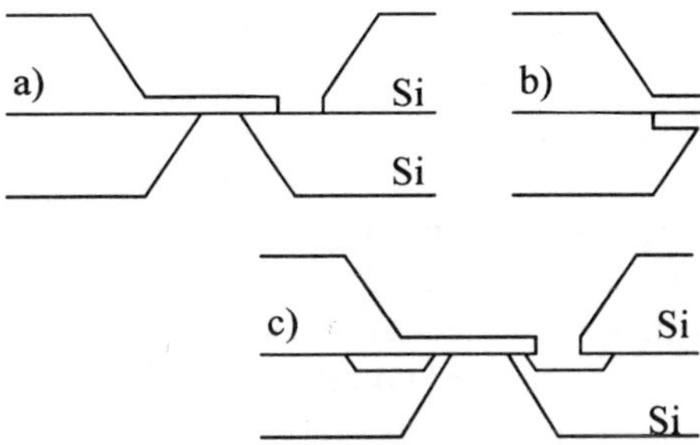

Figure 6.11: Cantilever valves: a) simple cantilever valve [28], b) recessed cantilever valve [44], c) cantilever valve with rim in the valve seat [43]

can be chosen with reference to figure 5.14. The oxide and nitride layers are now replaced by another oxide and nitride layer (see Design 1, figure 6.12.b), which are then patterned for the KOH etch. This etch has to be performed at 70°C and 10% wt in order to comply with optimum etch-stop conditions (section 5.3.1). The cantilever is formed at diffused places with a rough surface finish on the etched side and a smooth surface finish on the operating side. The duct is produced where boron diffusion was prevented. Figure 6.12.c depicts the etched devices.

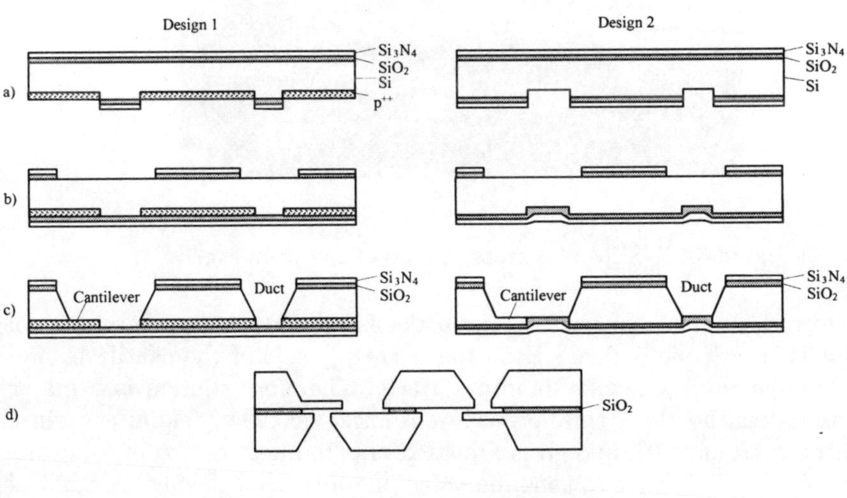

Figure 6.12: Cantilever valve fabrication steps

Instead of using a complex diffusion process for the cantilever, a simple and short anisotropic etch of 10 μm with an inverse mask on the front of the wafer allows a visually controlled etch stop with a 70°C, 30% KOH solution from the back (Design 2, figure 6.12.a – c). This technique gives a smoother surface finish at the etched side, but the thickness is not that easily controlled.

If one wafer is now turned and placed above another one, the valve can be completed (figure 6.12.d). Alignment of both is needed to position the cantilever on top of the duct. This is done by sawing two perpendicular new flats along the outer scribelines of the wafer. After oxidation of one wafer and RCA cleaning of both, room temperature bonding is performed on a jig, where the perpendicular flats are aligned. The intermediate oxide (500nm) is needed to give good bonding quality and to allow the release of the cantilever. The wafers are now annealed in oxygen at 1100°C to yield a permanent bond. Note that cantilever and duct are bonded directly together without the depression used by other authors [44]. This is done to avoid leakage. After sawing the individual chips, the intermediate oxide is etched with buffered HF to release the cantilever.

Figure 6.13: SEM of a cross section of the cantilever valve

Before drying, a sequential rinse in deionised (DI) water, pure alcohol and finally in toluene is done. Both the operating side of the cantilever and the valve-seat have a rather smooth surface. Therefore stiction is a major problem unless the above rinse procedure is executed. The yield after a rinse in toluene is around 80% and gives similar results to freeze-drying or n-hexane techniques [91]. Without toluene, no valve operates after drying. A SEM of a cross section of the finished cantilever valve is shown in figure 6.13

Measurements of the flow rate through the valve can be taken by measuring the travel speed of an air-liquid interface in the inlet tube connected to the valve. The minimum detectable flow rate with this apparatus can be estimated from 1 mm minimum detectable travel distance in 15 minutes. This gives approximately 0.3 μl min^{-1} for a hose diameter of 2.5 mm.

The flow behaviour for passive valves with three different cantilever dimensions is depicted in figure 6.14. The lateral side length of the square shaped cantilever was varied from 500 to 1000 μm. The duct was always positioned centrally below the cantilever. An 850 μm-sized square was used for anisotropically etching the duct. This gives an exit dimension of around 130 μm for a wafer of approximately 500 μm thickness. Allowing for 20 μm misalignment the edge of the cantilever is then at least 165 μm away from the duct, being enough for sealing the duct in case of a reverse pressure gradient. DI water was used as the test liquid.

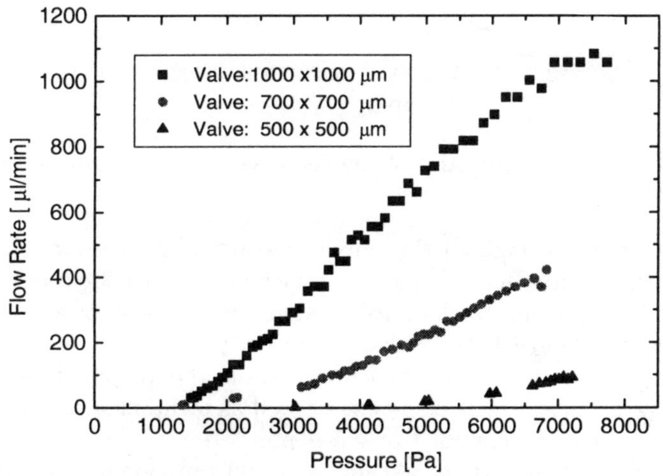

Figure 6.14: Flow characteristics of the boron doped bulk micromachined valves

All three valves show a rather similar behaviour over the pressure range. Below a threshold pressure of 1000 Pa (2000 Pa, 3000 Pa) no flow could be observed. Beyond the threshold pressure, the flow rate increases steadily before a linear region is reached at 2500 Pa (4000 Pa, 6000 Pa). Only for the valve design with 1000 μm sidelength could a sublinear range be observed, and then only for pressures greater than 6500 Pa. It is obvious from the data that the flow rate is higher for bigger cantilevers. The main pressure application to the cantilever is in the duct area. This area is further away from the cantilever root for larger cantilevers. Therefore, higher displacements and subsequent higher flow rates can be achieved. A leakage rate for reverse operation was not measurable for all three cantilever valves. Taking the minimum detectable flow, this gives a flow rate of less than or equal to 0.3 μl min^{-1}.

A similar behaviour but larger flowrates were observed for the fluid flow of the valves with the optical etch stop (figure 6.15). Both valves with the 500 μm-long and the 700 μm-long cantilever show a similar threshold pressure of 1000 Pa, below which the flow rate is very small. The 500 μm valve has

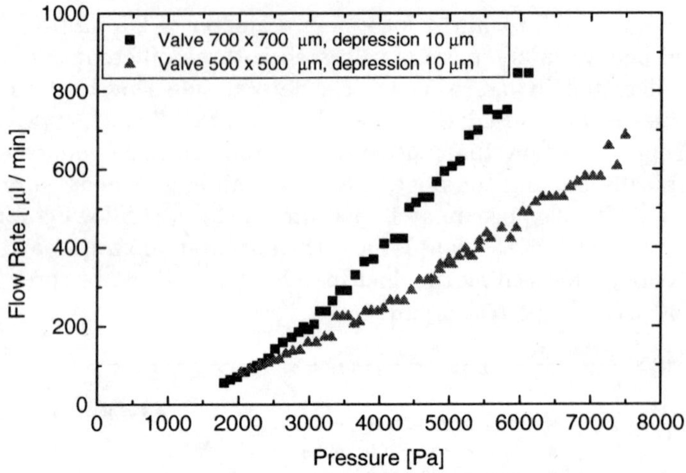

Figure 6.15: Flow characteristics of the depressed bulk micromachined valves

only a linear region throughout the whole pressure range, whereas the 700 μm valve shows first a superlinear behaviour which changes to linear at a pressure of about 3000 Pa. Owing to the mask design, a depression of approximately 10 μm is created beneath the valve, separating valve seat and cantilever. This distance is responsible for larger flow rates in the forward direction than achieved with the boron etch stop type as well as a rather high leakage rate. Reverse flow rates of 15 μl min^{-1} and 5 μl min^{-1} for a pressure difference of 1.5 kPa across the valves with 500 μm and 700 μm sidelength respectively have been measured.

6.3.1.2 Diaphragm Valves

The second type of valve incorporates diaphragms for controlling the flow. The basic principle is identical to the cantilever valve. The diaphragm is either opposite to a rim (figure 6.16.a [43]) or the rim is attached to the diaphragm (figure 6.16.b [29, 124]). Unlike most other valves, the latter is based on silicon-Pyrex anodic bonding. In order to prevent the rim sticking to the Pyrex substrate, an oxide-spacer is deposited on the rim (selective bonding). As a small rim is used for both valves, stiction is only a small problem, but the fabrication process is rather complicated compared to the simple cantilever valve.

Similar diaphragm valves can be made by using surface micromachining (figure 6.17): The nozzle is anisotropically etched from the back, while the front is covered with a sacrificial oxide and a structured poly-Si layer on top forms the actual valve. By removing the oxide, a free-standing flap is obtained and the valve is completed [123]. Apart from poly-Si, the flap can also be made out of metals like copper and titanium [134, 149] or single crystalline

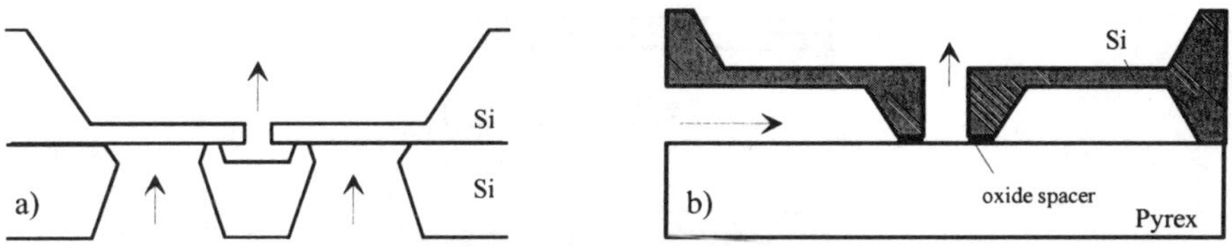

Figure 6.16: Diaphragm valves: a) simple diaphragm valve [43], b) diaphragm valve where the membrane includes the rim [124]

silicon bonded to Pyrex and etched back [150]. The advantage of surface micromachining is that only one wafer is necessary whereas two wafers are required for each bulk micromachined valve. However, the process is rather complicated and stiction is one of the major problems.

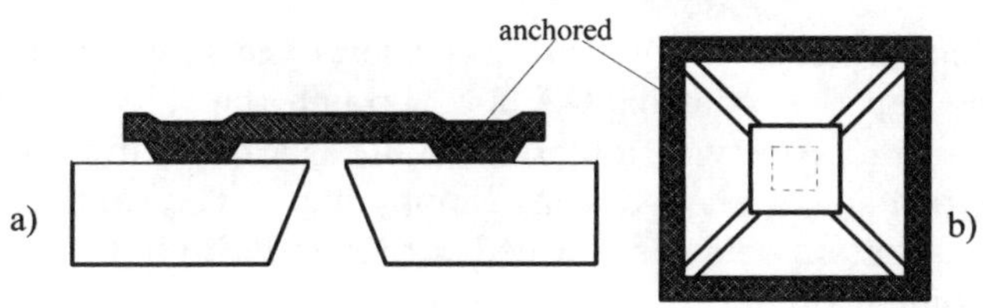

Figure 6.17: Diaphragm valve with surface micromachining technology [123]: a) cross section, b) top view

The process procedure for making a surface micromachined diaphragm valve is now described in more detail. The fabrication starts with a double-sided alignment step. Because of its high etch rate in buffered HF [151], a 1 μm-thick phosphosilicate glass (PSG) of approximately 8% boron and 4% phosphorus content was used as the sacrificial layer. The glass can be deposited in an APCVD reactor at atmospheric pressure and 400°C with a gas flow of 40 sccm silane, 20 sccm PH_3 in 99% Ar and 40 sccm B_2H_6 in 99% Ar.The sacrificial layer is etched away in the area where the polysilicon is to be anchored to the silicon substrate (figure 6.18.a). Standard photolithography and dry etching in an RIE etcher can be employed.

After this, polysilicon can be deposited in several ways:

- 2.0 μm polysilicon deposited at 610°C with a gas flow of 50 sccm silane subsequently doped with phosphorus tiles at 1000°C for 1.5 h in N_2 atmosphere to reduce its intrinsic stress [87].

- 1.5 μm amorphous silicon deposited at 560°C with a gas flow of 50 sccm silane subsequently annealed in a nitrogen atmosphere at 1100°C for 1 h to generate polysilicon.

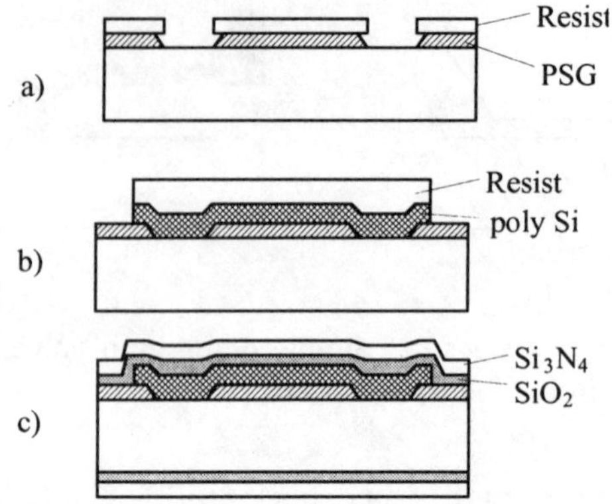

Figure 6.18: Fabrication of the valve in polysilicon (part I)

The surface micromachining process allows the fabrication of different designs of the flap that controls the flow through the valve (cf. figure 6.19): for example, a cantilever type, a plate with linear arms (stiff flap) and a plate with long perpendicular shaped arms (floppy flap). The cantilever and both plates with suspension arms are formed with standard photolithography and poly-Si dry etching (figure 6.18.b) in an RIE etcher.

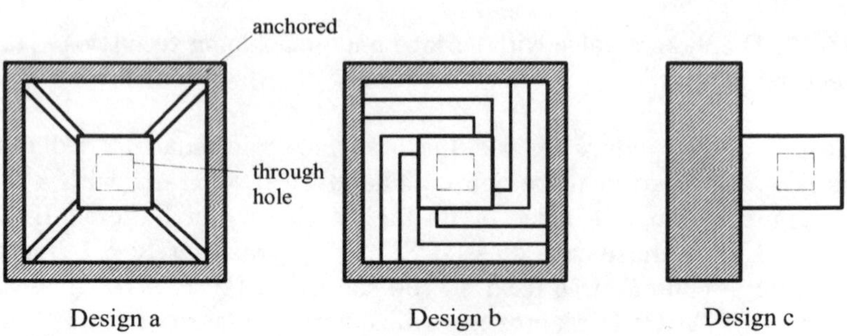

Figure 6.19: Different designs for surface micromachined valves: Design a) Stiff flap valve, Design b) Floppy flap valve, Design c) Cantilever valve

Now a 200 nm-thick LTO layer and a 160 nm-thick LPCVD nitride is necessary as a mask layer for the following KOH etch (figure 6.18.c). Structuring both layers on the back is done with dry etching (figure 6.20.d). This is followed by a through etch in 70°C, 30 % KOH, stopping at the PSG layer (figure 6.20.e). Care has to be taken during the last part of the etch, as the PSG oxide is not very dense and shows a rather high etch rate in KOH. After removing the silicon nitride in orthophosphoric acid the wafer is protected with

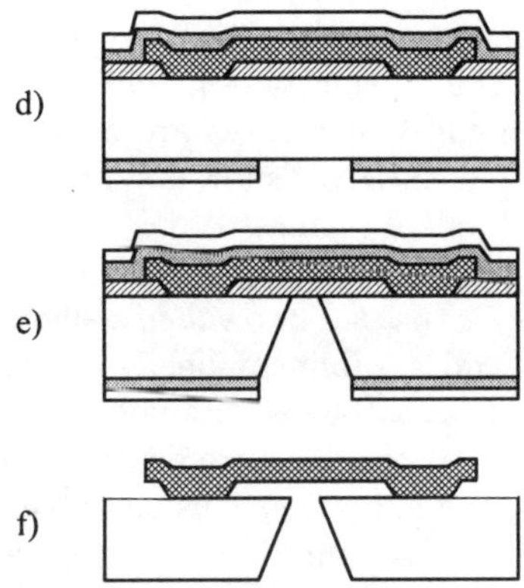

Figure 6.20: Fabrication of the valve in polysilicon (part II)

resist and diced into chips. In a similar way to the bulk micromachined valve, a buffered HF step taking several hours and a subsequent rinse in DI-water, ethanol and toluene is carried out to release the polysilicon structures and to reduce stiction (figure 6.20.f). By forming poly-Si structures on both sides, two counter-opposed valves, as needed for a micropump, can be produced.

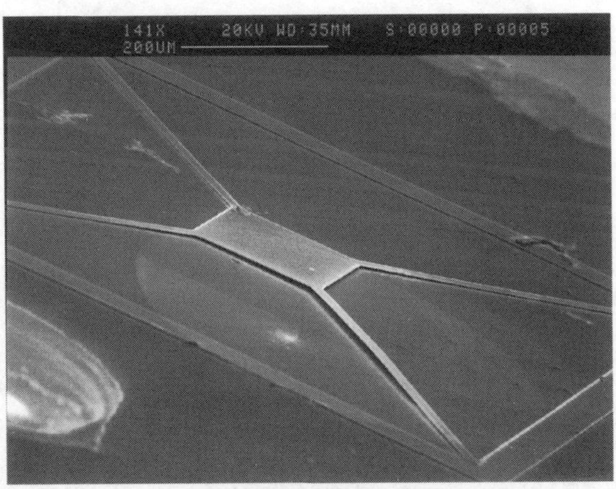

Figure 6.21: SEM of a finished surface micromachined valve with a stiff flap

A SEM of a finished device is shown in figure 6.21. The stiff flap valve has been released with HF and treated in toluene as described above. As can be seen the stress in the structure is slightly too large which induces buckling in one of the beams.

The flow rates of all three surface micromachined valve types are shown in figure 6.22. The polysilicon thickness is 2μm and the test liquid is water. It is obvious that the highest flow rates are achieved with the floppy flap. Both the stiff flap and the cantilever show a similar behaviour, leading to the conclusion that the mechanical stiffness of both valve types is similar.

It can be concluded from the graphs that the flow rates show a linear behaviour with pressure for the stiff flap valve, a slightly superlinear behaviour for the cantilever valve and a sublinear behaviour for the floppy suspension valve. A similar sublinear behaviour is obtained for a converging duct without any flap, which suggests that the pressure drop due to the nozzle is the main restriction to the flow in the floppy suspension valve. The leakage rate for all three valve types is less than the minimum detectable flow rate, i.e. less than or equal to 0.3 μl min^{-1}.

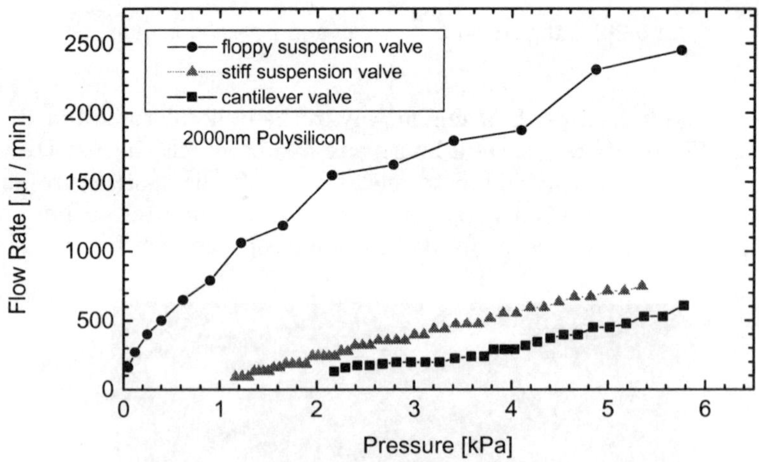

Figure 6.22: Flow characteristics of the polysilicon valves

A similar behaviour to the polysilicon valves is obtained for the valves with amorphous deposited silicon. Again, the floppy flap shows a far higher flow rate than the two other designs (figure 6.23). However, the flow rates of these valves are smaller than for the polysilicon valves. It seems that the mechanical stress in the amorphous deposited silicon is larger than in the polysilicon. This even seems to overcompensate the lower stiffness of a thinner beam. The leakage rate of all valve types is comparable with the one obtained from the polysilicon valves, i.e. less than or equal to 0.3 μl min^{-1}.

Diaphragm valves have also been fabricated from polyimide and titanium, as shown in figure 6.24. Here, a diaphragm of titanium and a seat of polyimide act together as a passive valve. The titanium layer is sputter deposited to achieve a 2.7 μm-thick membrane. By photolithography and etching in hydrofluoric acid, an opening of 180 μm diameter is created in the titanium.

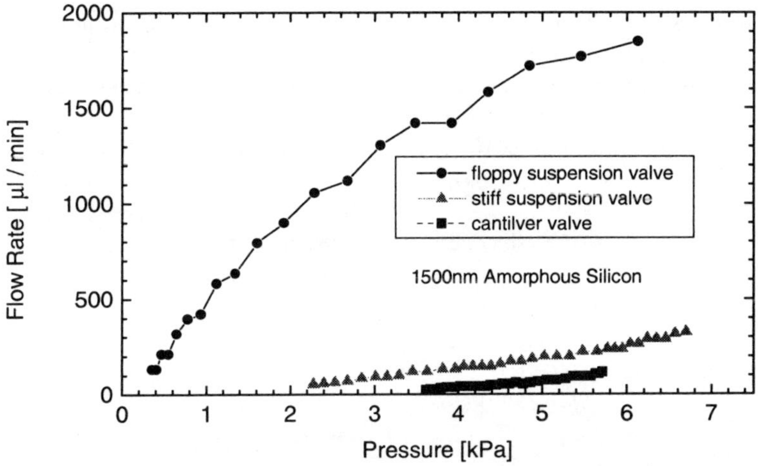

Figure 6.23: Flow characteristics of the amorphous deposited silicon valves

Then copper is deposited with a thickness of 1 μm to act as a sacrificial layer for the following polyimide deposition. The thickness of the active layer is 2.15 μm and is achieved by spin coating. As the polyimide is photosensitive, the layer is structured with lithography. Finally, the sacrificial layer is etched in 30 % HNO_3 and a surface-active substance. A rigid frame is then glued onto the front, and the titanium membrane is separated from the wafer. The finished device is depicted in figure 6.24. The leakage rate of such valves has been measured at less than 6 μl s^{-1} for helium gas at pressures between 1 and 10 kPa. The forward flow rate was up to 60 ml s^{-1} and thus a factor of 10^4 larger than the leakage.

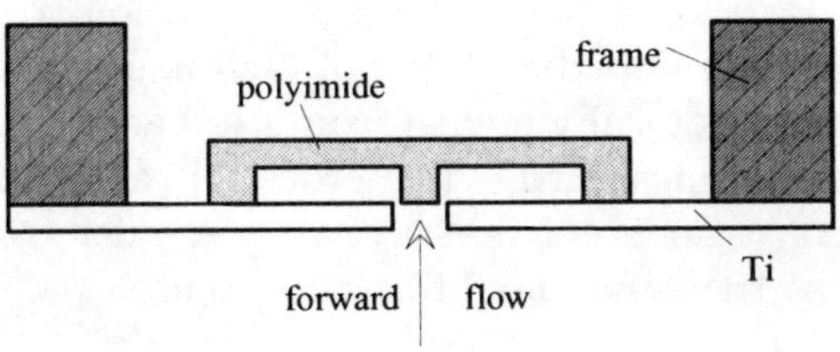

Figure 6.24: Diaphragm valves on titanium membranes [152]

Another diaphragm valve has been described by Shoji [153]. It is based on a silicone filling in a silicon through-hole and gives the extremely good forward to reverse ratios necessary in a high-pressure fluid pump. The fabrication starts with the etching of a through-hole in silicon with KOH and leaves a thin silicon nitride membrane at the bottom of the hole. Then a thin layer of positive photoresist is deposited on the sidewalls of the hole and the through-hole is filled with silicone using a micro-pipette. Again, another layer of positive photoresist is deposited as a sacrificial layer. Finally, a negative resist is deposited and exposed to produce the connection between the silicone stopper and the substrate. In the last stages, the silicon nitride membrane is etched and the positive resist is removed with acetone. A picture of the finished device can be seen in figure 6.25.

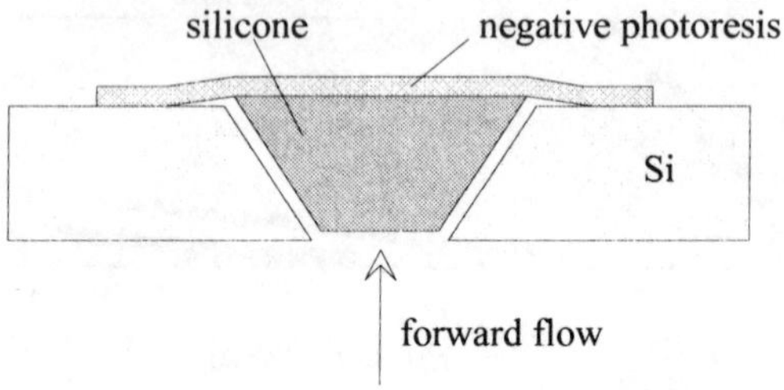

Figure 6.25: High pressure check valve developed by Shoji [153]

Forward flow rates of several 1000 μl min^{-1} have been achieved with DI water as a test liquid and pressures up to 10 kPa. The reverse leakage did not exceed 1.5 μl min^{-1}for 2 kPa and gets as low as 0.01 μl min^{-1} for pressures up to 100 kPa.

6.3.1.3 Bivalvular Valves

A different one-way valve has been described by Yang [154]. This device is a bivalvular microvalve composed of bulk micromachined p^+ doped silicon diaphragms. The shape can be seen in figure 6.26. The process is started with the formation of V-grooves followed by a p^+ diffusion so that the sidewalls of the groove are heavily doped and do not etch in KOH. Next, the wafer is isotropically and anisotropically etched from the back to create the two flap parts of the bivalvular microvalve. The process itself is rather complicated, but results in the construction of the valve on one wafer. No stiction problem has been reported so the valve should be easy to use.

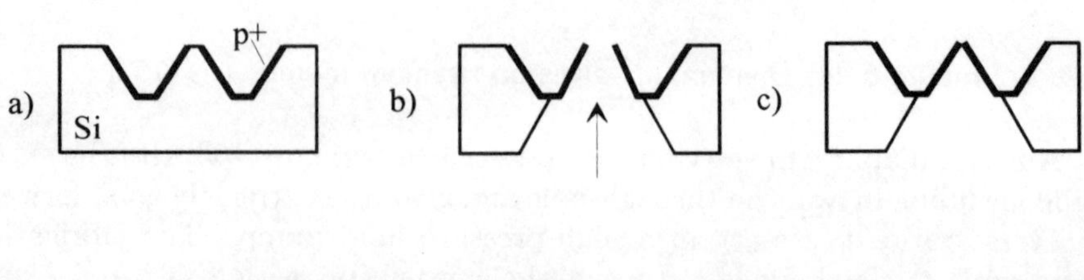

Figure 6.26: Bivalvular microvalve [154]: a) p^+ doped V-groove, b) finished device open, c) finished device closed

The valve shows an almost linear behaviour of the flow rate with respect to applied pressure from 0 to 4 kPa. The maximum flow rate measured at 4 kPa was around 1.6 ml min^{-1}. However, the leakage rate is rather large as well. For a backpressure of 4 kPa the leakage is around 50 μl min^{-1}, offering a forward to reverse ratio of 32 : 1.

6.3.1.4 Diffuser/Nozzle Valve

A completely different valve type is the dynamic diffuser/nozzle valve. Unlike all other valves, the diffuser/nozzle element does not offer a large difference in the flow rate between forward and reverse directions. The forward to reverse flow rates are typically 10:9 and are small compared to cantilever valves (better than 100:1).

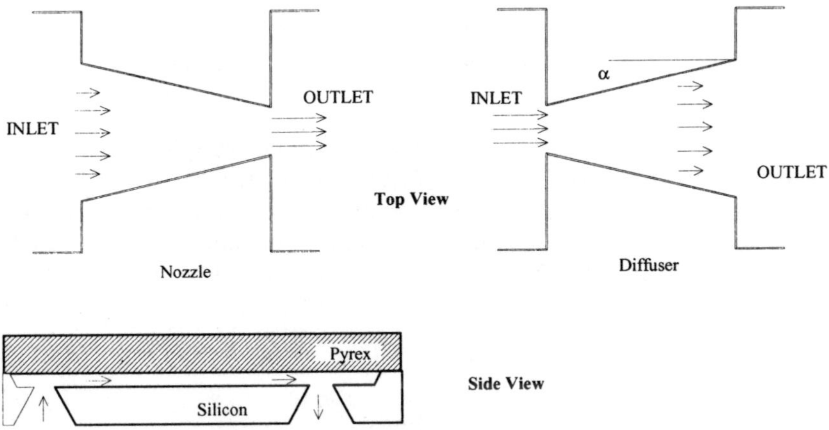

Figure 6.27: Diffuser/nozzle element with small angles (top view and side view)

Two different flow regimes are possible for diffuser/nozzle valves, as described earlier in section 2.1.4. If the taper angle is small (around 3 degrees) the diffuser direction is preferred compared to the nozzle direction [8, 155]. If the angle is enlarged, the flow domain changes and the nozzle direction is the preferred flow direction [125]. Diffuser/nozzle elements with small angles are usually formed by dry etching in silicon and sealing with Pyrex (see figure 6.27). Dynamic valves with large angles (figure 6.28) make use of the anisotropic etching behaviour of silicon in KOH. This allows only one angle (35.3 degree) but has the advantage of a very simple process.

The fabrication sequence of a dynamic diffuser/nozzle valve with large angles is given as an example: After dry etching to form the double-sided alignment marks on both sides of the wafer, the wafer is oxidised in a dry oxygen atmosphere. Subsequently a silicon nitride layer is deposited. The thicknesses are 160 nm (silicon nitride) and 80 nm (silicon dioxide), respec-

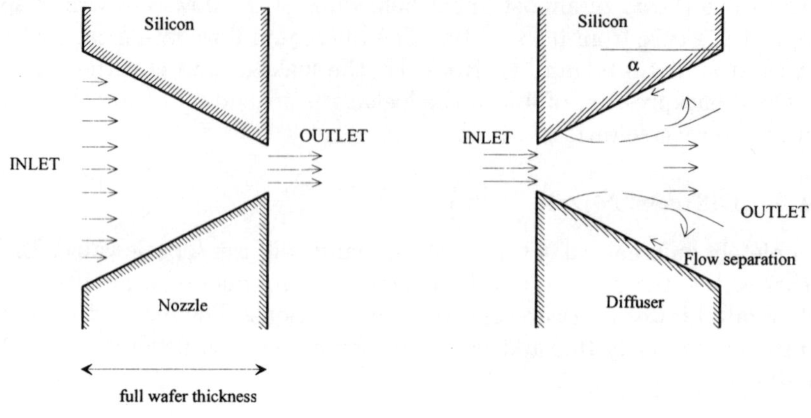

Figure 6.28: Dynamic valve with large angles (side view)

tively. Now the mask with the duct area is photolithographically printed on one side of the wafer and both layers (SiO_2 and Si_3N_4) are dry etched. This is then repeated for the other side to create two counter-opposed valves as is required for micropumps. Thus two displaced openings are generated on opposite sides of the chip. Now the wafer is etched in 30% KOH at 70°C to create both valves at the same time. After stripping the silicon nitride and the silicon dioxide, the chip is finished (figure 6.29).

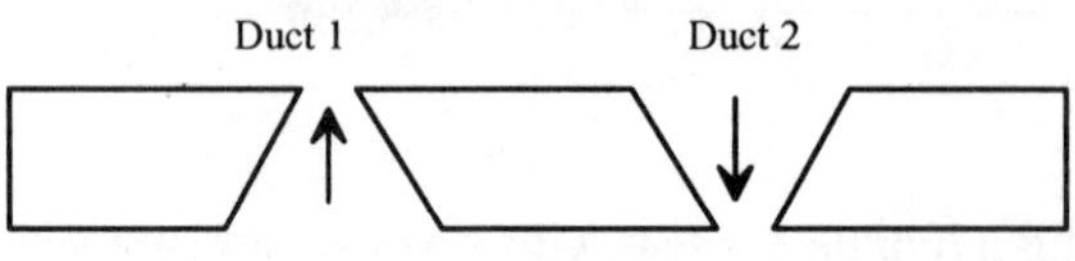

Figure 6.29: Diffuser/nozzle valve for a micropump

The flow rate measurement of the dynamic diffuser/nozzle valve has already been presented in figure 2.6 but is given again in figure 6.30. Over the whole pressure range, the flow through the diffuser is lower than the flow through the nozzle, as expected from theory. The test liquid for the measurements is DI water, the size of the small opening of the valve is 130 × 130 μm and the wafer thickness is 500 μm.

The other fabrication technique of diffuser/nozzle valves has the taper angle of the structure as a degree of freedom allowing better optimisation of forward/reverse flowrate ratio. Olsson [155] and Jiang [156] produce this valve by dry etching and anodic bonding of a Pyrex wafer on top. This gives a taper on only two (etched walls) rather than all four sides. An improvement is possible with laser assisted RIE [8] where the third side, i.e. the bottom

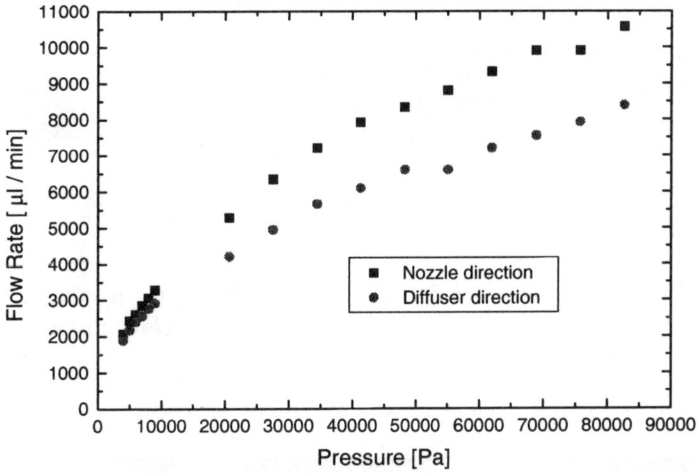

Figure 6.30: Measured flow rates for a dynamic passive valve

surface, can be tapered as well. As mentioned before, the flow behaviour changes for small taper angles preferring the diffuser direction. Depending on the intended average volume flow, the optimum taper angle α varies between 2 and 4°, with the smaller angles favoured at flow rates above 350 μl min^{-1} [8].

6.3.2 Actuation Techniques for Valves

More complicated structures are needed for active microvalves. Although different actuation mechanisms have been described in the literature, the basic principle is similar: a flap controls the flow by actively opening or closing the passage for the fluid. For example, the valve can be activated by the thermal actuation principle of a bimetallic element [157, 158]. The principle of bimetallic actuation has already been described in section 3.2.2 and is considered again in section 6.4.1.4. In the valve designed by Jerman [158], the flow path is either normally closed or normally open depending on the position of the bimetal flap (figure 6.31). The valve operates with gas flows in the range of 0 to 150 ml min^{-1} and pressures between 6.9 and 344 kPa. The switching times are around 100 ms. Lisec [157] employs a valve that is based on the buckling effect of a heated pure silicon bridge structure. The achieved flow rates range from 0 to 700 ml min^{-1} for gases at pressures up to 100 kPa with a switching time of 15 ms.

Another active valve has been presented by Trah [159] where the flow is steered by a bimorph cantilever beneath or above a bossed membrane using the Coanda effect (figure 6.32). If the flow is deflected below the membrane, the kinetic energy is partially transformed into potential energy, lifting the membrane and allowing a flow through the outlet. The valve was designed for water flow rates of about 75 ml min^{-1} and system pressures of 600 kPa.

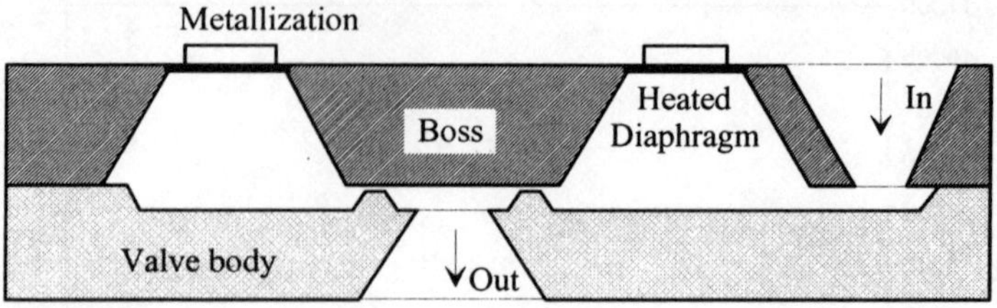

Figure 6.31: Bimetallic actuator for a microvalve [158]. Both metal-silicon diaphragms deflect the central boss towards the outlet and close the valve.

The maximum switching frequency was measured as 1 kHz.

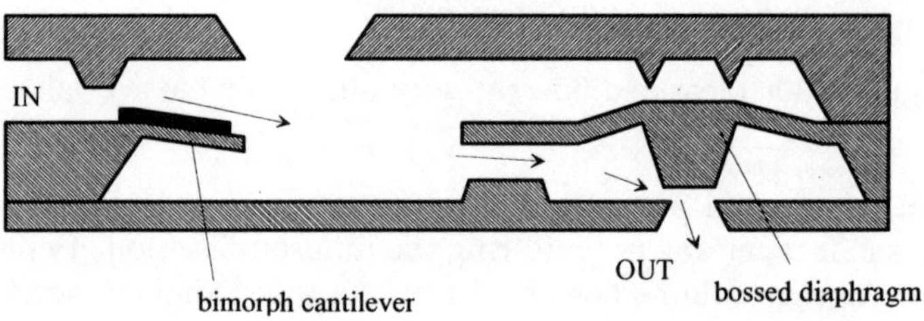

Figure 6.32: Schematic view of the bimorph cantilever valve with beam steering and hydraulically actuated diaphragm by Trah [159]

A second actuation principle is of electrostatic nature [160], as described in section 3.2.4. This actuation technique was introduced by Bosch [161]. A flap, which sits on top of a duct, forms the first electrode (figure 6.33), whereas the counter electrode is embedded in the valve seat. When an electric field is applied, the flap is deflected towards the second wafer and closes the passage. The maximum pressure for which the valve is fully operating was measured as 16 kPa, and a gas flow of up to 3 ml min^{-1} can be controlled at a switching frequency of greater than 2.5 kHz.

A normally open flap valve has been created using a similar idea [162]. Cr is deposited on silicon dioxide and curls the Cr/SiO_2 cantilever upwards (figure 6.34). This effect is due to the Cr layer being under tensile stress and resting on silicon dioxide, which is under compressive stress. If an electric field is applied between the Cr electrode and the lower electrode, the cantilever is bent and closes the underlying orifice. No measurements have been presented so far to quantify the control function of the valve.

A similar technology to that described by Bosch has been employed to fabricate an electromagnetically driven microvalve [163]. A gold coil is electroplated on top of a polysilicon diaphragm (figure 6.35) and a permanent magnet is fixed on the bottom to achieve the actuation of the membrane with

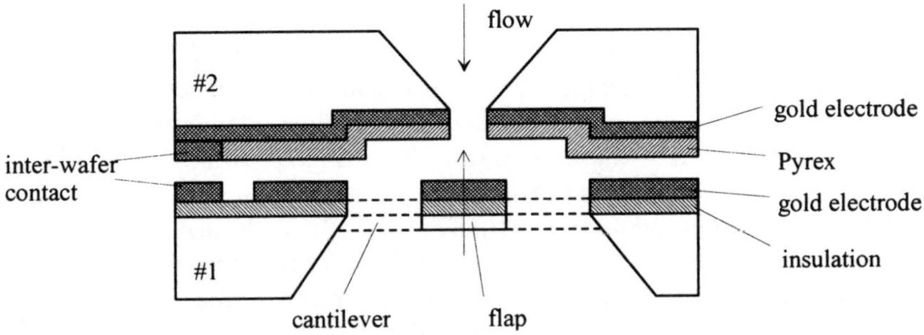

Figure 6.33: Electrostatically actuated microvalve [161]

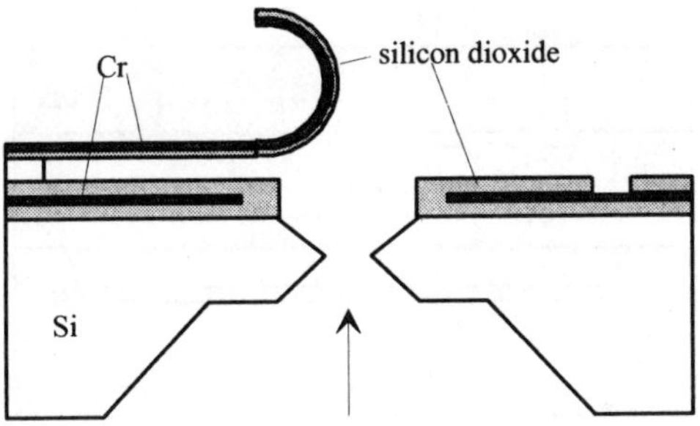

Figure 6.34: Microvalve with curled-up actuator [162]

the coil. In order to reduce the danger of sticking and improve the seal for gases, a silicone rubber is screen-printed onto the valve seat. The valve has been designed for working pressures up to 50 kPa and gas flows between 2 and 20 ml min^{-1}, as required for a miniature air gas analyser. However, no measurements have been presented to date.

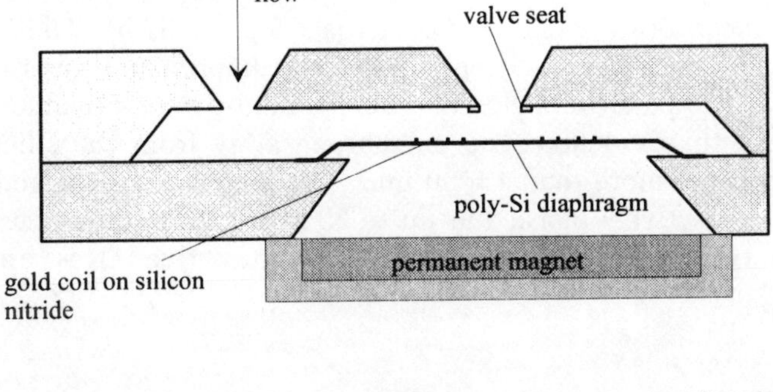

Figure 6.35: Electromagnetically actuated valve [163]

Thermopneumatic actuation can be used for the control of liquids with active valves [164]. In this case a heated volume is bordered on one side by a flexible silicon diaphragm of 50 μm thickness (figure 6.36). When the volume is heated, the diaphragm presses against the opening of an underlying through-hole and closes the flow path. The heated volume is made from Pyrex by the use of ultrasonic drilling and is filled with Fluorinert^{TM} liquid optimised for thermopneumatic actuation. It also includes a protection for overpressure to avoid the membrane breaking at large pressures in the flowing fluid. The fabrication of the other part is also quite simple using KOH and a fillet etch (HF : nitric : acetic acid). For assembly, the silicon layers are fusion bonded and then anodically bonded to the Pyrex layers. With this valve, differential pressures of up to 1 MPa can be controlled with flow rates between 0 and 10 gs^{-1} for the refrigerant R-134a.

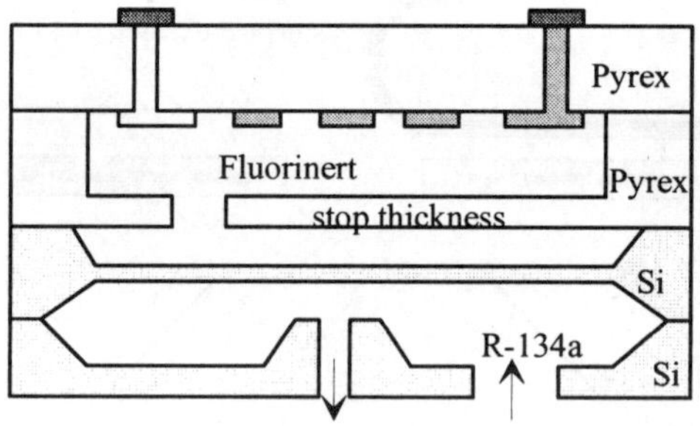

Figure 6.36: Thermopneumatical actuated microvalve by Henning [164]

Finally, there are piezoelectrically actuated active valves. The first valve of this type has been described by Esashi [165], where a stacked microactuator deflects a bossed diaphragm towards or away from the second wafer (figure 6.37.a). This opens or closes a passage between the two wafers and thus controls the fluid flow. The maximum gas flow that could be controlled was 15 ml min^{-1} at 70 kPa. A different type of valve using piezoelectric control was described by Stehr [166]. An elastic diaphragm is glued on top of a silicon wafer incorporating inlet and outlet (see figure 6.37.b). The diaphragm itself is actuated by a piezo-bimorph. When the diaphragm is pressed against the valve seat the passage is closed, whereas fluid flow is possible when the piezoelectric actuator deflects the membrane away from the silicon wafer. Water flow rates of more than 1.5 ml min^{-1} at 2 kPa were achieved, and the fluid could be effectively blocked at up to 20 kPa in both directions.

A summary of device properties for some of the active valves can be found in table 6.1.

a)

b)

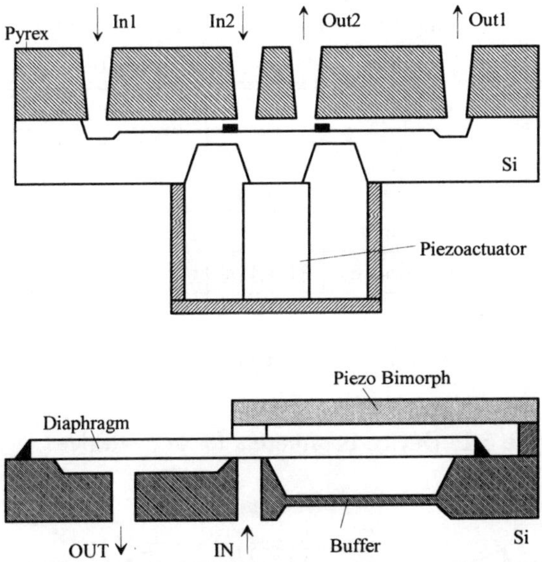

Figure 6.37: a) Active four port injection valve by Esashi [165], b) Microvalve with bi-directional pump effect by Stehr [166]

6.4 Micromachined Pumps

Micropump research is one of the biggest fields in microfluidics. The pumps are usually either mechanical of the membrane type or alternatively the flow is induced by an electric field. Membrane pumps have been described by Zengerle [28], who used electrostatic actuation of the membrane and passive cantilever valves. Gerlach [125] and Olsson [10] both employed dynamic passive valves and membrane actuation produced by surface mounted piezoelectrics. Pumps with electric field induced flow have been of the electrohydrodynamic type as described by Richter [168] or electroosmotic type as described by Jacobson [169].

6.4.1 Membrane-Actuated Pumps

Membrane pumps work on the principle of an oscillating membrane creating a varying chamber volume (figure 6.38). When the chamber volume increases, liquid flows through an inlet valve into the chamber. When the chamber volume decreases, the fluid is forced out through an outlet valve. A difference in the flow resistance between the inlet and the outlet for different flow directions is essential for operation. Valves can be either of the cantilever type, membrane type or diffuser/nozzle type.

Author	actuation mechanism	fluid	min. switching time [10^{-3}s]	max. pressure
Jerman [158]	bimetal	gas	100	344 kPa
Lisec [157]	bimetal	gas	15	100 kPa
Trah [159]	bimetal	water	1	600 kPa
Bosch [161]	electrostatic	gas	0.4	16 kPa
Henning [164]	thermopneum.	R-134a	–	1 MPa
Esashi [165]	piezoel.	gas	–	70 kPa
Stehr [167]	piezoel.	water	–	20 kPa

Table 6.1: Device parameters for active valves

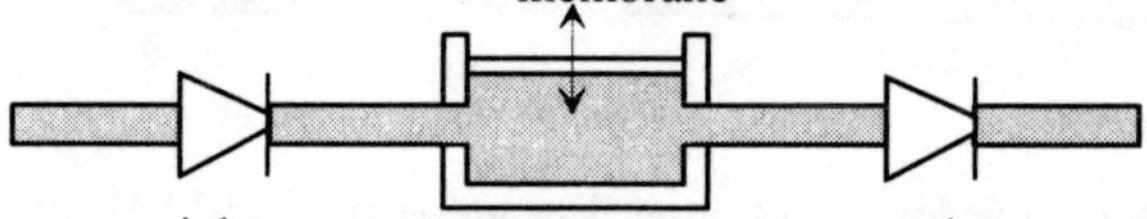

Figure 6.38: Schematic membrane pump

6.4.1.1 Overview

Actuation of the membrane can be carried out in several ways:

- Pneumatic [43]: An applied pressure from an external pump (figure 6.39.a) on one side deflects the membrane and thus changes the chamber volume on the other side. For this technique, an external pump source is needed and therefore further integration is not possible.

- Thermopneumatic [29]: A heated volume (figure 6.39.b) deflects the membrane in one direction, whereas cooling releases the membrane tension. If the heated volume is small enough reasonable pump frequencies can be achieved. The heater and the chamber can be fabricated in silicon technology, using bonding of Si-wafers and a heater resistor [29].

- Piezoelectric [10, 103, 124–128]: A piezoelectric layer is attached to the membrane (figure 6.39.c), such that a change in layer tension leads to a deflection of the membrane sandwich. The piezoelectric plate can either be surface mounted with conductive epoxy or screen-printed as described in section 5.4.1. Also, possibilities exist for either sol-gel fabrication [99, 170], where an organic fluid containing PZT is spun on the wafer and fired at high temperatures, or sputter deposition [171], where a PZT film is created by co-sputtering the correct composition onto silicon.

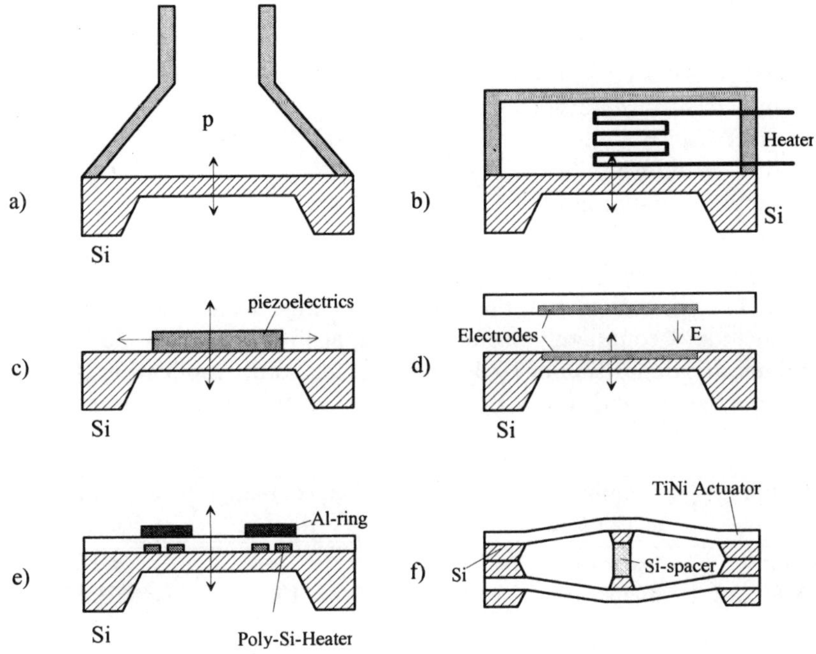

Figure 6.39: Actuation principles for micropumps: a) pneumatic, b) thermopneumatic, c) piezoelectric, d) electrostatic, e) bimetallic, f) shape-memory

- Electrostatic [43]: An electric field between the membrane and a fixed counter electrode (figure 6.39.d) attracts the membrane and thus deflects it. Both electrodes can be fabricated using silicon techniques. Bonding of the two wafers with a predefined spacer for the two electrodes is possible.

- Bimetallic [172]: Thermal heat dissipation occurs when a voltage is applied to a poly-Si resistor (figure 6.39.e). Coupled with the aluminium-silicon bimetal effect the heat dissipation in an aluminium ring causes a deflection of the whole membrane. The reason for this is a difference in the thermal expansion coefficients between aluminium and silicon.

- Shape-memory effect [27]: TiNi metal forms a membrane on two sides of an actuator, deflected in the centre by a spacer (see figure 6.39.f). When the top TiNi actuator is heated with resistive heat dissipation, the top diaphragm straightens and deflects the lower diaphragm further. Then the current is switched off and the bottom diaphragm is heated, so that an opposite movement of the bottom membrane is achieved. Attached to an inlet and outlet valve system, this actuator can be used for driving fluid.

A comparison of maximum pump rate and backpressure for most micropumps published so far is given in tables 6.2 and 6.3.
Key: smm = surface micromachining,
act. = active,
diff. = diffuser,
diaph = diaphragm,
ext = external,
mem = membrane.

Some other fabrication methods for micropumps have also been reported. These include a completely surface micromachined pump by Judy [89] and organometallic deposition of PZT on silicon for pump applications by Brooks [102]. However, in both cases, no further report of a working device has been published.

Author	pump mechanism	valve type; fluid	max. pumprate [μl min^{-1}]	max. backpressure
v. Lintel 1988 [124]	piezoel. membrane	diaphragm; water	• 8 100 V_{pp} 1 Hz	• 10 kPa 100 V_{pp} 1 Hz
Esashi 1989 [123]	piezoel. (ext) membrane	smm diaph; water	• 15 90 V_{pp} 30 Hz	• 8 kPa V_{pp} 30 Hz
v.d. Pol 1990 [29]	thermo-pneumatic membrane	diaphragm; water	• 34 6 V_{pp} 0.85 Hz	• 3.5 kPa 6 V_{pp} 0.5 Hz
Shoji 1990 [127]	piezoel. (ext) membrane	smm diaph; water	• 20 100 V_{pp} 20-40 Hz	• 15 kPa 100 V_{pp} 25 Hz
Smits 1990 [128]	peristaltic (piezoel.)	act. diaph; water	• 100 100 V_{pp} 15 Hz	• 6 kPa 100 V_{pp} 15 Hz
Stemme 1993 [126]	piezoel. brass mem.	diff./nozzle; water	• 4000–16000 20 V_{pp} 110 Hz	• 1–2 kPa 20 V_{pp} 110 Hz
Rapp 1994 [107]	pneum. (ext) LIGA mem.	diaphragm; air	• 82 50 kPa_{pp} 5 Hz	• 4.5 kPa 50 kPa_{pp} 5 Hz

Table 6.2: Comparison of all micropumps published up to date (part I)

Author	pump mechanism	valve type; fluid	max. pumprate [μl min^{-1}]	max. backpressure
Schomburg 1994 [173]	thermo-pneum. LIGA	diaphragm; air	• 44 15 V_{pp} 5 Hz	• 2.8 kPa 15 V_{pp} 5 Hz
Zengerle 1995 [47]	electrostatic membrane	cantilever; water	• 250 200 V_{pp} 830 Hz • -350 200V_{pp} 3400 Hz	• 27 kPa 200 V_{pp} 10-100 Hz • 7 kPa 200V_{pp} 3400 Hz
Zengerle 1995 [174]	electrostatic membrane	diaphragm; water	• 800 200 V_{pp} 900 Hz	• 35 kPa 200 V_{pp} 1-900 Hz
Gerlach 1995 [175]	piezoel. membrane	diff./nozzle; water	• 400 50 V_{pp} 1 kHz	• 3 kPa 50 V_{pp} 3 kHz
Olsson 1995 [10]	piezoel. membrane	diff./nozzle; water	• 16000 80 V_{pp} 540 Hz	• 16.5 kPa 80 V_{pp} 540 Hz
Carrozza 1995 [176]	piezoel. membrane	ball valve; water	• 2700 300 V_{pp} 70 Hz	• 25 kPa 300 V_{pp} 70 Hz
Zhan 1996 [172]	bimetal membrane	cantilever; water	• 44 5.5 V_{pp} 0.5 Hz	• 11 kPa 5.5 V_{pp} 0.5 Hz
Stehr 1996 [166]	piezoel. (ext) membrane	act. valve; water	• ± 1000 150 V_{pp} 60 Hz (+) 150 V_{pp} 140 Hz (-)	• 17 kPa 200 V_{pp} 190 Hz
Benard 1997 [27]	shape-memory membrane	diaphragm; water	• 50 0.6 V_{pp}, 0.95 Hz 0.95 Hz (+)	-
Koch 1997 [103]	thick-film PZT membrane	cantilever; ethanol	• 122 600 V_{pp} 100 Hz	• 2 kPa 600 V_{pp} 200 Hz
Koch 1997 [104]	thick-film PZT membrane	diff./nozzle; ethanol	• 155 600 V_{pp} 2.2 kHz	• 1 kPa 600 V_{pp} 2.2 kHz

Table 6.3: Comparison of all micropumps published up to date (part II)

6.4.1.2 Thermopneumatic Pumps

A thermopneumatic pump has been described by van de Pol [29]. As shown in figure 6.40, the actuating part comprises a cavity filled with air, a square silicon pump membrane and a built-in aluminium meander which serves as a heater resistor. The resistor is located on a sheet of silicon suspended by four silicon beams, which provide thermal isolation of the central region. The flexible membrane at the bottom end of the heated chamber can deflect the pump chamber and thus lead to pumping. The valves are of the diaphragm type where the diaphragm is in the middle of the wafer and the closing is against the Pyrex ground plane.

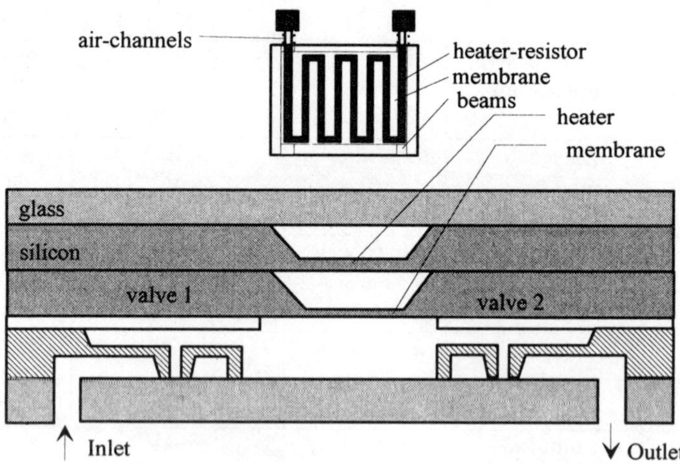

Figure 6.40: Thermopneumatic actuated micropump by van de Pol [29]

Due to downscaling effects, the generated mechanical power of a thermopneumatically actuated pump is good and the voltage can be kept in the range of 6 – 13 V. Because of a thin pump membrane, the pump stroke is quite large, i.e. 23 μm central displacement or 0.5 μl for a membrane geometry of 7.2 $\times$ 7.2 mm. The full amplitude is reached after 25 seconds. Due to this low cycle time, the maximum operating frequency is limited to a few cycles per second. The full size of the pump is around 30 $\times$ 15 $\times$ 2 mm.

A thermopneumatic micropump fabricated with the LIGA technique has been presented by Schomburg [173]. The working principle is identical to the one described above and a schematic of the device is depicted in figure 6.41. The pump case consists of polysulphone (PSU) and the membrane for actuation and valve purposes is made from a 2.5 μm-thick polyimide. The overall size of the pump is 10 $\times$ 10 $\times$ 7 mm. For the thermopneumatic actuation, a titanium wire is attached to the membrane and heated by an electrical current. This device pumps both liquids and gases, as the central membrane deflection is 100 μm, giving a large displaced volume. The maximum achievable pump rates are 44 μl min^{-1} and the maximum backpressure 2.8 kPa. The test fluid in both cases was air and the pump was operated at 15 V and a frequency of 5 Hz.

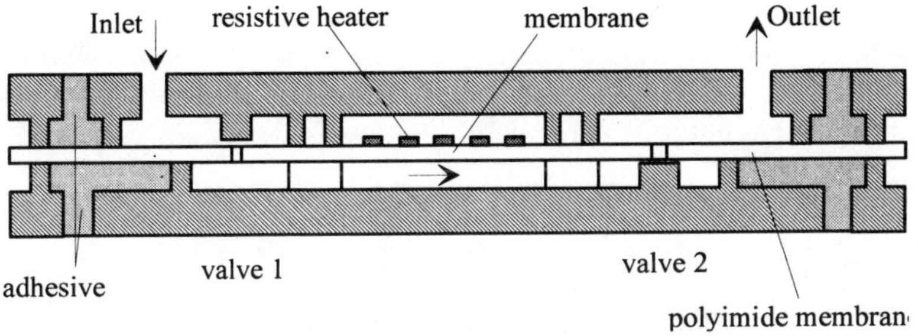

Figure 6.41: Thermopneumatic actuated micropump by Schomburg [173]

6.4.1.3 Piezoelectric Pumps

Quite a few pumps have been described in which the piezoelectric effect drives the pump membrane. The first micropump was developed by van Lintel [124]. The pump uses the same diaphragm valves as the pump by van de Pol. Instead of thermopneumatic actuation, this pump employs a surface mounted piezoelectric disc of 10 mm diameter and 200 μm thickness. The full size of the pump is 30 × 15 × 2 mm.

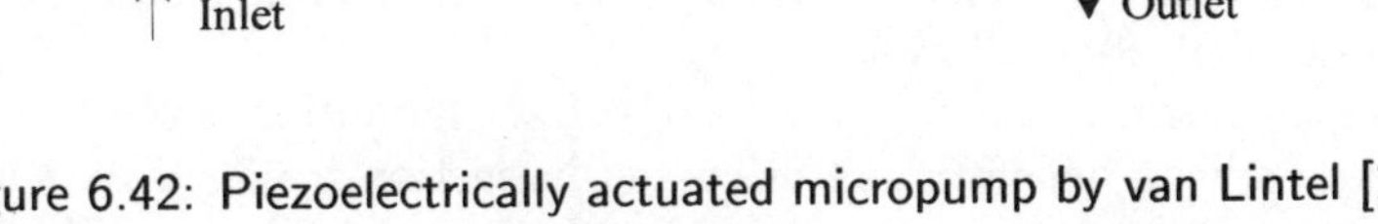

Figure 6.42: Piezoelectrically actuated micropump by van Lintel [124]

Piezoelectric actuation requires rather large voltages. Thus, a voltage of 100 V is necessary to achieve a volume stroke of 0.21 μl for a membrane of 12.5 mm diameter. Although the driving frequency of piezoelectrically actuated membranes is theoretically not restricted, all measurements of the pump were taken at frequencies less than 1 Hz.

A newer model of this pump has been presented by Gass [130], who integrated a flowsensor on the pump to achieve a pressure insensitive pump rate. Backpressure-insensitive operation has been demonstrated for pressures up to 9 kPa and flow rates up to 100 μl min^{-1}. The operating frequency of this pump has been taken up to 80 Hz but larger pump rates are obtained at half of this value due to dynamic effects in the valves.

The next piezoelectric micropump was presented by Shoji, who uses external piezoelectric actuation and surface micromachined diaphragm valves. A diagram of this pump is depicted in figure 6.43.

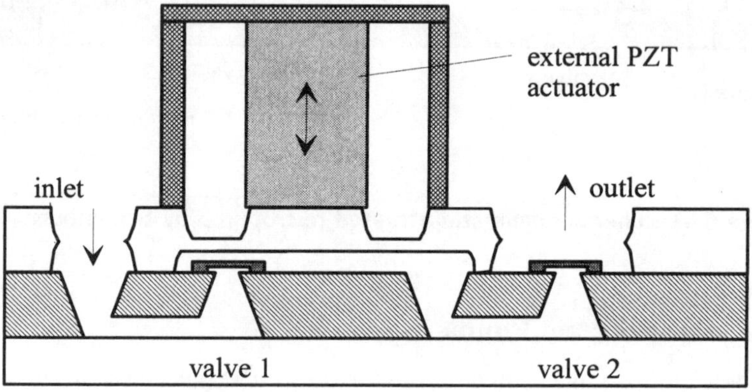

Figure 6.43: Piezoelectrically actuated micropump with surface micromachined valves by Shoji [127]

This pump consists of a silicon wafer between two glass plates. In order to avoid ripples occurring during the pump actuation, a dual pump operating at complementary periodic cycles as well as a buffer pump with a synchronous buffer connected in series to the pump have been constructed. This is particularly important in chemical analysing systems. With a voltage of 100 V and a frequency of up to 100 Hz, both pump systems give flow rates up to 30 μl min^{-1}. The external piezoelectric actuator is connected to the silicon membrane with epoxy, which makes the whole pump rather big. The overall size is about 20 × 20 × 10 mm.

An improvement to the fabrication of surface mounting is the technique of thick-film-on-silicon actuators as discussed in section 5.4.1. Bulk silicon micromachined cantilever valves were used with the thick-film-on-silicon hybrid actuator and compared with a conventional surface mounted actuator. The fabrication of these components has been already described in the previous sections and will only be reviewed briefly. The passive cantilever valves are based on a B^+ anisotropic etch stop layer for the definition of the cantilever and the creation of a conical duct on the same wafer. For ease of construction, two identical chips are placed on top of each other to complete both inlet and outlet valves. Bulk PZT material of 6.25 × 3.25 × 0.25 mm was glued with conductive epoxy to the silicon membrane on the actuator chips. The second type of actuator uses screen-printed PZT with 5% lead zinc borosilicate as a binder, and cermet gold top and bottom electrodes, as this combination gives good performance. A sketch of this micropump is shown in figure 6.44.

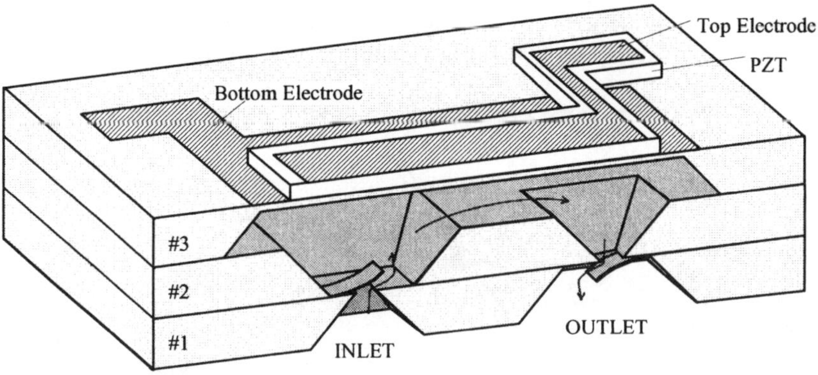

Figure 6.44: Schematic of a piezoelectric micropump with cantilever valves [103]

Thin metallic wires are glued on the contact pads of the actuator chip with conductive epoxy to allow simple electrical access. The epoxy is cured at 130°C for 30 minutes.

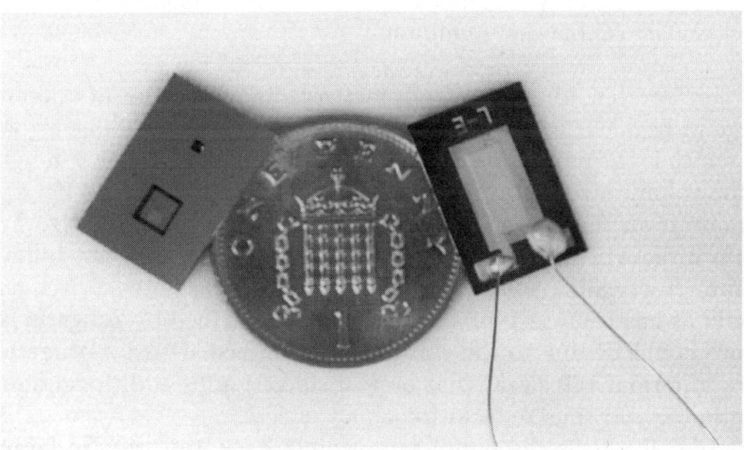

Figure 6.45: Valve chip (left) and screen-printed actuator (right)

For the surface mounted devices, the bulk piezoelectric plate is glued with the same conductive epoxy to the substrate. Electrical contacts are made to the cured epoxy on the substrate and to the top of the plate. Figure 6.45

shows both the actuator and the valve chip. In order to give an idea of the size of the pump (15 × 10 × 2 mm) a British penny is put next to the devices.

The micropump with surface mounted piezoelectric actuators coupled with the cantilever valve of 1000 μm side length was measured for reference purposes. A 6.25 × 3.25 × 0.25 mm-sized actuator was employed. The actuation voltage was kept constant at 200 V_{pp} and the frequency was varied between 10 Hz and 3 kHz. The pump rates at the different frequencies and small backpressures up to 290 Pa are given in figure 6.46.

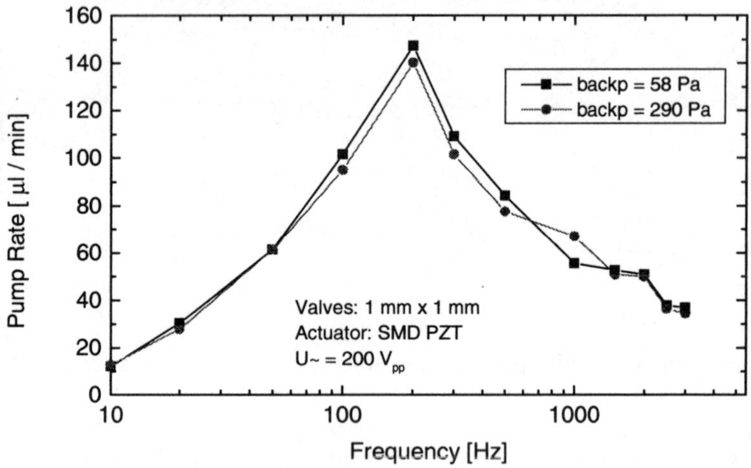

Figure 6.46: Frequency dependence of a surface mounted micropump with 1000 μm-square cantilevers: pump rate

As can be seen, the pump rate increases by one order of magnitude for a change in frequency from 10 to 200 Hz. The effective volume stroke of the pump, which is the pump rate divided by the frequency, is shown in figure 6.47. Here, the volume stroke stays between 22 nl and 15 nl up to 100 Hz, and starts to decrease from then on with the steepest part just after 200 Hz. The small variation in backpressure between 58 Pa and 290 Pa does not influence the pump rate drastically. No reverse pumping action was noted at frequencies up to 20 kHz as has been seen in a similar pump described by Zengerle [47]. The difference could be due to the rim at the valve seat used by Zengerle (figure 6.11.c). A normal valve seat (flat area of silicon) adds additional damping to the cantilever and thus reduces frequency effects.

Figure 6.48 shows the pump rate against backpressure. No large dependence on the backpressure up to 3 kPa is noted for small frequencies up to 50 Hz. A linear fit has been made for all curves from which the maximum achievable backpressure has been calculated. A backpressure of 11 to 13 kPa is possible for frequencies of 20 and 50 Hz. The behaviour changes between the frequencies of 100 and 200 Hz, where a drop in the pump rate is noted

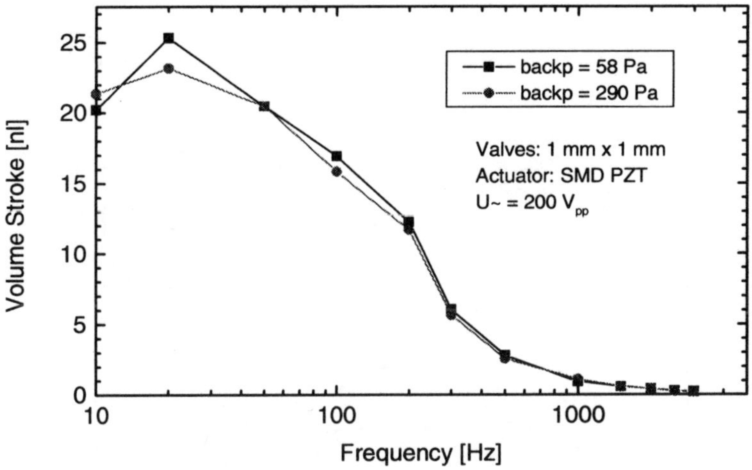

Figure 6.47: Frequency dependence of a surface mounted micropump with 1000 μm square cantilevers: volume stroke

with respect to the backpressure. In this case values of 4.96 and 3.52 kPa were calculated as the maximum achievable backpressure for 100 and 200 Hz respectively.

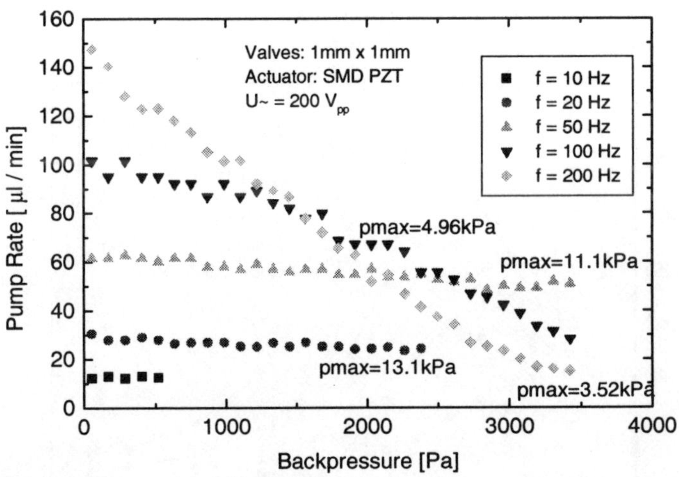

Figure 6.48: Backpressure dependence of a surface mounted micropump with 1000 μm cantilevers at frequencies up to 200 Hz

Figure 6.49 shows the pump rate with respect to the backpressure for frequencies greater than 200 Hz. An almost linear drop of the pump rate was observed when increasing the backpressure and the lines for various frequencies seem to be almost parallel to each other. Thus, for increasing frequencies, the maximum achievable backpressure decreases down to 1500 Pa at 3 kHz.

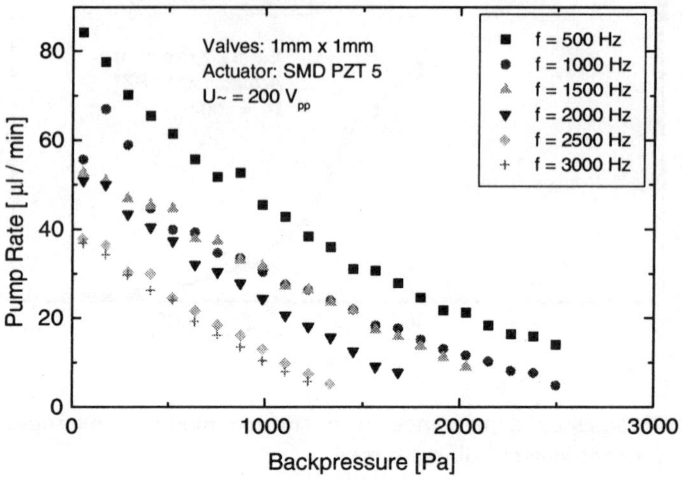

Figure 6.49: Backpressure dependence of a surface mounted micropump with 1000 μm cantilevers at frequencies beyond 200 Hz

A similar picture is obtained for cantilever valves of 700 μm side length. The same pump rate as with 1000 μm-long cantilevers is achieved at 10 Hz and 200 V_{pp} driving voltage (figure 6.50).

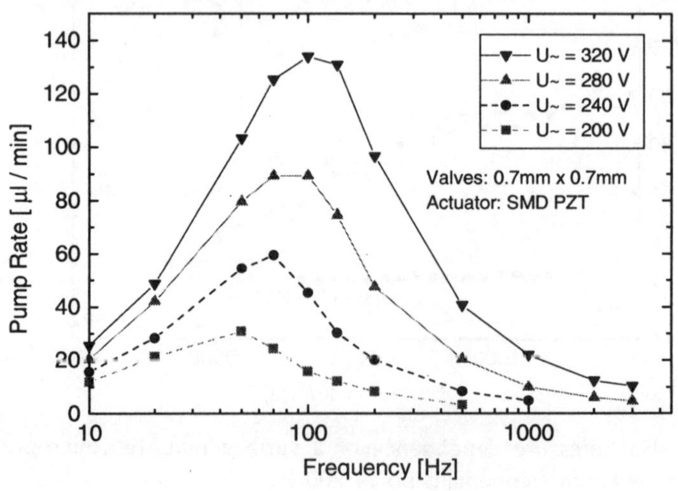

Figure 6.50: Frequency dependence of a surface mounted micropump with 700 μm cantilevers: pump rate

This gives confidence that the pump behaves in the quasi-static mode at very low frequencies and the valve is only needed to direct the flow. The maximum pump rate is now achieved at 70 Hz and 200 V_{pp}, increasing slightly to 135 Hz at 320 V_{pp} actuation frequency. This shows the influence of the valve at frequencies as low as 100 Hz. The value of the maximum pump rate has decreased from 140 $\mu l \, min^{-1}$ to 20 $\mu l \, min^{-1}$ for a change of the cantilever side length from 1000 μm to 700 μm. For higher frequencies, the pump rate as well as the volume stroke drop significantly to volume flows of less than 20 $\mu l \, min^{-1}$ and volume strokes of less than 1 nl above 1 kHz (figure 6.51).

The thick-film piezoelectric actuated micropump has also been examined for frequency and backpressure dependence. In order to achieve a reasonable driving voltage for the screen-printed piezoelectrics a transformer is placed between the output of the signal generator and the actuator. Unfortunately, this increases the minimum frequency at which the pump can be tested. The size of the piezoelectric layer is 700 × 700 × 0.1 mm and the length of the cantilever is 1000 μm.

The frequency dependence of the pump rate at zero backpressure is shown in figure 6.52 for two different actuation voltages (450 V_{pp} and 600 V_{pp}). A maximum pump rate of 61 μl min^{-1} and 122 μl min^{-1}, respectively, is achieved at frequencies of 100 Hz and 200 Hz.

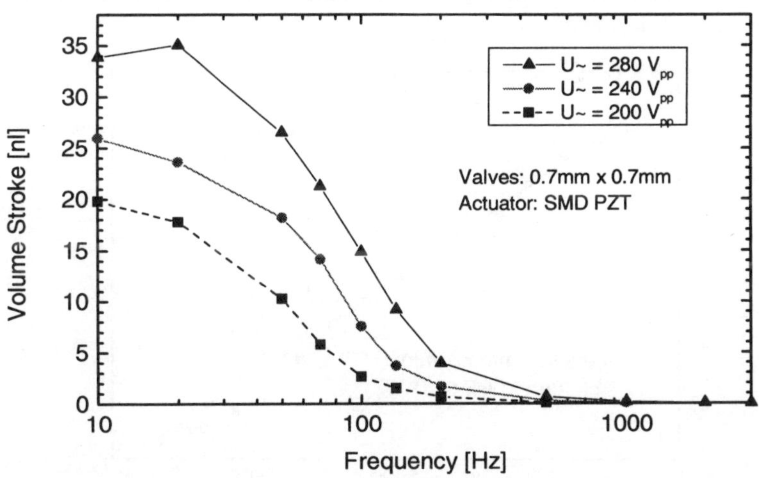

Figure 6.51: Frequency dependence of a surface mounted micropump with 700 μm cantilevers: volume stroke

The backpressure dependence is depicted in figure 6.53 for 600 V_{pp} and 200 Hz, where the maximum flow rate is achieved at zero backpressure. An almost linear drop can be observed and a maximum backpressure of 2 kPa is achievable. The backpressure dependence of the pump rate for 450 V_{pp} with the frequency as a variable is given in figure 6.54. Again, a change in the behaviour with respect to frequency can be seen. For lower frequencies, the

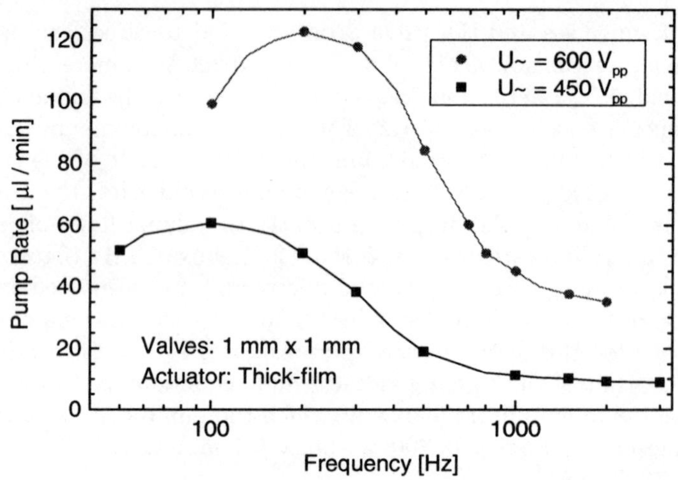

Figure 6.52: Frequency dependent behaviour of the micropump with thick-film-on-silicon actuator

achievable backpressure is usually higher. This tendency is identical with the surface mounted actuator.

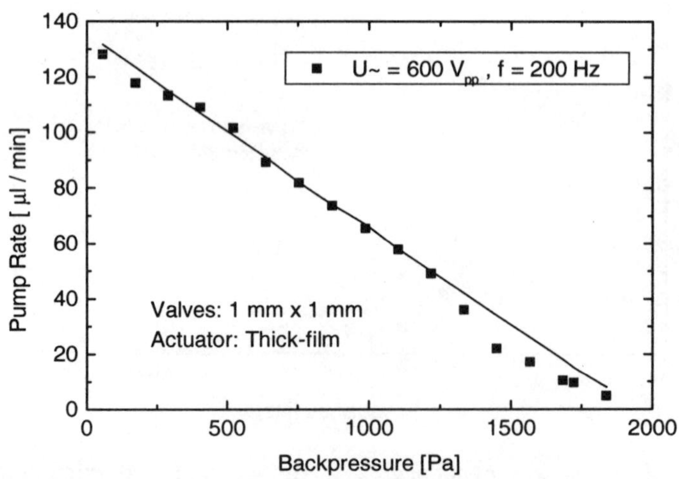

Figure 6.53: Backpressure dependent behaviour of the thick-film-on-silicon micropump at 600 V_{pp}

In comparison, surface mounted actuator equipped micropumps with 1 mm-long cantilevers achieved a maximum pump rate of 150 μl min^{-1} with 200 V_{pp} at 200 Hz. The maximum backpressure at this actuation was between 3.52 and 4 kPa. Assuming a linear behaviour of the pump rate with voltage, this is a factor of 6 larger than the thick-film-on-silicon micropump.

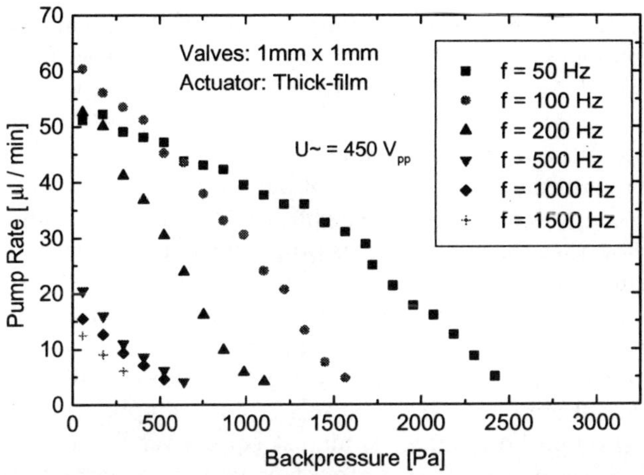

Figure 6.54: Backpressure dependent behaviour of the thick-film-on-silicon micropump at 450 V_{pp}

A different micropump mechanism was used by Stehr [166,167]. The first approach uses an elastic diaphragm connected to a bimorph cantilever beam and a membrane buffer. A picture of this pump type is shown in figure 6.55.

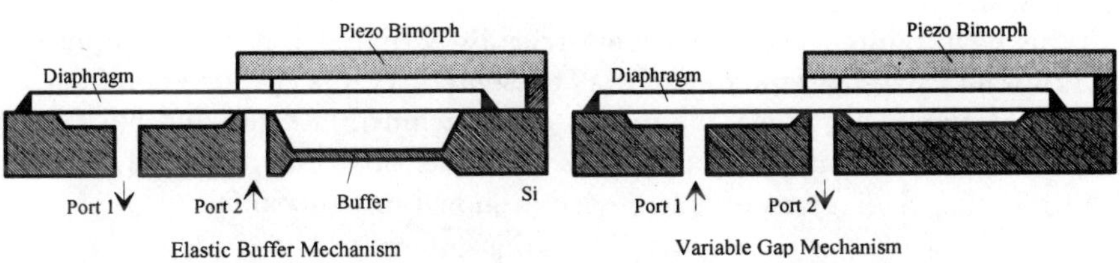

Figure 6.55: Microvalve with bi-directional pump effect by Stehr [166, 177]

The operation of the pump with elastic buffer mechanism is described as follows: The diaphragm is moved upwards leading to an upwards movement of the buffer membrane and a fluid flow through port 1 and port 2 to equalise the membrane movement. When the diaphragm has reached the maximum deflection, the buffer membrane relaxes to its resting position. Now the diaphragm moves downwards to close port 2 leading to a downward movement of the membrane and some flow through both ports. Once the diaphragm has closed port 2, the buffer membrane is still deflected downwards due to an excess pressure inside the pump chamber. The relaxation of the buffer membrane then leads to a fluid flow through port 1 as port 2 has been closed.

The pump has been shown to pump air at 100 V driving voltage and frequencies between 800 and 1300 Hz from port 2 to port 1 with flow rates of up to 8.5 ml min^{-1}. At frequencies above 1500 Hz the direction reverses to a flow from port 1 to port 2 with flow rates of up to 8 ml min^{-1}. This is due to dynamic effects of the buffer membrane.

The other operating principle of the pump does not need the buffer mechanism but relies on the difference in height of the diaphragm from fluid ports 1 and 2. The flow rate through port 2 is proportional to the product of chamber pressure and the cubed deflection of the diaphragm. For square wave actuation, the pressure is large when the diaphragm moves upwards but, due to a much smaller gap between the diaphragm and port 2 as compared to port 1, most of the fluid is sucked through port 1 into the displaced volume beneath the diaphragm. At large distances of the diaphragm from port 2, the fluid flow through both valves is almost equal. When the diaphragm is then moved downwards, the flow is almost equally large through both ports as a large pressure is obtained at the beginning of the opposite cycle. This pressure then drops, as does the height, leading to small flow rates through port 1 and even less through port 2. Integrating the whole flow over one cycle, a net flow from port 1 to port 2 is obtained and thus pumping is achieved.

This pump mechanism allows bi-directional pumping, due to dynamic effects [40] at different frequencies for water, at pump rates up to 1000 μl min^{-1} from port 1 to port 2 (50 Hz) as well as 700 μl min^{-1} from port 2 to port 1 ($>$ 100 Hz). At 190 Hz actuation frequency and 200 V, the maximum achievable backpressure is 17 kPa.

The next pump type is piezoelectrically actuated but uses simple diffuser/nozzle valves. Stemme [126] has presented the first prototype of such a pump for MEMS. However, the design was machined in brass and has rather large device dimensions, with a chamber diameter of 19 mm and a valve length of 3 to 4 mm. A diagram of this device is shown in figure 6.56.

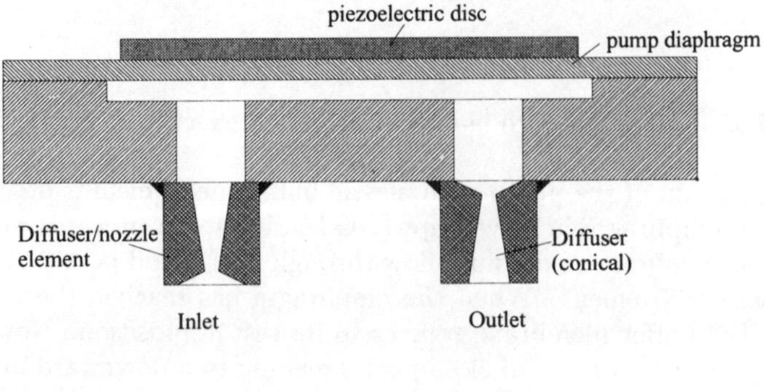

Figure 6.56: Piezoelectrically actuated pump with dynamic diffuser/nozzle valves by Stemme [126]

The vibrational amplitude of the piezoelectrically actuated membrane was 37 μm for a voltage of 100 V. Unlike most previous pumps, this type allows both water and air to be pumped at reasonable flow rates and backpressures. Air flow rates of up to 35 ml min^{-1} and backpressures of 2.5 kPa are possible, and the comparable figures for water are 4 to 16 ml min^{-1} and 2 kPa. However, it has to be remembered that this pump is conventionally machined as a prototype.

From the same group at the Royal Institute of Technology in Sweden comes a similar diffuser/nozzle pump based on microfabrication techniques. The valves of this pump are isotropically etched into silicon and the piezoelectric actuator is surface mounted onto the Pyrex plate, which also serves as the sealing of pump chamber and valves [10]. A schematic diagram of this pump can be found in figure 6.57.

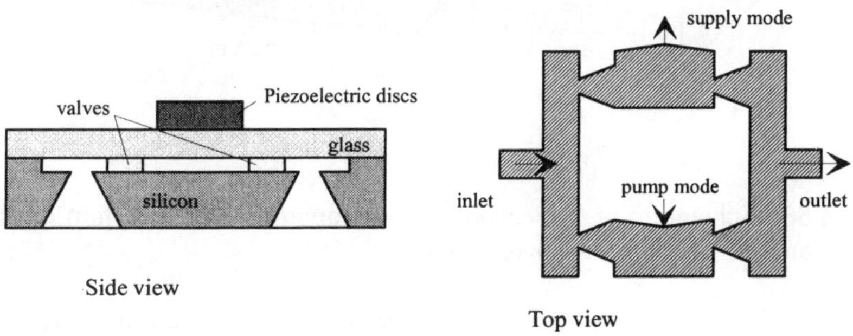

Figure 6.57: Piezoelectrically actuated pump with dynamic diffuser/nozzle valves by Olsson [10]

A dual pump reducing the effect of ripples during the pump strokes has been presented recently by the same Swedish group. It uses an identical fabrication technique as described above and the shape of the valves has been optimised by the use of deep RIE (cf. section 5.3.4) [178]. Optimum performance was obtained for a neck geometry of the valve of 80 × 80 μm, a diffuser length of 1073 μm and an angle of 3.5° for water or 6.5° for methanol. The pump achieves methanol pump rates of 3500 μl min^{-1} at backpressures of up to 60 kPa and water flow rates of 2200 μl min^{-1} at backpressures of 75 kPa.

A piezoelectric actuated membrane pump with KOH etched diffuser/nozzle valves has been presented by Gerlach [175]. Again, the piezoelectric plate was glued onto a Pyrex membrane. The maximum pump rate achievable with this pump is 400 μl min^{-1} and the backpressure is 7 kPa. An improved version with several valves for inlet and outlet and a reduced geometry of the chamber volume has been presented recently [179]. When the actuator of this pump is driven at resonance frequency, gases can be pumped at up to 7.5 ml min^{-1} and backpressure up to 2.8 kPa.

Again, an improvement on the fabrication of the actuation part can be achieved with the thick-film-on-silicon technique. The actuator fabrication has already been described in section 5.4.1.2. Ordinary KOH etched diffuser/nozzle structures have been employed as valves and both actuator and valve chip were combined with an intermediate silicone layer. A picture of the whole device is shown in figure 6.58.

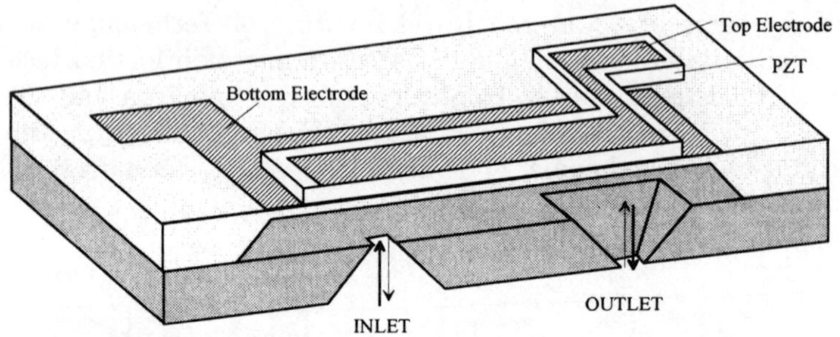

Figure 6.58: Schematic of a piezoelectric micropump with thick-film-on-silicon actuator and diffuser/nozzle valves [104]

The first experiments were done with the surface mounted actuator as a reference. The size of the actuator was 6.25 × 3.25 × 0.25 mm and the membrane geometry was 8 × 4 × 0.07 mm. Because of a possible influence on the dynamic behaviour of the pump by the silicone washer between actuator and valve chip, a blue-tape washer was tested as well. The height of the blue tape was only 0.1 mm, which is one tenth of the silicone seal. The frequency dependence of both pumps is given in figure 6.59. Although the pump with silicone washer shows a slightly higher pump rate up to 2 kHz actuation, the general behaviour of both pumps seems to be similar over the whole frequency range. The maximum pump rate with silicone washer was measured as 114 μl min^{-1} at 2.2 kHz, which is less than with blue-tape washer (184 μl min^{-1} at 2.8 kHz).

Between 2 and 7 kHz, the pump with blue tape exceeds the pump rate of the pump with silicone seal slightly but the general behaviour is still similar. This changes at around 11 kHz, when a small peak of the pump with silicone washer is transformed into a resonance behaviour for the pump with blue tape. At high frequencies (around 9 kHz with silicone and beyond 17 kHz with blue tape) the pump reverses its direction and achieves negative pump rates. It can be concluded from the measurements that the silicone seal does influence the pump behaviour but the general behaviour shown in the pump rate as a function of frequency is of the same form.

The last measurement on pumps with surface mounted actuators was the

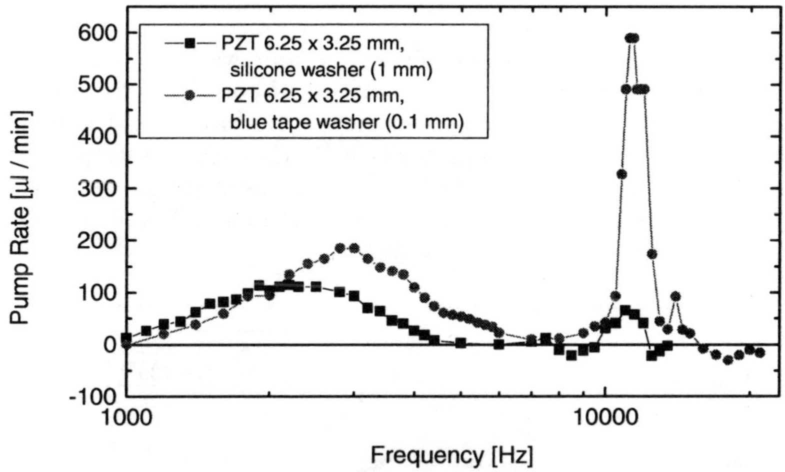

Figure 6.59: Comparison of the pump rate for different washer types

dependence of the pump rate on the backpressure that was applied at the outlet as shown in figure 6.60. The driving frequency of the actuator was kept constant at 2.2 kHz and 4 kHz respectively.

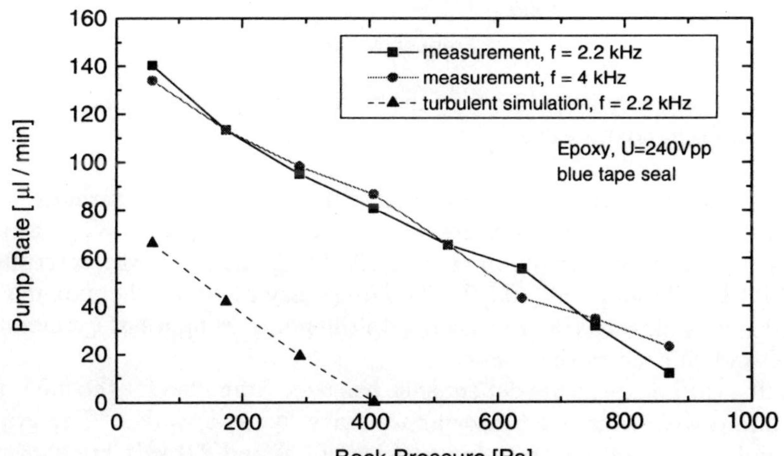

Figure 6.60: Backpressure dependence of the micropump with a surface mounted piezoelectric actuator

In order to achieve its backpressure, the pump was operating continuously, pumping liquid from the inlet pipe to the outlet. The average pressure difference between inlet and outlet as well as the flow rate were measured and plotted in figure 6.60. The maximum backpressure is rather low, lying around

1 kPa.

The novel hybrid thick-film-on-silicon actuator was employed with the passive dynamic valves as well. A combination of both leads to a very simple device from the processing point of view. The valves only require one anisotropic etch step and the actuator is built up with anisotropic etching and a subsequent screen-printing of electrodes and piezoelectric material. The geometry of the tested actuator was 7 × 3 × 0.1 mm. Silicone was used as the seal between the valve chip and actuator.

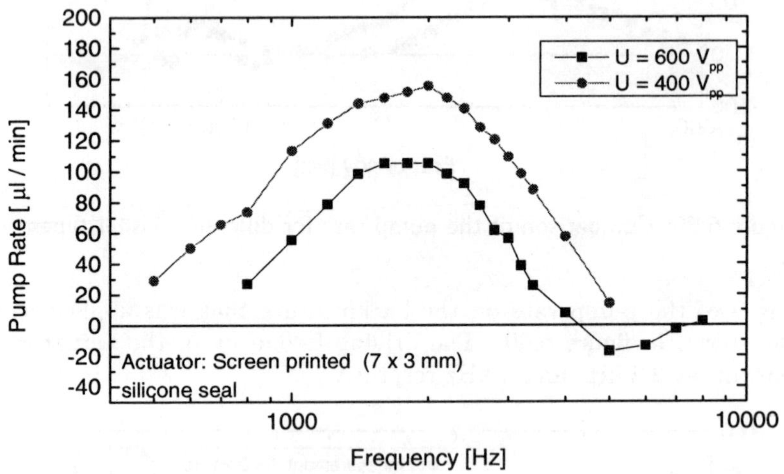

Figure 6.61: Frequency dependence of the thick-film-on-silicon actuated micropump with diffuser/nozzle valves

A graph of the frequency dependence of the micropump is given in figure 6.61. Two different driving voltages were used for testing (400 V_{pp} and 600 V_{pp}). A maximum in the pump rate of 105 (155) μl min^{-1} was determined at 1800 (2000) Hz and 400 (600) V. The frequency at which the maxima are obtained is very close to that obtained with the pump using a surface mounted actuator and silicone washer.

Finally, the backpressure dependence has been examined for the micromachined pump with hybrid actuator and passive dynamic valves. The graphs at 400 and 600 V and constant frequencies of 1.4 and 2.0 kHz are shown in figure 6.62. The maximum achievable backpressure for 600 (400) V and 2 kHz is 1 kPa (550 Pa). These values are similar to the maximum achievable backpressure with the surface mounted actuators. However, the voltage applied to the bulk piezoelectrics was a factor of 3 less than for the screen-printed devices.

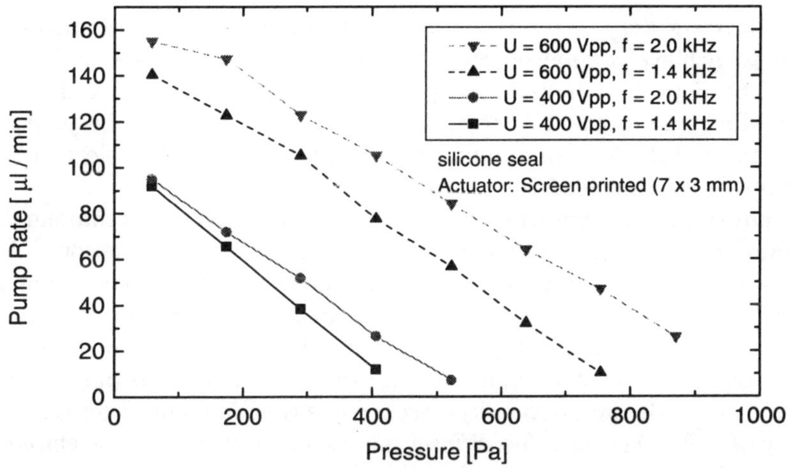

Figure 6.62: Backpressure dependence of the micropump with screen-printed actuator

6.4.1.4 Bimetallic actuated Pumps

An alternative to thermopneumatic and piezoelectric actuation of a membrane pump is to use the aluminium-silicon bimetal effect. This makes use of the difference in thermal expansion between the silicon substrate and a deposited aluminium layer.

A diagram of such a pump is given in figure 6.63 [172]. It consists of two passive cantilever valves of the type presented by Zengerle [43] where stiction is greatly reduced by the use of a small area of the valve seat on which the cantilever rests. The actuation is achieved with a 10 μm-thick layer of aluminium on top of a silicon membrane of around 20 μm thickness. The outer dimensions of the pump are $6 \times 6 \times 1$ mm^3.

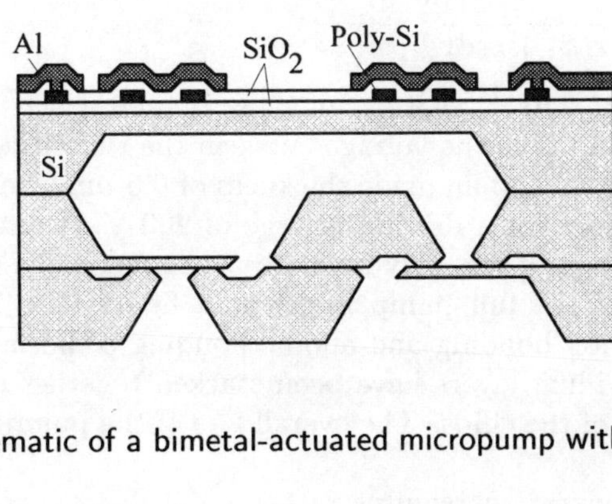

Figure 6.63: Schematic of a bimetal-actuated micropump with passive cantilever valves [172]

Most process steps of the actuator are compatible with MOS processing. In order to achieve a small electrical resistance of the poly-Si heater, the layer can be deposited in the amorphous state and then transformed into polycrystalline material at 1000°C with a solid source doping of phosphorus at the same time. The thick aluminium layer can be evaporated or electroplated after depositing a thin seed layer.

The advantage of bimetallic actuation is the fact that this technique does not require high voltages as compared with piezoelectric actuation. At a driving voltage of 5 V, the centre of a bimetallic membrane of 4 × 4 mm size could be deflected by 27 μm, which is very versatile for micropump technology. However, the disadvantage of this actuation principle is the low maximum frequency of operation. Although a pump rate of 45 $\mu\text{l min}^{-1}$ is possible with 5.5 V and 0.5 Hz, the pump rate decreases to less than 5 $\mu\text{l min}^{-1}$ for a driving frequency of 2 Hz. Therefore no diffuser/nozzle based valves can be employed with this technique. For microfluidic systems, the maximum backpressure of 12 kPa achievable with a bimetallic actuated pump at 5.5 V is reasonable.

6.4.1.5 Electrostatically actuated Micropumps

Another interesting pump type has been fabricated by Zengerle [28]. This pump is based on passive cantilever valves and electrostatic actuation of the pump membrane. The actuation principle has been described in section 3.2.4 and works by application of an electric field between two electrodes, one of which is fixed. The other electrode is a flexible diaphragm, which can be attracted towards the fixed electrode. For large dimensions this technique is very inefficient, as it does not generate a big force. However, at the micron scale, the force is large due to the small spacing between the two electrodes. For example, the generated electrostatic pressure for the ideal case of two flat electrodes with air of thickness d_1 and oxide of thickness d_2 acting as dielectric is [28]:

$$P_{el} = \frac{\epsilon_0 \epsilon_1}{2} \left(\frac{\epsilon_2 d_1}{\epsilon_1 d_2 + \epsilon_2 d1} \right)^2 \frac{U^2}{d_1^2} \tag{6.1}$$

where $\epsilon_0 = 8.85 \cdot 10^{-12}$ A s $\text{V}^{-1}\text{m}^{-1}$, ϵ_i is the dielectric constant of air or silicon dioxide and U is the voltage between the two electrodes. Assuming 4 μm separation in air and an oxide thickness of 0.5 μm, the created pressure will be almost 10 kPa for a driving voltage of 200 V. A more exact solution for the flexed diaphragm can only be obtained numerically.

A schematic of the full pump is given in figure 6.64. Although other techniques like fusion bonding and anodic bonding could have been used, in this case the individual layers have been stacked together and bonded with glue at the border of the chips. The overall size of the pump is approximately 7 × 7 × 2 mm.

Electrostatic operation requires rather large driving voltages of around 150 – 200 V, which is in the range used for piezoelectric actuation. With

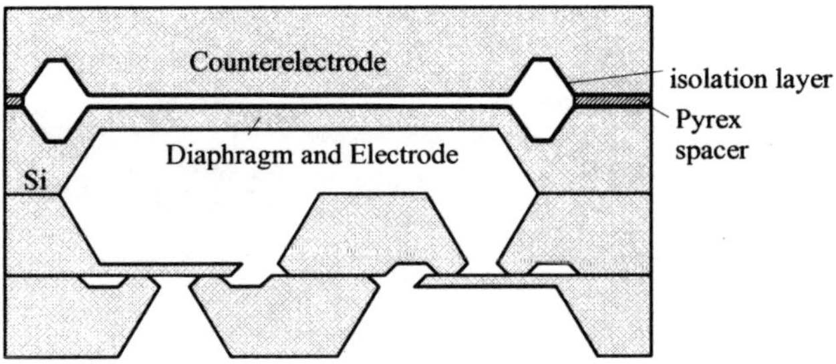

Figure 6.64: Schematic of an electrostatic actuated micropump with passive cantilever valves [47]

these voltages, a maximum pump rate of 300 μl min^{-1} is possible at around 1 kHz driving frequency. This device is also capable of bidirectional pumping due to the inertia of the passive cantilever valves in the liquid. At high actuation frequencies the cantilever valves cannot follow the diaphragm actuator. Thus the inlet valve is still open when the diaphragm reduces the chamber volume. At around 3.5 kHz, the pump achieves a negative pump rate of 380 μl min^{-1}which is reduced to almost zero at 10 kHz.

The pressure during a pump cycle has also been measured for square wave voltage actuation. The pressure signal during the supply mode is dominated by oscillations. These oscillations are typically influenced by the inertia of the fluid in the tubes connected with the micropump. Also, the pressure peaks for supply and pump mode are quite different. This is due to the electrostatic drive mechanism.

The behaviour of the pump can be more easily understood with a fluid-mechanical analogy. The driving diaphragm can be seen as a spring with a damping element emulating the attached fluid. The resistance of the valves acts as a non-linear damping element. If this resistance is low enough, the system can be set into damped oscillations.

A second interesting electrostatically actuated pump has been described by Judy (figure 6.65 [89]). It has been built completely with silicon surface micromachining and uses four silicon nitride layers as well as two poly layers. The channel has been created by etching PSG glass in buffered HF (5:1) and rinsing in methanol and DI water.

The operation principle of this pump is as follows: At the beginning, all moveable electrodes in poly #2 are at 50 V and therefore attracted to the other electrode in poly #1 (grounded), closing all three passages. The first step is to open the input by grounding its electrode and then opening the pump chamber. Now, the input is closed again and the output is opened thereafter. If the pump chamber is closed at this stage, some fluid is pushed through the outlet. Finally, the outlet is closed and the cycle can start anew. Regarding test results, only the valves have been reported to be operating.

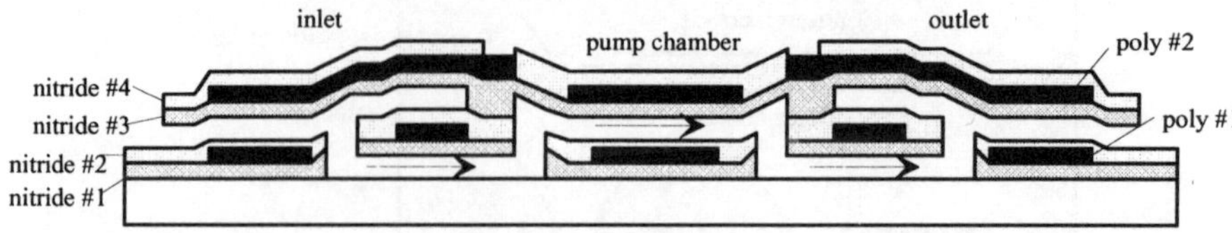

Figure 6.65: Schematic of a surface micromachined pump with electrostatic actuation [89]

No reports have yet been presented on a fully working pump. A great advantage of such a pump would be its fabrication process being compatible with any microelectronic processing. The disadvantage, however, is the small pump volume per stroke, making it only attractive for applications where low volumes have to be moved.

6.4.1.6 Pumps with the Shape-Memory Effect

The last of the pumps with membrane actuation is a pump operated through the use of the shape-memory effect. It has been described recently by Benard [27] and uses TiNi for the shape-memory effect as well as polyimide cantilever valves.

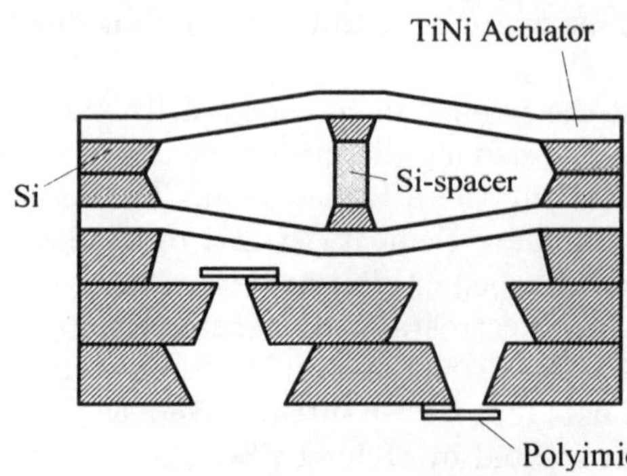

Figure 6.66: Schematic of a micropump with shape-memory-effect actuator [27]

The fabrication of the actuator is as follows: The silicon is covered with a 1.5 μm-thick oxide and patterned to generate three posts as seen in the top layers of figure 6.66. Then the wafer is etched in KOH to leave a small silicon membrane, covered with the oxide. On the front, the oxide is stripped and replaced by a 3 μm-thick layer of sputter deposited TiNi. The deposited material is amorphous and needs to be annealed to form a polycrystalline layer that displays the shape-memory effect. Now the silicon membranes are etched completely in EDP leaving only the bossed TiNi layer supported by the silicon frame. Two TiNi actuators are bonded together with a silicon spacer in the centre to form the actuator as depicted in the top part of figure 6.66.

The valves are fabricated in a similar way. The silicon wafers are covered with an oxide of 1.5μm thickness. Two holes are then etched anisotropically with KOH into the silicon to leave thin silicon membranes on the front. The oxide is then replaced by a sputter deposited Al layer of 4 μm thickness which is structured in the areas of the anchor. Now the rest of the silicon membranes are etched away. After this, polyimide is deposited, patterned with a photolithography step, and the final structure is hard baked. The diaphragm valve suspended on beams is then released with an aluminium etch and the valve part completed. The forward to reverse flow rate ratio of these valves is around 500 : 1 which is sufficient for good pump performance.

Passing a current through the upper or the lower TiNi layer, the relevant diaphragm is stretched, increasing or decreasing the chamber volume. The excitation voltage for the pump with a shape-memory-effect actuator is quite low at around 0.6 V, but a current of 0.9 amp is required resulting in a power consumption of 0.54 Watt. With a frequency of around 1 Hz, pump rates of 55 μl min^{-1} are possible. Due to a low maximum frequency of this actuator type, passive valves are required that cannot be replaced by the simpler diffuser/nozzle valves. Although the power consumption is large compared to piezoelectric actuation, the pump has the advantage of a low supply voltage.

6.4.2 Electric Field Actuated Pumps

The second mechanism used for micromachined pumps is the flow induced by electric fields. Electric field induced flow pumps are quite efficient on the micron scale. The big advantage of these pumps is that they do not require any moveable parts as compared with membrane pumps. However, large voltages are necessary to produce reasonable flow rates. It is the force produced by an electric field on charged particles that produces flow or pump action. If the number of charged particles within the fluid is large, the whole fluid is dragged along due to frictional forces. The pumps are categorised by the method of charge generation. A general overview of the pump performance of these pumps can be found in table 6.4.

6.4.2.1 Electrohydrodynamic Pumps

Electrohydrodynamic pumps require fluids with a conductivity between 10^{-8} and 10^{-10} S cm^{-1} [168] in order to function. In general, the forces acting on a fluid are the electric field force)$q \cdot E$, the dielectrophoretic force, the dielectric

Author	pump mechanism	valve type; fluid	max. pump rate [μl min^{-1}]	max. back-pressure	electrode separation
Richter 1991 [180]	electro-hydro-dynamic	valveless; ethanol	14000, 800 V (dc)	2.5 kPa, 700 V (dc)	350 μm
Fuhr 1994 [181]	electro-hydro-dynamic	valveless; water	0.45, 32 V_{pp}, 200 kHz	-	10 μm
Jacobson 1994 [169]	electro-osmotic	valveless; water	0.016 2.67 kV (dc)	-	165 mm

Table 6.4: Comparison of micropumps operated with the electric field

force and the electrostrictive force [182]:

$$\vec{F} = q \cdot \vec{E} + \vec{P} \cdot \nabla \vec{E} - \frac{1}{2} E^2 \nabla \epsilon + \nabla \left(\frac{1}{2} \rho \frac{\delta \epsilon}{\delta \rho} E^2 \right) \tag{6.2}$$

where E is the electric field, ϵ the material permittivity, $\vec{P}$ the polarisation vector and ρ the mass density of the fluid. The last term, i.e. the electrostrictive force, is only important if the fluid is compressible. Also, the dielectrophoretic force does not lead to a continuous flow for d.c. fields and is considerably smaller than the electric field force for a.c. fields.

Conventional EHD pumps are based on the electric field force and can be classified into injection EHD and induction EHD according to the generation of the charges. Figure 6.67 illustrates the principle of EHD pumping.

The onset of injection EHD pumping starts with the formation of space charges through injection or electrolytic processes at the boundary layers between the electrode and the liquid at high electric field strengths [180]. In the case of intermediate field strength (20 kV cm^{-1}), the charge creation within the liquid at one or both electrodes is achieved by dissociation or electrolytic processes within the liquid itself. The charges generated by electrolytic processes are of both polarities and are moved in opposite directions in the fluid. For high fields (around 100 kV cm^{-1}) the liquid is ionised at the electrode and ions then injected into the bulk liquid. The injected charges at high fields are of the same polarity as the electrode from which they are injected. Both low and high electric fields can be employed for EHD pumping where the applied voltage can be a.c. as well as d.c. Due to a different electrochemical behaviour of the liquid at the two electrodes, one electrode is favoured for the generation of ionic charges. This permits directional flow and thus pumping of the liquid.

Moving of the ionic charges and friction processes within the liquid then moves the whole liquid. The space charge around the electrode varies not only with the fluid but also with the distance between the two electrodes.

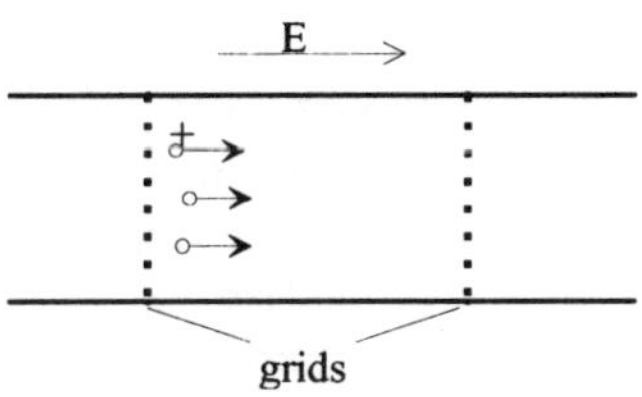

Figure 6.67: Principle of electrohydrodynamic pumping

The electrohydrodynamic pump described by Richter [168] is fabricated from two single crystalline silicon wafers. The individual wafers are etched from the front and the back of the wafer with KOH to create the grid. The electrode contact is made via deposition of a metal on the back side of each wafer. On the upper wafer, the grid is etched back by 10 – 60 μm to define the grid distance and Pyrex is sputter deposited on the lower wafer to allow anodic bonding of the two devices. Finally, the upper wafer is placed upside down onto the lower wafer and bonding achieves permanent contact. Alternatively, the grid distance can be increased to 350 μm by placing the upper chip directly onto the lower chip (figure 6.68). Ethanol pump rates of 2000 μl min^{-1} [180] and pressures of 1.2 kPa [168] have been reported for a pump with a grid area of 3 $\times$ 3 mm, 350 μm spacing of the grid and 300 V supply voltage. The pumped volume increases approximately as the square of the voltage and inversely as the square of the electrode separation. Methanol, acetone and propanol can also be pumped with this technique. DI water, however, leads to outgassing (gas production due to hydrolysis) at the electrodes and thus to the creation of bubbles and a reduction in the pump performance over time. The pump measures 7 $\times$ 7 $\times$ 1 mm but can be reduced to 4 $\times$ 4 $\times$ 1 mm size.

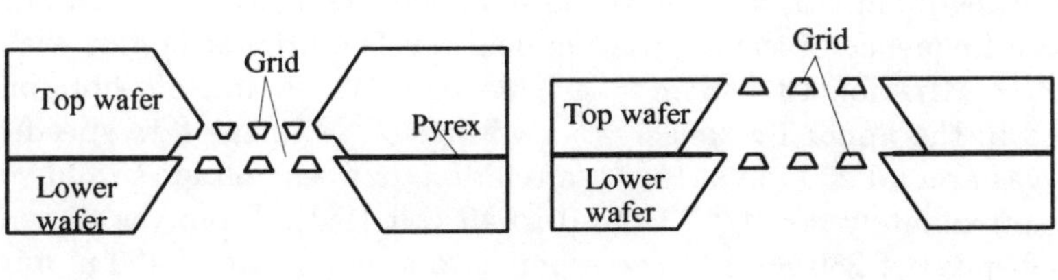

Figure 6.68: Electrohydrodynamic pump by Richter [168, 180]

In induction EHD, the charges are induced in the fluid. Melcher [183] has shown that a gradient in conductivity is obtained if a temperature gradient is realised in a slightly conducting fluid. According to Maxwell's equations, a gradient in permittivity or conductivity is necessary to have free charges existing within the liquid. The free charges, and thus the whole fluid, can then be transported in a constant electric field. In order to avoid the additional heat generation within a micropump, a second mechanism allows the induction of charges within the fluid. Waves of electric fields travelling at right angles to a liquid/solid interface induce free electric image charge in the liquid at the interface. These charges travel in synchronism with the potential waves but lag behind because of charge relaxation effects. If the propagation velocity of the travelling wave is too slow, i.e. the dominant charge relaxation time too long, then very little charge is induced. If the propagation velocity is too fast, the image charge is only slightly displaced from the potential waves. Thus there is an optimum drive velocity which produces maximum pumping.

A planar travelling-wave EHD pump has been fabricated by Bart [184]. It consists of poly-Si bus bars, which are 10 μm wide, 1.75 μm thick and 10 μm spaced. Every third electrode is connected together, and three phase drive is applied to produce a travelling electrostatic wave in a direction normal to the poly-Si bars. Pumping has been observed visually for silicone oil ($\sigma = 10^{-12}$ S m^{-1}) with drive frequencies from 1 Hz to 2 kHz and the greatest pumping at lower frequencies. However, no figures have been given for the velocity.

More recently, a device was described which enabled electrohydrodynamic pumping of water [185]. Unlike the fully planar device by Bart, an encapsulated planar travelling-wave pump was described by Fuhr [181]. The electrodes were formed by electroplating 1 μm-thick gold layers onto a sputtered and photolithographically structured Cr/Au layer. Gold or platinum is necessary as the electrodes to avoid any electrolyte electrode reactions in aqueous salt solution. For the generation of the travelling wave, a four phase electrode system was used with 90° phase shift between each (figure 6.69). The surface of the planar chip was then covered with a glass plate. A channel of 50 μm depth, 200 – 600 μm width and 4 mm length was cut into the glass plate before assembly.

Voltages of up to 50 V have been applied and it was found that the optimum frequency of the alternating field was 200 kHz for liquids with 10^{-3} S m^{-1}, 2 MHz for 10^{-2} S m^{-1} and so on. Water and alcohol could be pumped in the upper Hz range. At a voltage of 35 V, the flow speed for DI water was around 250 μm s^{-1} with a relation between voltage U and velocity v of approximately $v \sim U^{2.5}$ from 0 to 40 volt [181]. From the dimensions, the flow speed of 250 μm s^{-1} corresponds to a pump rate of 450 nl min^{-1}.

6.4.2.2 Electroosmotic Pumps

If there is a double layer of charge between two materials, of which one is a fluid, pumping can be achieved by applying an electric field [186] (figure 6.70).

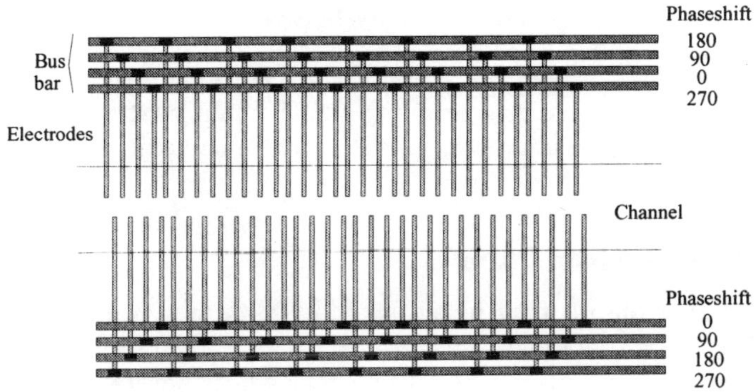

Figure 6.69: Electrode arrangement for a micromachined electrohydrodynamic pump [181]

The difference between electroosmotic and electrohydrodynamic pumps is in the charge generation. As the charge density (ionic or electronic) normally differs between two materials, equilibrium will be established at the interface. Therefore, the materials are oppositely charged within the electrical double layer at the interface. The layer in the fluid is called the *Helmholtz layer* and is located near to the usually negatively charged walls of the capillary. With an external electric field, these charges are then dragged along the fluid channel.

Unlike the normal Hagen-Poisseuille flow profile (chapter 2), micropumps using the electroosmotic principle achieve an almost flat flow profile over the whole cross section. Also, the viscous resistance to mass transfer is two times higher in the Hagen-Poisseuille flow profile, making this technique particularly suitable for use at the micron scale.

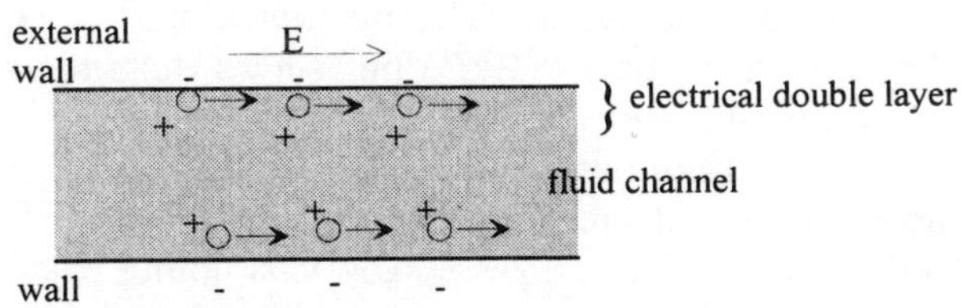

Figure 6.70: Principle of electroosmotic pumping

For high electric field strengths (27 – 163 $\mathrm{V\,cm^{-1}}$) over a 165 mm-long capillary, a moderate speed of 0.13 – 0.78 mm $\mathrm{s^{-1}}$ and flow rates of 3 – 18 nl $\mathrm{min^{-1}}$ have been achieved [169]. The dimensions of the channel cross section were 5.6 μm × 66 μm.

6.4.3 Other Actuation Techniques

In addition to the mechanisms described above, some other techniques are available for micropumps. A gear pump that is mainly fabricated with LIGA technology and an ultrasonic pump using surface acoustic waves will now be described.

6.4.3.1 In-Line Gear Pumps

Microfluidic gear pumps may be successfully fabricated using LIGA technology [187,188]. Due to its high aspect ratio, LIGA is ideal for the fabrication of such devices. Gear pumps consist of two gear wheels intersecting each other. If they are rotated, fluid is moved around the edges of the two gear wheels (see figure 6.71). Unlike the other methods, this technique is particularly good for pumping very viscous fluids. For example, glycerin-water has been pumped with flow rates up to 200 μl min^{-1} and pressures in excess of 100 kPa [187].

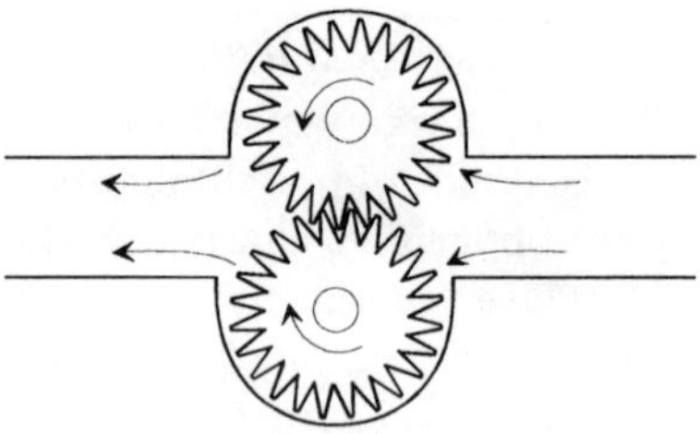

Figure 6.71: Micro gear pump for viscous fluids [187]

The only disadvantage is the actuation technique, which requires an external motor to drive the gear wheels. Also the pump performance for water, i.e. a low viscous medium, is not as good. Up to 7 kPa backpressure is possible with water as the test liquid. The maximum backpressure for gases decreases further and yields only 2.5 kPa but shows that micro gear pumps are self-priming and bubble resistant.

A similar pump has been presented by Dewa [188] based on LIGA technology. The big advantage of this device is that a Ni/Fe bar has been placed on one of the gear wheels. This allows magnetic coupling from an actuation source from outside and does not require any direct connection to the gear wheels as in [187]. The total thickness of the pump is 600 μm and the outside diameter of the pump body is 3.175 mm. With the non-axial placing of the coupled gear wheel, a nearly uniform magnetically coupled torque is produced. This pump produces a water flow of up to 350 μl min^{-1} and backpressure of up to 14 kPa [188]. Lifetime tests of the pumps suggest that they can be operated periodically over 7 months ($2 \cdot 10^7$ revolutions) and do not show any degradation in the performance. As before, this pump allows the pumping of air, which makes it a self-priming pump that is not affected by air bubbles during the pumping.

6.4.3.2 Travelling Flexural Wave (Ultrasonic)

The last principle used for micropumps is an ultrasonically induced movement of the liquid. Unlike the transportation of the previous pumps, this device is mainly intended for moving fluids along a surface and not through pipes and whole systems of channels and microfluidic devices. The principle is described as follows: An interdigital finger structure is positioned on top of a piezoelectric material which is itself again on a silicon nitride membrane. By application of a suitable voltage, a surface acoustic wave is generated that travels along the membrane. These surface acoustic waves induce movement in the fluid boundary close to the surface and thus transport the fluid along. The basic device is depicted in figure 6.72.

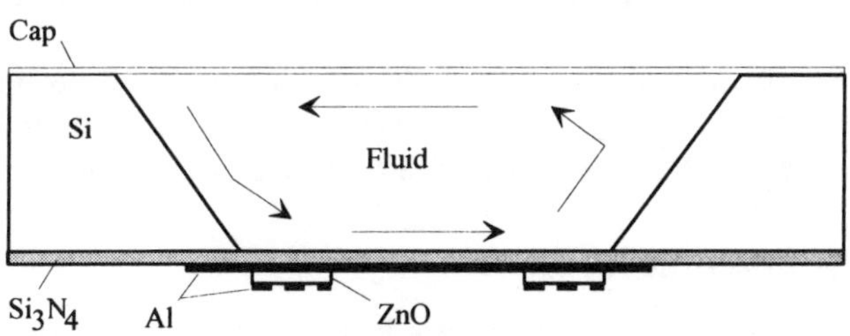

Figure 6.72: Travelling flexural wave type ultrasonic micropump [189]

The fabrication of such a transducer is quite simple and consists only of KOH etching of the silicon to leave a silicon nitride membrane. On top of the membrane, a sandwich of aluminium, zinc oxide and aluminium is sputter deposited. For efficient generation of surface acoustic waves, the top aluminium layer needs to consist of interdigitated fingers. The transducer couples most effectively when driven so that its spatial period (typically 100 μm) equals the ultrasonic wavelength. The device is thus driven at 1 to 5 MHz to generate flexural plate waves with a velocity in the range of 100 to 500 m s^{-1}. With a driving voltage of 15 V, the fluid velocity was about 100 μm s^{-1}. Also, a test has been performed with diluted whole blood, where the red blood cells were moving with the fluid in the direction of wave propagation without damage.

The device is quite interesting as it also allows trapping and spinning of coliform bacteria by a standing wave, i.e. where two interdigital finger actuators at both ends of the membrane are actuated synchronously. Also, the mixing action of a fluid loaded membrane is another interesting application, where the fluid close to the membrane is stirred ultrasonically.

6.5 Physical Sensors for Microfluidics

The following section moves away from the directing of fluids by nozzles, valves or pumps and deals with the physical sensing techniques used in microfluidics. The most important physical quantities in microfluidics are the flow velocity and the pressure. Sensors in these two areas are reviewed in this section.

6.5.1 Flowsensors

Silicon micromachining techniques have been widely used for the realisation of microstructures for microflowsensors. Different physical principles may be used for the measurement of a fluid flow: the thermal measurement method which relies on the thermal interaction between the sensing element and the flowing fluid; the drag-force method, based on the piezoresistive effect of the silicon material and the bending of a cantilever due to the dragging of the fluid; and the pressure difference method, which depends on the drop in pressure due to friction processes.

A summary of some of the flowsensors which have been presented to date can be found in table 6.5. Unless otherwise stated, the measurements of the flow range and other characteristics of the devices were done with a liquid.

6.5.1.1 Drag-force Flowsensors

Several micromachined flowsensors based on the principle of drag-force have been presented to date [130, 196, 197]. Figure 6.73 shows such a cantilever-based flowsensor. The flowsensor consists of a cantilever beam with an integrated strain gauge resistor. When the cantilever is immersed in a flowing fluid, a drag-force is exerted on the beam resulting in a deflection of the cantilever. This deflection creates a stress in the silicon and thus changes the resistance of the piezoresistive silicon material.

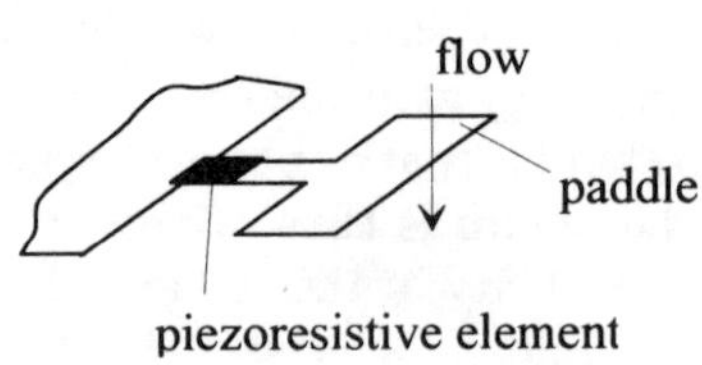

Figure 6.73: Flowsensor based on drag-force method

The drag-force F_d is proportional to the flow velocity if viscous forces are dominant and proportional to the square of the flow velocity if inertial forces are dominant. Gass [130] for example reported a piezoresistive rectangular cantilever of 3 mm length and 10 μm thickness for flow sensing. With this device, water flow rates of 5 to 500 μl min^{-1} have been measured at a sensi-

Type	chip size	flow range	response time [s]	sensitivity
drag-force integrated [130]	0.01 × 1 × 3 mm	5 – 500 $\mu l\,min^{-1}$	–	4.3 $\frac{\mu V\,V^{-1}}{\mu l\,min^{-1}}$
floating element shear stress [190]	0.18 × 0.14 × 5 mm	1 – 100 kPa	–	–
capacitive pres. (gas) [129]	9.7 × 3 mm	0.001 – 4 Torr	–	200 $\frac{ppm}{mTorr}$
thermal (gas) on cantil. [191]	1.6 × 0.4 × 0.03 mm	0 – 30 $m\,s^{-1}$	0.45	0.01 – 0.5 $\frac{mW\,mW^{-1}}{m\,s^{-1}}$
microbridge (gas) [192]	3 × 0.3 × 0.004 mm	0 – 10 $ml\,min^{-1}$	–	30 – 100 $\frac{mV}{ml\,min^{-1}}$
vibrating cantilever [193]	7 × 2 × 0.05 mm	0 – 40 $m\,s^{-1}$	0.075	–
thermal transit time mode [194]	9 × 5 mm	0.1 – 10 $ml\,min^{-1}$	1	–
thermal const. Temp. [48]	–	0 – 60 $\mu l\,min^{-1}$	–	2 – 8 · 10^3 $\frac{K}{m\,s^{-1}}$
thermal const. Temp. [195]	–	1 – 4 $ml\,min^{-1}$ 10 – 3000 sccm (gas)	–	–

Table 6.5: Characteristics of micromachined flowsensors

tivity of 4.3 $\frac{\mu V\,V^{-1}}{\mu l\,min^{-1}}$. The two flow rates correspond to fluid velocities of 33.5 mm s^{-1} and 3.35 m s^{-1}.

Another example is a cantilever with an integrated buried resistor, which has been obtained by high voltage ion implantation [196]. This sensor is reported to be less sensitive to surface electrical fields and moisture and more stable than conventional strain gauge resistors. The disadvantage is the low sensitivity per unit deflection due to the maximum stress being on the surface and not in the volume. The non-amplified sensitivity has been determined as 1.5 $\frac{\mu V\,V^{-1}}{\mu l\,min^{-1}}$ with water as the test liquid.

Flow sensing based on a change in the vibrating amplitude of a cantilever has been reported by Hetrick [193]. In this sensor, the mechanical cantilever is driven at its resonance frequency by piezoelectric actuators. Due to the damping effect in the high velocity region (> 5 m s^{-1}) the maximum amplitude of the cantilever is inversely proportional to the gas flow velocity and the gas density.

6.5.1.2 Differential Pressure Method

The next flow sensing principle relies on the measurement of the differential pressure of a flowing liquid (figure 6.74). When a liquid flows through an aperture, a pressure loss occurs. If a membrane is mounted across the aperture, the pressure difference can be measured and the flow velocity can be determined. The measurement window for the sensor by Cho [129] was between 0.001 and 4 Torr with a sensitivity of 200 ppm per mTorr.

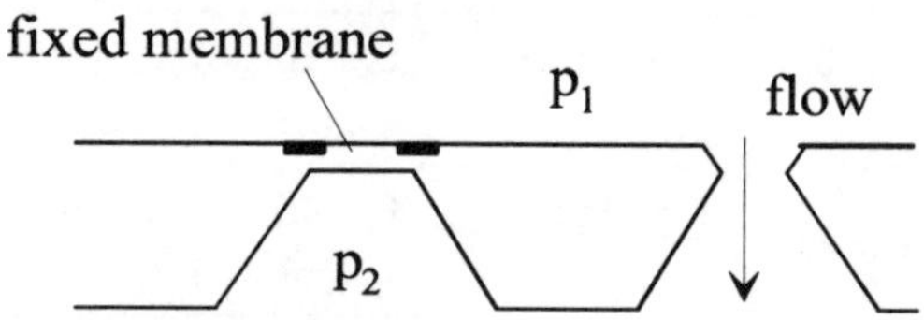

Figure 6.74: Flowsensor based on the differential pressure method

A second device in this category has been presented by Oosterbroek [198] using two absolute pressure sensors at both sides of a small channel with known fluidic impedance. The pressure sensors are of the capacitive type using a thin silicon membrane as the moveable diaphragm and a fixed electrode on a Pyrex wafer anodically bonded to the silicon wafer. The channel is defined with a second Pyrex wafer on the bottom of the silicon wafer. This system permits not only the determination of the flow rate but also the absolute pressure of the fluid at the outlet of the sensor. Flow rates between 0 and 270 μl min^{-1} have been measured with a capacitance change from 2.63 to 2.67 pF.

6.5.1.3 Thermal Flowsensors

Thermal flow sensing relies on the thermal interaction between the sensor and the flowing fluid. Depending on the underlying physical principle, there are three main thermal flow sensing techniques:

- Constant temperature mode [131, 191]: When cooling a hot object, the heat transferred from a heated element to a fluid stream depends on the flow velocity. Under steady state conditions, there is a balance between the input electrical power to the chip P_E and the sum of the losses due to convective cooling P_V by the gas and heat conduction P_C through the silicon beam:

$$P_E = P_V + P_C = h\,A_1\,(T_{\text{beam}} - T_{\text{gas}}) + K\frac{A_2}{l}\,(T_{\text{beam}} - T_{\text{chip}}) \quad (6.3)$$

where h is the convective heat transfer coefficient of the fluid and K is the thermal conductivity of the beam material. A_1 is the surface area of the beam that is in contact with the gas, A_2 is the cross section area of the beam and l is the length of the beam. If the temperature difference between the beam and the gas, $T_{\text{beam}} - T_{\text{gas}}$, is kept constant with the help of an external temperature regulator, and the thermal conductivity, density and viscosity are constant, the power dissipation P_E can be written as follows:

$$P_E = C_1\sqrt{v} + P_C \quad (6.4)$$

where v is the velocity of the fluid and C_1 is a constant.

Stemme for example reported a flowsensor based on this principle with a measurement window of 0.8 – 30 m s^{-1}, a response time of 0.45 s and a sensitivity of 0.01 – 0.5 $\frac{\text{mW(mW)}^{-1}}{\text{m s}^{-1}}$ [191].

A second example of a sensor in constant temperature mode is a hot film anemometer, where the flowsensor consists of a pair of Ti micro-heaters [192]. Titanium offers not only a high thermal coefficient of resistance (approximately 0.5 % K^{-1}) but also excellent adhesion to the underlying SiO_2 and has therefore ideal characteristics for the temperature sensor. The output voltage signal of the device versus the gas flow rate is reported to be linear in the range between 0 – 1 ml min^{-1}, deviating from the linear form above 1 ml min^{-1}. The measurable flow range is between 0 and 10 ml min^{-1} which corresponds to gas flow velocities of 2.4 cm s^{-1} to 2.4 m s^{-1}.

- Time of flight mode [131]: A heater is mounted in a liquid stream and is continually pulsed with a certain amount of electrical energy. Further downstream of the heater, a temperature sensor is fixed which is used to measure the time delay between heat source and heat detection. A simple device is obtained if both heater and temperature sensor use the same resistive films. This mode of operation is used when the heat is transported faster by convection rather than by diffusion, i.e. at large flow velocities.

- Constant power mode [131]: In the constant power mode, the power for heat generation is kept constant and the temperature is measured at or close to the heater. The optimum operation condition for this technique is a moderate flow velocity and the major advantage is its very fast response time to changes in the flow rate.

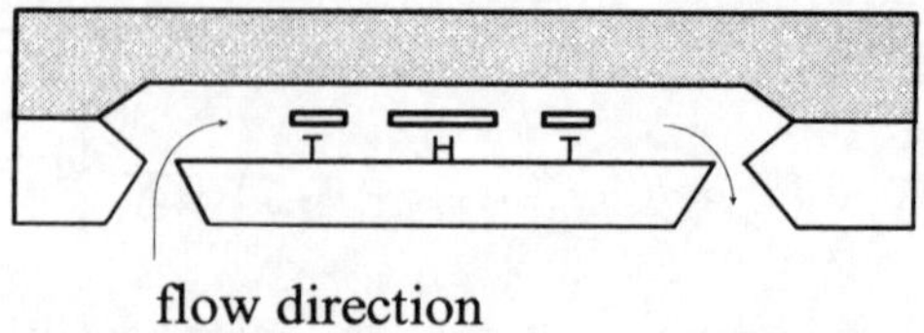

Figure 6.75: Flowsensor based on the thermal principle of constant power

An insulating bridge on silicon with heaters and temperature sensors of Pt has been presented as a flowsensor operating at constant power mode (figure 6.75). The modelling of this device was discussed in detail in section 4.6.3. Here a heater (H) is mounted between two temperature sensors (T). Conductive heat transfer is modified by the flow of the fluid. Therefore the sensor mounted downstream receives more heat than the one upstream. The resulting temperature difference is related to the flow velocity of the fluid. The measurement range for the use of water in this device has been determined as $0 - 2 \cdot 10^{-3}\ \mathrm{m\,s^{-1}}$ and $0 - 60\ \mu\mathrm{l\,min^{-1}}$ [48]. A very similar device has been presented by Kohl, giving the measurement range of $1 - 4000\ \mathrm{ml\ h^{-1}}$ for liquids and 10 – 3000 sccm for gases [195].

6.5.1.4 Electrohydrodynamic Flowsensors

As with electrohydrodynamic pumping, the principle of such a sensor is injection of charges at one electrode grid and subsequent detection of a charge pulse at the second grid. These charges are carried along in ionic species (figure 6.76). The measurement range for DI water was $8 - 1000\ \mu\mathrm{l\,min^{-1}}$ [168]. A disadvantage of this principle is its restriction to liquids with high resistivity. Another problem of EHD flow measurement is that the injected charges themselves can cause pumping and thus could falsify the measurement. In order to overcome this, an injection electrode with two detection electrodes (upstream and downstream) is suitable, as the two generated flows will compensate each other.

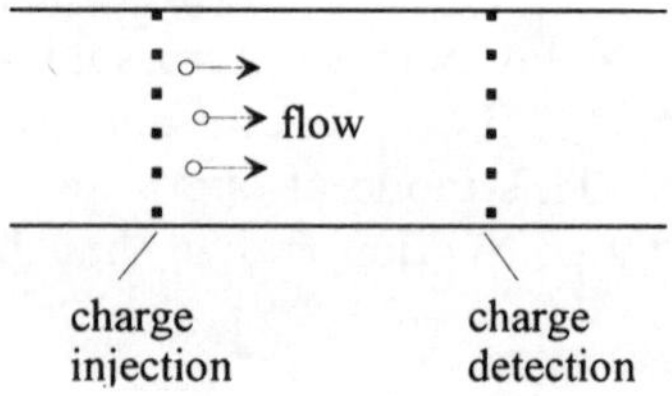

Figure 6.76: Electrohydrodynamic flow sensing principle

6.5.1.5 High-speed X-Ray Micro-Imaging

A somewhat different technique for measuring the flow has been presented by Leu [199] using high-speed X-ray micro-imaging of steady state flow in micro pipes. The setup for the measurement consists of an X-ray monochromator and a high frame rate CCD camera. Flow images of vortices are collected and the flow pattern is reconstructed with the help of a flow recovery algorithm. This makes it possible to plot velocity profiles of micro channels and complicated two-dimensional flow fields.

6.5.2 Pressure Sensors

The first bulk micromachined pressure sensor was fabricated by Honeywell in 1962. Since then quite a number of both bulk micromachined and surface micromachined pressure sensors have been developed. Micromachined pressure sensors are readily available commercially, driven largely by the automotive sector. A major advantage of silicon micromachined sensors is the possibility of integrating electronics on the same chip. This is useful for temperature compensation as well as for calibration purposes and provides short distances from the transducer to the amplification unit and therefore large signal to noise ratios. Applications or potential applications of silicon pressure sensors are mainly found in automotive, water resources, aeronautics and food production where the markets are large.

Pressure sensors on silicon may operate with resonant structures, piezoresistive elements on membranes or utilise the capacitive transducer principle. Fibre optic pressure sensors are also possible. In this case the optical path length in a fibre is altered due to bending of a membrane to which the fibre is attached. When used in a Mach-Zehnder interferometer the constructive and destructive optical interference is determined with a lateral a-Si:H p-i-n photodiode [200] and enables the stress or pressure to be determined.

Here, only a short overview of the mainstream silicon pressure sensors will be given.

6.5.2.1 Piezoresistive Pressure Sensors

A piezoresistive silicon pressure sensor is built on a silicon diaphragm. The diaphragm can be either surface micromachined or bulk micromachined. Resistors are implanted or diffused at the edge of the diaphragm, where the largest stresses occur (see figure 6.77). The piezoresistive sensing technique relies on a change in electrical resistance due to stress and strain in the resistor. Usually, a bridge of four resistors at each edge of the diaphragm is used to electrically detect any stress in the diaphragm. The output of the bridge circuit is nearly linear with the applied pressure which deflects the diaphragm.

Although most surface micromachined pressure sensors use poly Si and only a minority use epitaxial single crystalline silicon, a differentiation of the

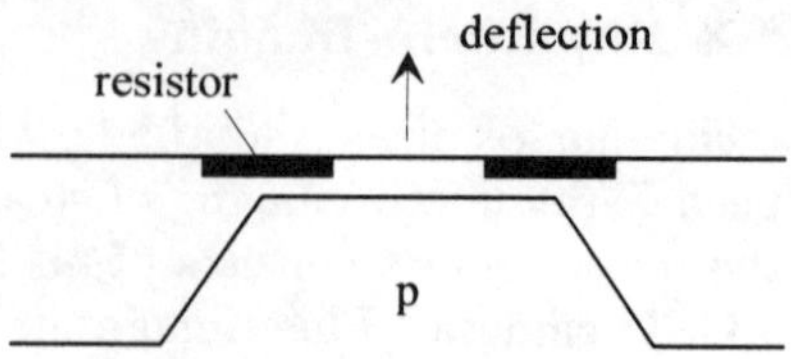

Figure 6.77: Bulk micromachined piezoresistive pressure sensor

sensors can also be made by the material used for the piezoresistive element. This can be single crystal silicon or polycrystalline silicon. The polysilicon strain gauge is less sensitive to stress than single crystal silicon but allows smaller chips to be fabricated.

Some typical figures of merit achievable for piezoresistive silicon pressure sensors are summarised in table 6.6.

Figure of merit	Single Crystalline Si	Poly Crystalline Si
pressure range	-600 – 1000 mbar [201] 0 – 100 mbar [203] up to 500 bar [205]	0 – 1 bar [202] up to 172 bar [204]
overpressure	500 % of full scale [206] mechanical blockage possible [201]	300 % of full scale [206]
sensitivity	1.83 $\frac{\mathrm{mV\,V^{-1}}}{\mathrm{bar}}$ [207]	1 $\frac{\mathrm{mV\,V^{-1}}}{\mathrm{bar}}$ [208]
sensitivity drift	< 0.1 % for 45 h at 200°C [207] < 0.1 % for 2 month [209] self test possible with actuation [201]	virtually no changes after 1 million test cycles [204]
T range	-40 – 120°C [206] up to 300°C with SOI [207]	-40 – 180°C [206] up to 300°C with SOI [207]
power	6.5 W [203]	–

Table 6.6: Properties of some piezoresistive silicon pressure sensors in bulk and surface micromachining

6.5.2.2 Capacitive Pressure Sensors

Capacitance sensors generally sense the movement of a diaphragm by measuring the capacitance between the diaphragm and a second electrode (figure 6.78). Most of the recent developments in this area have used micromachined silicon structures for diaphragm and electrode. The capacitance is generally of the order of a few pF or less which is why the measuring circuit is often integrated on the same silicon die as the sensor or on a separate die to make a small hybrid structure. Linearity of response may be obtained by using shaped electrodes or by allowing the diaphragm to touch the lower electrode (with a suitable insulating layer). The touch technique also gives a sensor which will show good resistance to overpressure, but will only start to operate properly if the pressure is high enough to cause the electrodes to touch.

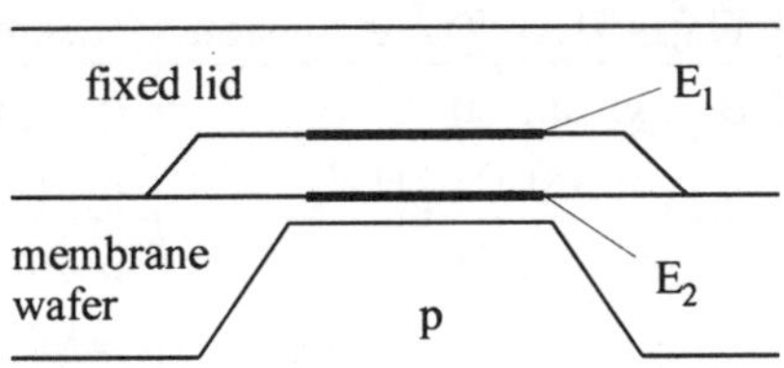

Figure 6.78: Capacitive pressure sensor

A great advantage of the capacitive sensor technique is that the basic sensing mechanism is stable and relatively free from interference from other measurands, e.g. temperature. Also of particular interest is that the measurement circuit may be designed for very low power applications. The properties of some capacitive pressure sensors can be found in table 6.7.

6.5.2.3 Resonating Pressure Sensors

Several silicon micromachined resonant sensors have been described to date. These typically employ a vibrating silicon structure fabricated on a silicon diaphragm where it acts as a strain gauge. As the diaphragm deflects, the resonator is stretched or compressed, thereby altering its stiffness and therefore its resonant frequency. The resonating element must be vacuum encapsulated and this can be done on chip, thereby isolating the resonator from the pressurised media, or by packaging the sensor within a vacuum.

Some sensors are reviewed according to their properties in table 6.7. The first and best known is the Druck "vibrating butterfly" pressure sensor [213, 218, 219], the second is a resonant differential pressure sensor developed by Yokogawa [211] and the last is a silicon resonant pressure sensor developed for aerospace applications by Solatron Weston [215].

Figure of merit	Capacitive Transduction	Resonant Structures
pressure range	0 – 35 bar smm [210] 10 – 100 kPa [212]	0 – 1300 mbar [211] up to 48 MPa [213]
overpressure	250 % full scale [214]	mechanical blockage possible
sensitivity	0.5 $\frac{fF}{Pa}$ [212]	10 ppm resolution at 1300 mbar [211]
stability	–	0.01 % of full scale for 12 month [215]
T range	0 – 70°C [210] -30 – 80°C [214]	-60 – 80°C [216] up to 450°C [215] with optical fibre
power	7.5 mW [217]	–

Table 6.7: Properties of capacitive and resonant pressure transducers

Chapter 7

Microfluidic Devices II: Particles and Chemistry

7.1 Introduction

This chapter concentrates on particle detection and handling as well as the whole area of chemical applications of microfluidics. The latter topic covers devices for chemical analysis, chemical reactors and sensors for biochemical substances.

7.2 Particle and Cell Handling with Microfluidics

An interesting application of microfluidics can be found in particle or cell handling. It has been shown that dielectrophoretic forces induced by high frequency alternating, rotating or travelling electric fields can be used to levitate and/or to move particles and cells in weak electrolyte solutions [220–223]. Also, it is possible to count particles by the change in the electric resistance of a conducting fluid if a particle of different conductivity passes.

7.2.1 Introduction to Dielectrophoresis

The general behaviour of the force of an electric field on a particle of net charge Q in a non-uniform field E_{el} is described by [224]:

$$\vec{F} = Q \cdot \vec{E_{el}} + (m \cdot \nabla)\, \vec{E_{el}} \tag{7.1}$$

where m is the dipole moment due to induced interface charges and for a spherical particle is given by:

$$m = 4\pi\epsilon_f\epsilon_0 \left(\frac{\epsilon_p^* - \epsilon_f^*}{\epsilon_p^* + 2\epsilon_f^*} \right) r^3 \cdot E_{el} \tag{7.2}$$

Here ϵ^* represents the complex polarisability of fluid f and particle p, ϵ_f is the relative permittivity of the fluid, ϵ_0 is the permittivity of free space ($8.854 \cdot 10^{-12}$ F m^{-1}) and r is the radius of the particle. The complex permittivity $\epsilon^* = \epsilon_r - j\frac{\sigma}{\omega}$ depends on both the real permittivity ϵ_r and the conductivity σ. Equation 7.2 implies that the dipole moment depends non-linearly on the frequency ω of the electric field.

In order to avoid electrolysis and electroosmosis in the fluid, it is important to employ a.c. fields although the electric field force is much larger than the dielectrophoretic force. Simplification of equation 7.1 with vector calculus and assuming a dielectric particle or high frequencies ω of the electric field, where the time-averaged Coulomb force on the particle is zero, leads to the following equation [224] of the dielectrophoretic force:

$$F_{\text{dep}} = \frac{1}{2\,E_{el}} \text{Re}\left[m(\omega)\right] \text{grad}\left(\left|E_{el}\right|^2\right) \tag{7.3}$$

where $|E_{el}|$ is the amplitude of the a.c. field. Depending on the polarisability of the particle with respect to the polarisability of the surrounding fluid, the particle moves to or from the highest field region. If the particle has a smaller permittivity than the fluid, negative dielectrophoresis occurs and the particle moves away from the high-field regions. Positive dielectrophoresis is obtained when the permittivity of the particle is larger than that of the fluid, and the particle is attracted to high-field regions. Because the dielectric force depends on the square of the electric field, both AC and DC fields produce a similar particle motion.

7.2.2 Particle/Cell Trapping

To date, several devices for handling particles with the dielectrophoretic force have been presented, of which the simplest is a quadrupole field generated by four electrodes [222, 223]. Depending on whether the dielectrophoresis is positive or negative, the cells move towards the electrodes or towards the centre. A picture of the electrode arrangement is given in figure 7.1.

The distribution of the absolute value of an electrical quadrupole field is plotted in figure 7.2. The size of the electrodes is 50 × 50 μm, the length of the structure is 100 μm and the distribution was simulated right above the ground. It is obvious that the electric field is almost zero in the centre of the arrangement, rising to larger values (arbitrary scale) in all directions away from the centre. Thus, a particle with negative dielectrophoresis will be trapped at the centre of the structure as the dielectrophoretic force will oppose any movement that is directed outwards (cf. equation 7.3).

7.2.3 Transportation and Separation of Particles

The principle of dielectrophoresis can also be employed to transport or separate different particles or cells. When a particle is moving in steady state

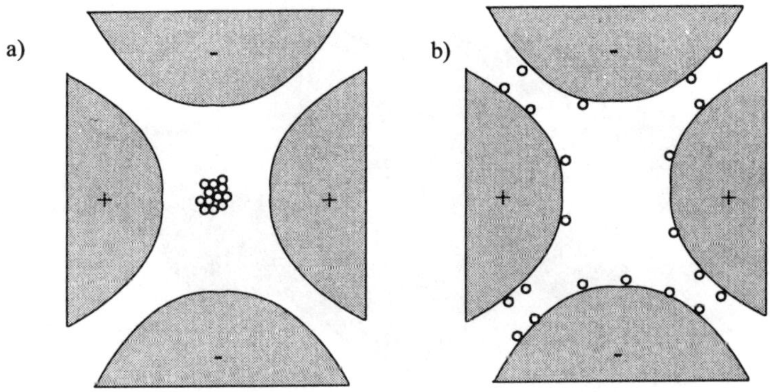

Figure 7.1: The effect of (a) negative and (b) positive dielectrophoresis on suspended particles with an electrical quadrupole field [222]

in a surrounding fluid by dielectrophoresis, the dielectric force and the fluid friction are in equilibrium ($F_{\text{dep}}(\omega) = F_{\text{fric}}(\omega)$):
• In an a.c. field, the dielectrophoretic force F_{dep} on a spherical particle of radius r_P and absolute complex permittivity ϵ_P^* submerged in a fluid of absolute complex permittivity ϵ_f^* is given by [225]:

$$\begin{aligned} &F_{\text{dep}}(\omega) = 2\pi\, r_P^3 \epsilon_f \epsilon_0 \operatorname{Re}\left[f\left(\epsilon_P^*, \epsilon_f^*\right)\right] \operatorname{grad}\left(E^2\right) \\ &\text{with } f\left(\epsilon_P^*, \epsilon_f^*\right) = \frac{\epsilon_P^* - \epsilon_f^*}{\epsilon_P^* + 2\,\epsilon_f^*} \end{aligned} \tag{7.4}$$

indicating that the dielectrophoretic force depends on the frequency ω of the electric field and that the real part is to be considered for the dielectrophoretic force. The absolute complex permittivity is given by $\epsilon^* = \epsilon - \frac{\sigma_{\text{el}}}{\omega} j$ with $j = \sqrt{-1}$ and σ_{el} being the conductivity of the liquid or particle.
• A second force acting on a particle submerged in a fluid is the frictional force due to viscosity of the fluid. The frictional force F_{fric} on a spherical particle which is pulled through a fluid with velocity v can be described by Stokes' law for laminar flow ($Re \ll 100$) [15]:

$$F_{\text{fric}} = 6\pi\, \mu\, r_P\, v \tag{7.5}$$

where μ is the viscosity of the fluid. If the electric field is moved within the liquid, a travelling wave is generated and particles are moved along with the electric field. Structures based on this have been presented by Fuhr [223,226], Huang [227] and Talary [228], with the electrodes arranged like railway tracks to achieve travelling electric fields (see figures 7.5 and 7.6).

In order to give the reader an idea of the field within the railway track system, the effect of just two horizontal electrodes (E_1, E_2) has been modelled. The electric field was simulated with Ansys 5.4 for a potential difference of

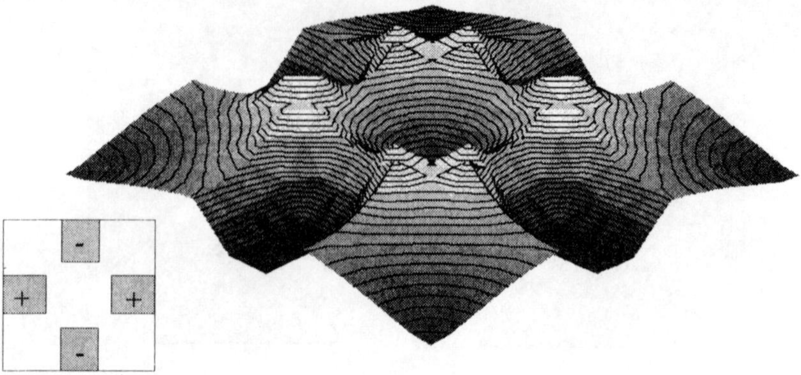

Figure 7.2: Electrical field distribution for four electrodes at the centre of the four sides

0.1 V between two electrodes of 5 μm width and 5 μm spacing. Both the square of the electric field and the gradient of the square of the electric field are plotted in figures 7.3 and 7.4 at a height of 2.5 μm above the electrodes. From the electric field, the value of $\text{grad}_x(E^2)$ has been calculated for the x-direction.

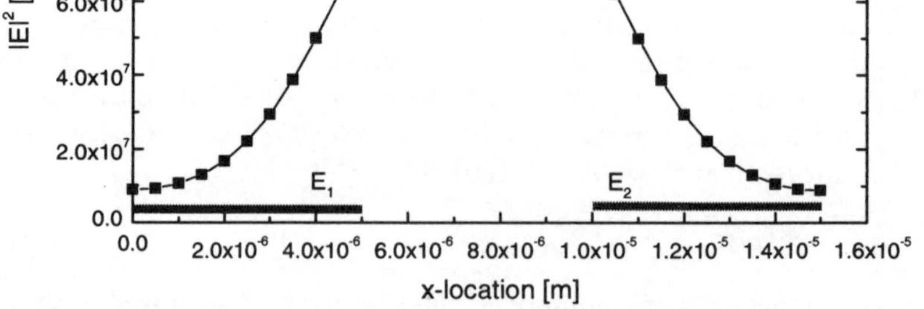

Figure 7.3: Electric field for two electrodes in a channel

Unless a particle of negative dielectrophoresis is exactly at the middle between the two electrodes, it is always deflected to one or the other electrode. A similar field distribution can be obtained with four electrodes at different potentials. By periodically varying the potential on the four electrodes, it

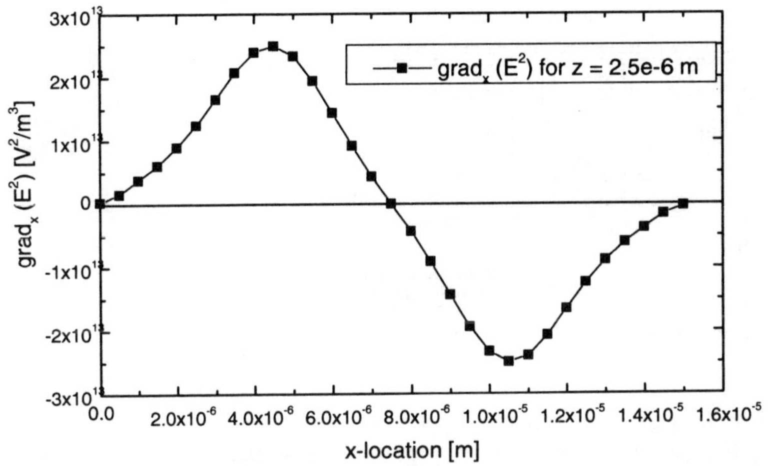

Figure 7.4: grad$_x(E^2)$ for two electrodes in a channel

is then possible to create a travelling electric field pattern and thus move a particle steadily along the structure.

For his railway tracks, Fuhr [226] used a thin bottom CrAu layer to electroplate 0.5 μm-thick gold onto a glass substrate. The structuring was done by using ion beam milling. Channels were formed by sawing 100 μm-deep cuts into the cover glass plate. In order to move the particles, phase shifts between four electrodes are used (0, 90, 180 and 270 degree). As negative dielectrophoretic particles are steered to the lowest electric field values, this travelling wave can be used to transport particles. The operation of the device is described as follows: At the beginning, the whole system needs to be flushed with a fluid containing the particles. Then the travelling wave is applied to transport the particles into the centre of the snail-type electrode arrangement. There, the particle can be trapped within a quadrupole field for further investigations. The electrode design used by Fuhr does not include any higher layers of metal, making it slightly problematic for more complicated railway track electrodes.

Talary [228] uses a slightly different scheme for the electrode railway track, where two finger structures are slightly misaligned in the horizontal direction. Furthermore, the travelling wave of the lower four bus bars has an additional shift compared to the upper bus bar as this produces uniform travelling field motion which transports the particles [227]. Cells of negative dielectrophoresis are repelled from the electrode area and focused in the centre of the two finger structures. There, they are moved in the same direction as the travelling wave. Cells of positive dielectrophoresis are just trapped at the electrodes. This device has been successfully used for separating viable and nonviable yeast cells [228].

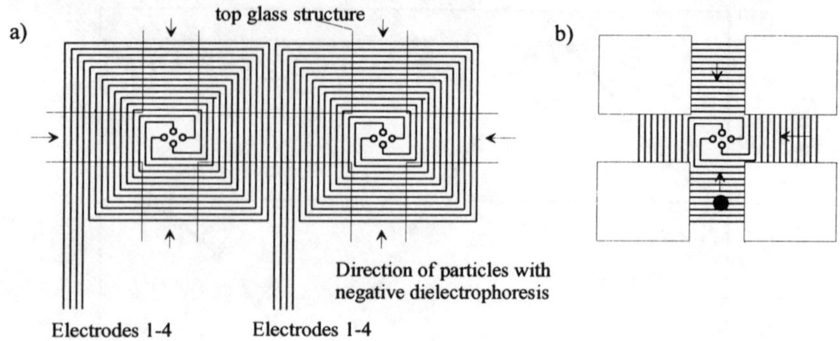

Figure 7.5: Layout of a particle micromanipulator, where particles with negative dielectrophoresis are transported on railway track electrodes with a travelling electric field: a) general layout of electrode system and top glass structure, b) electrode region that is used for the transport of particles [226]

In a similar way to Fuhr, the electrodes of Huang [227] and Talary [228] are fabricated using standard photolithography and subsequent etching of a chrome-gold layer on glass. For the bus bars a photosensitive insulator was deposited and structured before the deposition of a subsequent metal layer. The lift-off technique was employed for structuring the bus bars.

A somewhat different design for particle deflection using dielectrophoresis has been presented by Washizu [229]. The layout of the device is shown in figure 7.7. It consists of one inlet, two outlets and four electrodes at the sides of the channel. The electrodes on each side are separated by walls which project towards to the centre of the channel. In order to avoid permanent attachment of particles to the walls, an intermittent high-frequency voltage supply has been used (0.2 s active in every 0.8 s). To deflect a cell to one outlet, the electrodes on the side of the selected outlet are activated and the particle moves to this outlet due to the dielectrophoretic force.

Particle sorting can also be performed by microfluidic amplifiers as described in section 6.2.2 or by electroosmotic pumps with one inlet and two outlets. In these cases, the whole fluid is moved in a certain direction and the particle just follows the movement of the liquid.

7.2.4 Particle Counting

Counting of particles and especially red blood cells is a well-established method for macroscopic devices (see the book on particles by Allen [230]), but only a few designs of micromachined particle counters have been presented so far [231, 232]. One of the macroscopic devices in use is a Coulter Counter, where the change of the electrical resistance of a liquid filled capillary is mea-

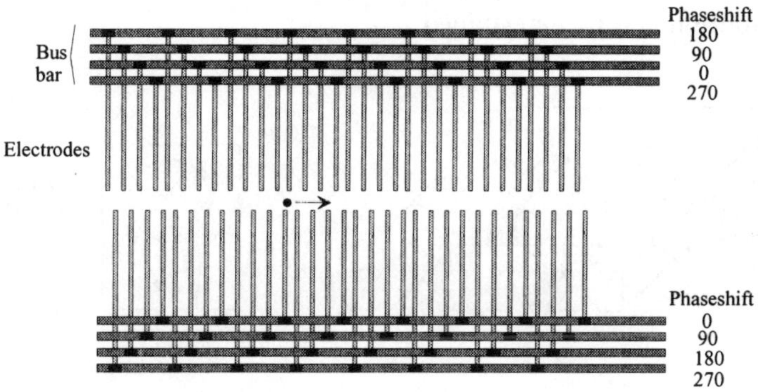

Figure 7.6: Electrode arrangement for separation of negative and positive dielectrophoretic cells [228]. A cell of negative dielectrophoresis moves to the right, which is the same direction as the travelling wave. A cell of positive dielectrophoresis is attracted towards the electrodes and remains there.

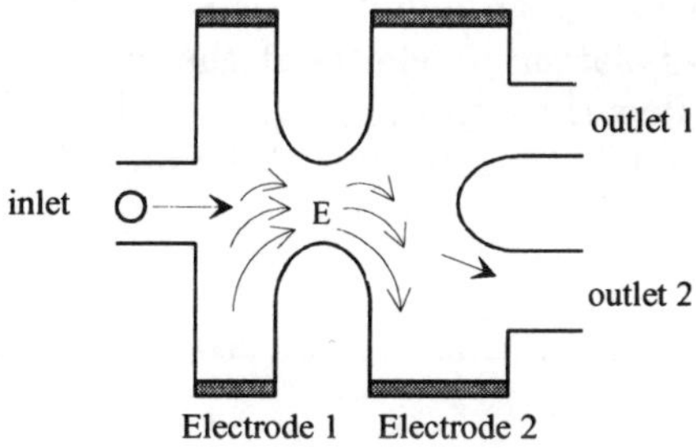

Figure 7.7: Deflection of particles with positive dielectrophoresis [229]

sured when a particle flows through. The simple construction of these devices allows analyses to be carried out rapidly with good reproducibility, and resolutions down to particle sizes of 0.6 μm diameter have been reported [230]. Such devices can be useful in quality control of high-tech fluids and characterisation of biological cells. For example, blood cell rheology has been studied by measuring the variation of transit time of cells through the capillary [233]. Diseased and aged cells are usually more rigid. This leads to an increased transit time which can be detected by the Coulter counter.

The Coulter principle is used to determine the number and size of particles in an electrically conducting fluid. The principle itself relies on particles passing through small capillaries fitted with an electrode at both ends. As a particle (P) passes, a change in the electrical resistance of the fluid between electrodes E_1 and E_2 within the capillary occurs and creates a pulse (see figure 7.8). By amplifying, sizing and counting these pulses, the number and size distribution can be determined.

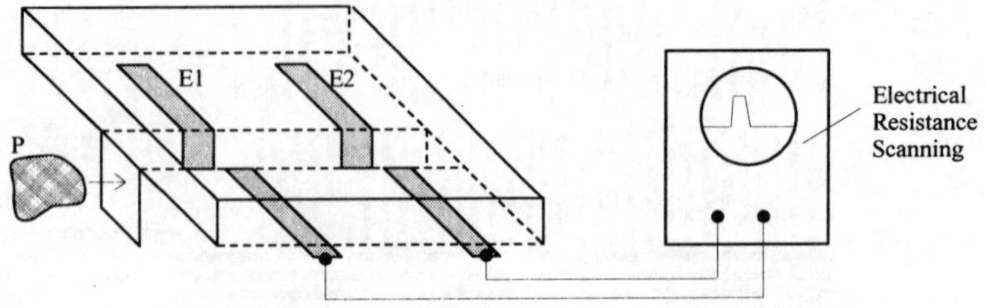

Figure 7.8: Schematic of the micromachined Coulter Counter

The basic assumption for the Coulter principle is the dependence of the electrical output signal on the particle volume (figure 7.9). Assuming that a particle in the capillary has an infinite resistivity (this is true even for conducting particles, as polarisation effects at the particle/electrolyte interface prevent any current flow through the particle itself), the change in the electrical resistance depends on the integral of the particle cross-section.

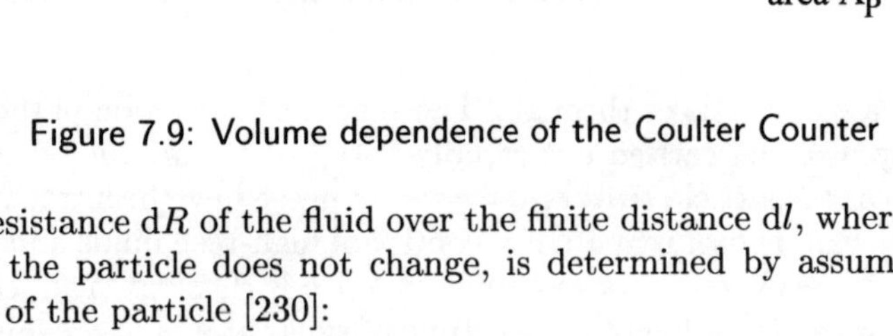

Figure 7.9: Volume dependence of the Coulter Counter

The resistance dR of the fluid over the finite distance dl, where the cross-section of the particle does not change, is determined by assuming infinite resistance of the particle [230]:

$$\mathrm{d}R = \rho_{el.f}\,\frac{\mathrm{d}l}{A_C - A_P} \tag{7.6}$$

where $\rho_{el.f}$ is the electrical resistivity of the fluid and A_C and A_P are the cross sectional areas of the capillary and the particle, respectively. The change in the electrical resistance over the length dl is given by:

$$\mathrm{d}(\Delta R) = \mathrm{d}R - \mathrm{d}R_0 = -\rho_{el.f}\,\mathrm{d}l\,\frac{A_P}{A_C^2}\left(1 - \frac{A_P}{A_C}\right)^{-1} \tag{7.7}$$

For a spherical particle of diameter $2r_P$, i.e. $A_p(l) = \pi\left(r_P^2 - l^2\right)$, the change in resistance becomes:

$$\Delta R = 2\rho_{el.f}\left(\frac{\arctan\left(\frac{r_P}{\sqrt{\frac{A_C}{\pi}-r_P^2}}\right)}{\pi\sqrt{\frac{A_C}{\pi}-r_P^2}} - \frac{r_P}{A_C}\right) \tag{7.8}$$

A particle counter has been presented recently by Larsen [231]. The device is completely micromachined (see figure 7.10). The channels were dry etched from the top to a depth of 50 μm and access holes were wet etched from the bottom. Finally, the dry etched trenches were sealed with a Pyrex cover plate. Tests have been performed with liquids and the resistivity has been measured as a function of the flowrate for a diluted KCl solution. The fluid to be examined is surrounded by a carrier fluid that does not mix with the test fluid due to the large velocity used. Electric resistance measurement was proposed with a four-point electrode arrangement, but no particle counting results have been presented so far.

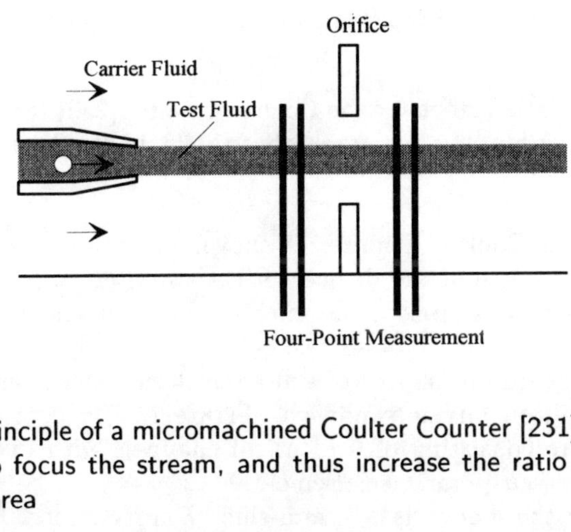

Figure 7.10: Principle of a micromachined Coulter Counter [231]. The orifice can be employed to focus the stream, and thus increase the ratio between particle area and fluid area

Another particle counter was described by the present authors [232]. The fabrication process of the Coulter Counter starts by etching double-sided alignment marks on the two sides of a silicon wafer. This is followed by trench etching (RIE) of 5 μm-deep channels on the front. Then the wafers are wet etched to a depth of 400 μm from the back to create the inlet and the outlet channels (figure 7.11.a). A 100 nm-thick titanium layer is now deposited by sputtering. This technique ensures a perfect side-wall coverage and therefore continuous electrical conduction of a crossing metal track. The titanium layer is then structured with photolithography and subsequent titanium etch using an optical endpoint detection (see figure 7.11.b). In order to protect the metal during the subsequent KOH etch a double layer of PECVD SiO_2 and Si_3N_4

is deposited. After this, all Si_3N_4 and SiO_2 layers are dry etched, to form the etch-mask for deepening the access channel. The access channels are etched further in KOH to a depth of 50 μm to connect with inlet and outlet (figure 7.11.c). As a last process for the silicon wafer, the plasma nitride and plasma oxide on top of the metal layer are removed by dry etching, where selectivity to titanium is very good.

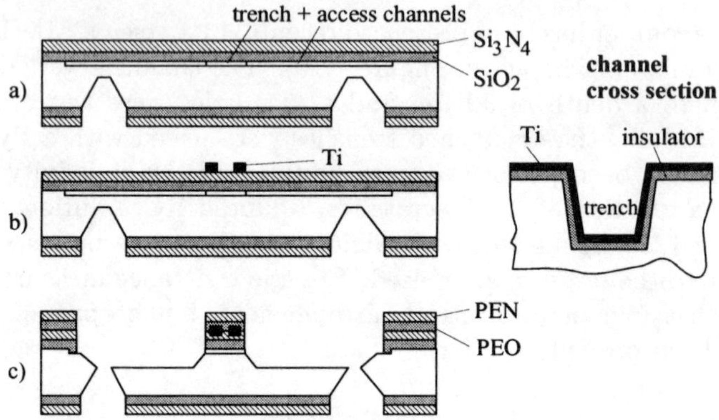

Figure 7.11: Fabrication steps for the Coulter Counter [232] (not to scale). The channels are not visible, but run from left to right under the electrodes in the plane of the diagram

A SEM of the Coulter Counter is shown in figure 7.12. The Coulter Counter is in the centre of the device with a pre-filter structure depicted to the right, the electrodes on top, the inlet to the right and the outlet to the left.

The electrodes do not show up well against the background as they are embedded in a silicon nitride sandwich. Figure 7.13 gives a close-up of the capillary structure consisting of 9 identical channels with two electrodes at each end to allow four-point measurements.

The cover of the trench is a 1 mm-thick Pyrex wafer. To allow access to the contact pad of the electrodes, the Pyrex wafer is recessed in this area by wet etching with 48 % HF using Cr/Au masking layers (cf. section 5.4.2). Before bonding, two perpendicular flats are sawn along the outer scribelines of both wafers for alignment purposes. Then both wafers are cleaned with fuming nitric and aligned on a hot plate using additional holders.

With a voltage of 800 V and a temperature of 390°C, a permanent chemical bond is achieved. The top Pyrex wafer is then sawn to the depth of the etched cavity, leaving the silicon wafer untouched, and removing only the glass in the contact area. After this, the chips are diced and the sensor is finished. Although electrical tests have not yet been performed, flow tests with latex spheres in DI water allowed optical tracing of the particles for supply

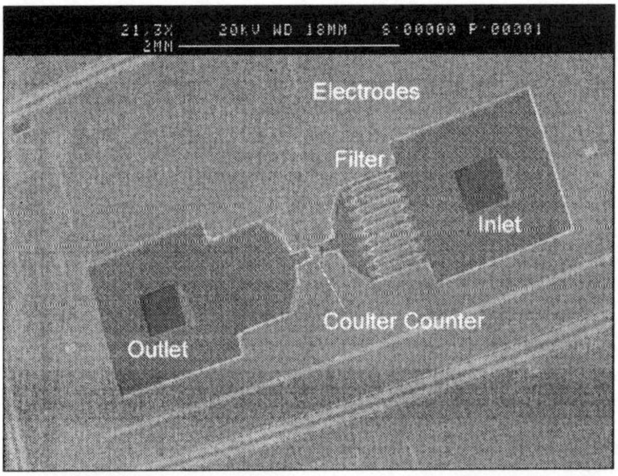

Figure 7.12: SEM image of the Coulter Counter

pressures up to 1 kPa.

A similar system has been used by Tracey [234], where the passage of red blood cells through micromachined capillaries was studied optically rather than electrically.

7.3 Mixing and Chemistry on the Micronscale

Microfluidics is beginning to find uses in chemistry. Some examples of applications are: measurement of the reaction gradient (stopped flow), screening of chemicals for drug discovery or DNA synthesis, enzyme & substrate reactions, high temperature and light induced reactions (Diels-Alder reactions or cyclo additions). Compared with conventional reactors several advantages are gained by moving to microsystems. Clean reactions are possible due to constant flushing out of the products, a very homogeneous temperature distribution is achievable for endothermic reactions, potentially hazardous reactions are made safer through the use of small quantities and larger volumes could be processed, if necessary, by using many microsystems in parallel.

7.3.1 Diffusive Mixing Principle

Mixing several fluids in chambers at the micron scale is not as easy as it might seem at first sight. The Reynolds number at these dimensions is usually rather small, and therefore no turbulence occurs. Thus liquid streamlines do not interfere with each other, resulting in zero mixing. However, over small distances mixing can be accomplished by diffusion. Alternating streams of

Figure 7.13: Coulter Counter: Magnification of the channels

small dimension can mix completely by this process. The reason for the diffusion is the large gradient of the concentration of fluid molecules which exists when two different liquids have a common interface. This behaviour can be described by the second law of Fick [15]:

$$\frac{\mathrm{d}n}{\mathrm{d}t} = D_c * \mathrm{div\ grad}\ n \tag{7.9}$$

where n is the concentration for a particular fluid molecule type and D_c is the solute diffusion constant. A measure of the mixing distance is the diffusion length L_D which is given by $L_D = \sqrt{D_c\, t}$, where t is the time of mixing.

In order to describe the efficiency of the mixing of two fluids next to each other, a dimensionless factor F_{mix} is introduced [235]:

$$F_{\mathrm{mix}} = \frac{D_c\, t}{l^2} \tag{7.10}$$

where t is the stream contact time allowed for mixing and l is the system dimension perpendicular to the flow direction. In general, the mixing process ranges from substantial up to almost complete for F_{mix} values in the range of 0.1 to 1. $F_{\mathrm{mix}} = 1$ corresponds to $l = L_D$.

Before designing a diffusive mixer, some preliminary calculations have to be done to ensure that mixing will be achieved with the proposed design. Two basic designs can be employed for mixing: either lateral mixing of several parallel streamlines or vertical mixing of several layers of fluid (figure 7.14).

For a mixer with streamlines next to each other (figure 7.14.a), a mixing length of $L = 2.4$ mm length with $l = 85$ μm central distance between two streamlines and $d = 5$ μm depth is taken as an example. The two liquids A and

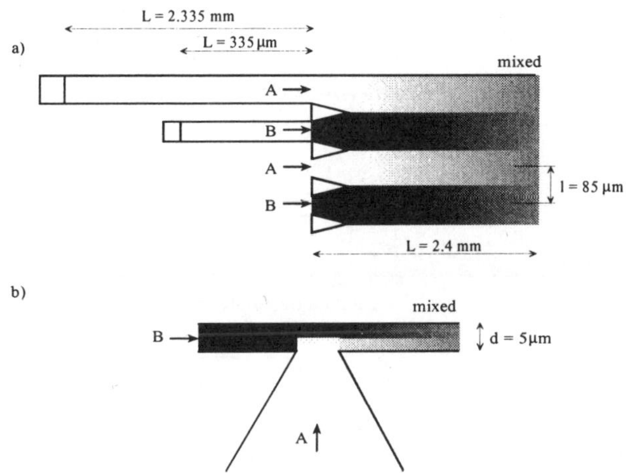

Figure 7.14: Principle of the micromixers: a) top view of lateral mixing (design 1), b) cross section of vertical mixing (design 2)

B are introduced as shown. The efficiency of mixing can now be calculated by using equation 7.10, with D_c being the solute diffusion constant for water (10^{-9} m^2 s^{-1} for a range of molecules in aqueous solution [236]). For a mixing time of $t = 1$ s and with water based liquids the ratio of mixing is $F_{\text{mix}} = 0.14$, i.e. good mixing. This implies a flow rate through a 2.21 mm-wide mixer of 1.707 μl min^{-1}. The volume flow / pressure relationship for laminar flow in long channels from table 2.1 can then be employed to calculate the pressure difference necessary to achieve such a throughput:

$$p = Q \cdot \frac{C_{fr}\, L\, \mu}{2\, A\, D_h^2} = 2.86 \text{ kPa} \tag{7.11}$$

Here, C_{fr} is the friction coefficient ($C_{fr} = 96$ for rectangular cross section where the width is much larger than the height), D_h is the hydraulic diameter and A is the cross sectional area. The resulting pressure is a figure which is easily achievable by most of the micropumps presented to date. Therefore microfluidic systems with a micromixer and micropumps are quite practical.

All micromixers presented in the literature are based on the diffusion principle, where improved mixing is obtained by reducing the channel dimensions. Two simple diffusive micromixers have been fabricated and tested by the authors [237]. These devices are shown schematically in figure 7.14. The fabrication process starts with the formation of 5 μm-deep trenches by dry etching, using a 1.4 μm-thick wet oxide as the mask. These trenches form the supply channels and the mixing area (figure 7.15.a) and are etched at the same time. Both inlets and the outlet are then wet etched with KOH at 70°C and 30% wt. to allow fluid connection from the back of the wafer (figure

7.15.b), using Si_3N_4 and SiO_2 as the etch mask. Finally, a wet etch is made from the front to allow fluid in from ports A and B and out at the outlet (figure 7.15.c). As before, a 1 mm-thick Pyrex wafer is anodically bonded to seal the mixing area and to allow viewing (figure 7.15.d).

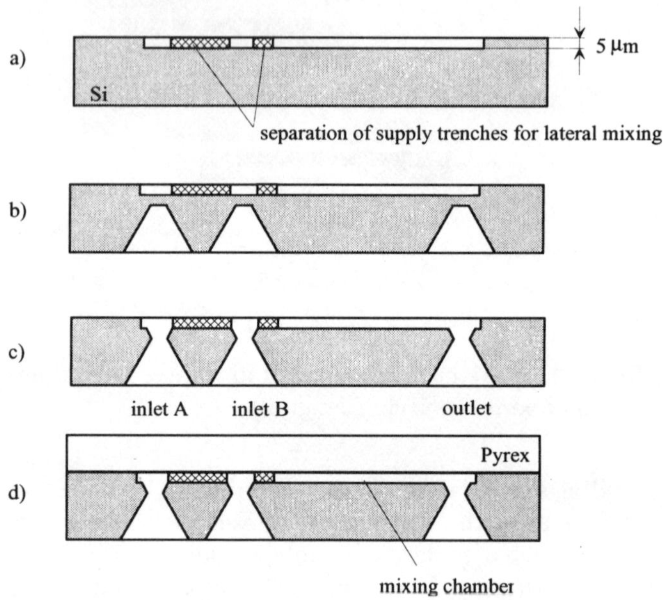

Figure 7.15: Micromachining of the diffusive mixers [237]

The micromixers were tested with liquids of characteristics similar to those of water (density, viscosity and solute diffusion constant). For the first pair of test fluids, an ammonia solution of 10% dissolved in water and a solution of phenolphthalein dissolved in ethanol and water (5g of phenolphthalein, 500 ml of ethanol, 500 ml of water) were used. When both liquids interact, the colourless liquids change to a very intensive pink colour. A test box made of perspex with a glass window on top was used to introduce small amounts of liquid into the mixers and to be able to view the mixing. Transparent tubing is advantageous as it allows the measurement of the flow rate by measurement of the travel speed of the air/liquid interface. Pressure can conveniently be applied by adjusting the height of a liquid column.

Figure 7.16 shows the mixing with phenolphthalein and ammonia solutions at 7 kPa for the first mixer [237]. The inlet holes are visible on the left side. The dark pink colour to the right indicates mixing of the two fluids. At 9 kPa the mixing in this mixer becomes good but is not complete. To achieve mixing (uniform change to pink as in figure 7.16) the pressure must be lower than 8 kPa.

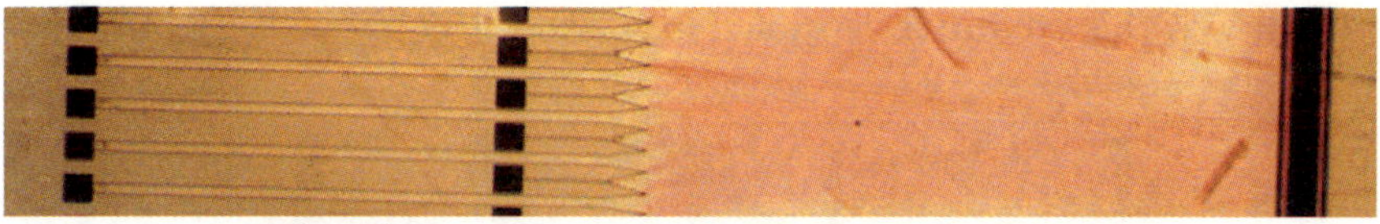

Figure 7.16: Mixing of phenolphthalein and ammonia in mixer design 1 at 7 kPa

Another interesting result can be obtained with commercially available red and green inks from Parker Ltd. [238]. Prior to the experiments the inks need to be mixed with alcohol and filtered through 0.2 μm pores to avoid handling problems. At a large pressure (>10 kPa) the liquids showed no mixing (figure 7.17) as the colours show a good contrast up to the outlet. At a pressure of 7.8 kPa (figure 7.18), the contrast at the entrance to the mixing chamber is still quite high, but the contrast of red and green near the outlet starts to get weaker.

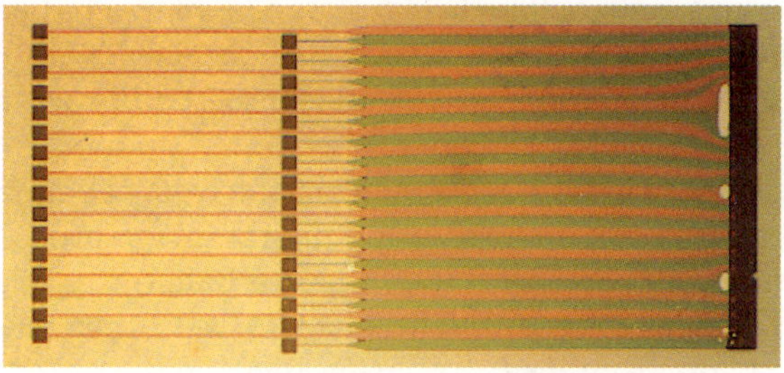

Figure 7.17: Mixing of red and green ink in mixer design 1 at $p > 10$ kPa

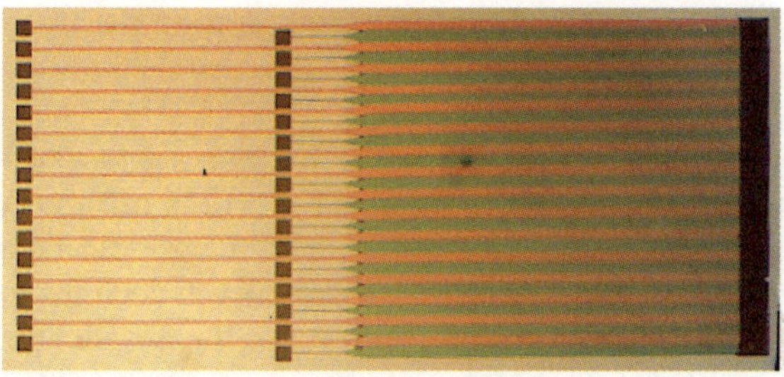

Figure 7.18: Mixing of red and green ink in mixer design 1 at $p = 7.8$ kPa

Perfect mixing of the two coloured inks is achieved at around 2 kPa, where a blue-grey mixing colour is formed after half the mixing length. However, there is still some weak contrast between the two liquids near the outlet. A photograph of this condition is shown in figure 7.19.

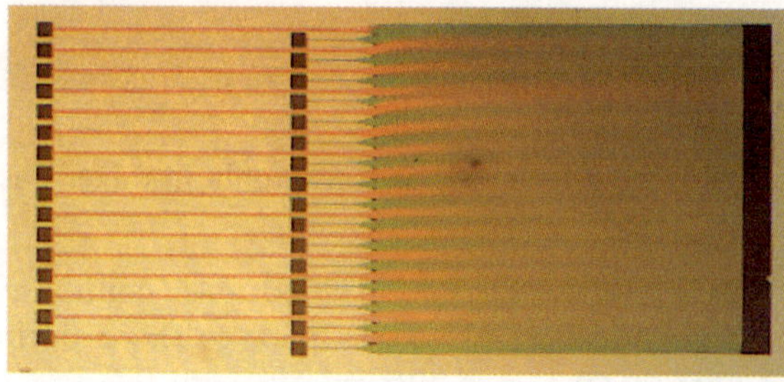

Figure 7.19: Mixing of red and green ink in mixer design 1 at $p = 2$ kPa (x-direction horizontally, y-direction vertically)

The inks show the mixing process more clearly than the phenolphtalein and acid system as the latter has a sharp transition for the pink indicator appearance, the same pink colour being obtained whether the mixture is 1:9 or 9:1. All pressures (2 – 7 kPa) are easily obtained with a micropump. Therefore, this type of micromixer fits well into the library of microfluidic devices.

In order to get more quantitative information from the mixing of the inks it is possible to analyse the pictures using digital image processing. In order to do this, clear liquid (9 parts DI water, 1 part ethanol) was mixed with green ink (9 parts ink, 1 part ethanol). The pictures were then scanned into the computer and the colour intensity variation registered as a grey-scale image. The grey-scale intensity was then plotted along a vertical line (y-direction) for several distances x from the entrance of the mixing chamber. In order to avoid any colour changes because of light intensity variation, a reference picture with only green ink was used and the grey-scale intensity of these line scans were subtracted from the mixing picture. A graph with several curves of the normalised grey-scale intensity is given in figure 7.20 for an inlet pressure of 2 kPa.

It can be seen from the plots that the mixing starts as soon as the two liquids are in contact, as no rectangular waveform is obtained for grey-scale values at 10 μm distance. All the curves show a periodic behaviour around the same centre line and the amplitude of the curves declines for larger travel distances and therefore longer contact times. As the amplitude is almost zero after 2.3 mm travel distance, an almost complete mixing can be assumed at this point. The image processing avoids any subjective influence in viewing the photographs and therefore gives a more quantitative measure of the diffusive mixing.

Mixing for the second mixer with phenolphthalein and ammonia solution [237] requires different pressures for both inlets to gain perfect mixing. The

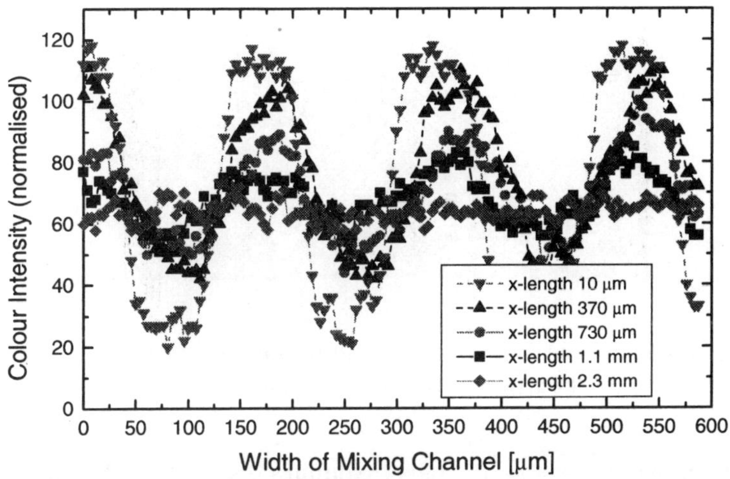

Figure 7.20: Grey-scale intensities at 2 kPa inlet pressure

first inlet (B in figure 7.14) needs 4 kPa inlet pressure, whereas 3 kPa is applied to the second inlet (A in figure 7.14). This is necessary in order to equalise the pressures at the point where the liquid through inlet A meets with liquid B. A pink colour corresponding to mixing could be observed under these conditions. The flowrate at this pressure was 0.9 μl min^{-1}.

The mixer has also been evaluated with red and blue ink. Liquid B comes from the far-left, whereas liquid A is "injected" in the middle of the mixer. At this point liquid B is on top of liquid A as indicated in figure 7.14. A photograph of the mixer at 3.9 kPa for inlet B and 3.6 kPa for inlet A is given in figure 7.21. In this case, the area from the second inlet A up to the outlet is mainly red. Between the first inlet A and the second inlet A, the colour indicates mixing of blue and red. Owing to an additional pressure drop up to the second inlet A, the blue liquid is at a lower pressure than the red liquid and therefore more of the red liquid enters. Only at the first inlet A are both pressures similar, which results in mixing.

A second picture was taken at pressures of 3.9 kPa for inlet B and 2.9 kPa for inlet A. In figure 7.22 no mixing colour can be observed between the first and the second inlet A. Thus, the pressure at the first inlet A seems to be too small for mixing. However, this time the pressure at the second inlet A is just enough to allow perfect mixing in these areas. The pressures to achieve proper mixing (2 – 4 kPa) with this mixer are also suitable for integrated microfluidic systems.

In both pictures, a difference in mixing can be observed for lateral mixing of the streamlines, and vertical mixing. Because the two fluids are viewed on top of each other, the exact amount of mixing cannot be deduced. However, we know that, because of the dimensions of the two mixers and equation 7.10,

Figure 7.21: Mixing of red (left - B inlet) and blue ink (centre rows - A inlets) at 3.9 kPa (blue) and 3.6 kPa (red) with mixer design 2

the lateral mixing between individual streamlines is slow ($l = 85\ \mu$m), whereas the vertical mixing is extremely fast ($l = 5\ \mu$m).

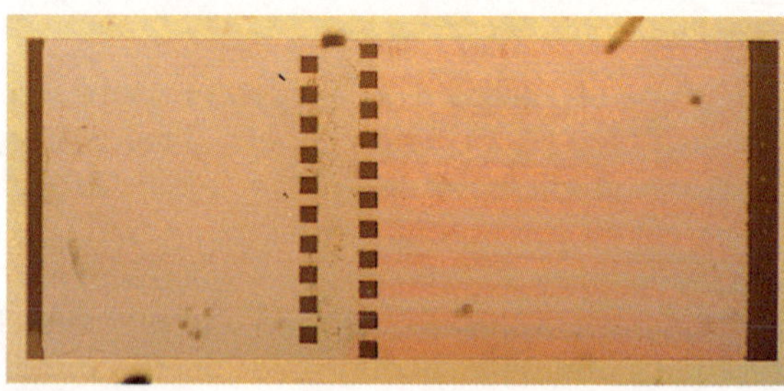

Figure 7.22: Mixing of red and blue ink at 3.9 kPa (blue) and 2.9 kPa (red) with mixer design 2

An improvement to the simple mixer (design 1) can be made by increasing the boundary surface area by laminating the fluids on top of each other, a technique which was for example developed by Schwesinger [134]. Two fluids are introduced on top of each other (figure 7.23.a). After this, the stream is divided laterally into two flows. The second flow is then steered on top of the first in order to reduce the dimensions and therefore the diffusion distance l. Fabrication of the Y-shaped mixer was by anisotropic etching of channels into a silicon wafer and isotropic etching of a fork into a Pyrex wafer. Both wafers are finally bonded by anodic bonding.

A similar principle was used by Branebjerg [239]. This system is realised by bonding together a lower silicon wafer and an upper glass wafer. One liquid flows in a channel made in the Pyrex and the other liquid comes from below from a channel in the silicon. At the intersection, the lower liquid meets with the upper liquid through underetched slits in a silicon plate. The fluids for the mixing experiments were phenol-red and an acid. At input flow rates in the order of 0.5 – 12 μl min^{-1} and pressures between 0.2 and 5 kPa, a mixing time of 0.1 – 0.3 s was obtained [239].

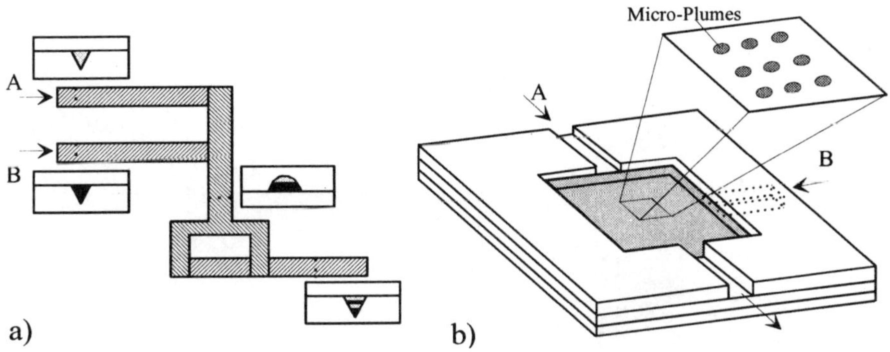

Figure 7.23: Schematic principle of a laminating mixer [134] (a) and an injection mixer [131] (b)

Another mixer with injection of one liquid into the other with microplumes was proposed by Elwenspoek [131]. The principle of this mixer is depicted in figure 7.23.b. During the operation, the mixing area is first filled with liquid A and then liquid B is injected through microplumes into the bottom area of the chamber. The contact surface is increased by using an array of nozzles with 100 μm spacing. The fabrication of the device is done on two wafers. On the front of the silicon wafer, V-grooves are etched prior to the etch of a membrane from the back. This forms the nozzle area. A Pyrex wafer with inlet ports from the side and a cover glass are anodically bonded on top of the silicon wafer and the device is finished. Mixing times of 1.2 s for water and uranine ($C_{20}H_{10}O_5Na_2$) were achieved with flow rates of 34.2 μl min^{-1}.

A simple mixer fabricated with the LIGA process has been described by Möbius [240]. In this design, two fluid streams are split into many small liquid sheets, which are then alternately combined into one stream perpendicularly to the incoming fluids. The device is made in photo-etchable glass (FOTURAN^{TM}) where the exposed parts of the glass are crystallised by heat treatment and can be etched in HF.

7.3.2 Chaotic Mixing Processes

This micromixer design uses pumps to actively improve the mixing of fluids. Flows with low Reynolds numbers ($Re \ll 1$), as obtained in microfluidic devices, are symmetric and reversible, so mixing is usually not possible with external stirring. However, by using a chaotic flow field generated by external pumps, the path and final position of fluid particles are extremely sensitive to their initial position. A chaotic flow field is achieved with two pumps connected via source and sink to the mixing chamber. The mixing chamber is initially filled with fluid A at the top of the chamber and fluid B at the bottom (figure 7.24.a). After several cycles of pumping both liquids through the chamber, complete mixing is achieved (figure 7.24.b). No test results have yet been presented for this novel mixer.

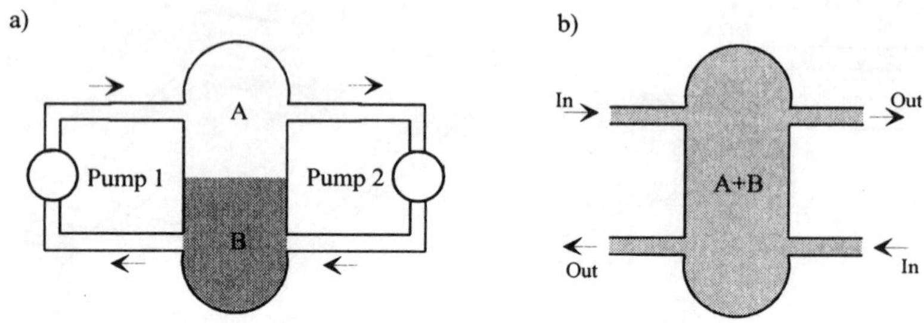

Figure 7.24: Chaotic Mixer [133]: a) Two liquids are introduced into the chamber; b) Complete mixing after several pump cycles

7.3.3 Microreactors and Chemistry on a Chip

Microreactors are micromixers in which the fluids that are mixed react chemically to form a desired product. Reactors have been designed for both liquids and gases. Due to the small scale, controlled reactions with intermediate products can be realised which would not be possible for large-scale reactions.

An example of such a controlled reaction is the following microreactor for a Pt-catalysed NH_3 oxidation, presented by Srinivasan [241]. Ammonia is oxidised to NO in a first reaction, but in subsequent reactions, NO is reduced to N_2. By optimising contact time, temperature and inlet composition, the reaction can be controlled so that NO is the main product. The device is a T-shaped silicon microreactor with two inlets and one outlet. The channels are closed from the top with a low-stress silicon nitride film and on the bottom with an Al base plate. NH_3 and O_2 are supplied to the reaction zone, which is partly coated with Pt acting as a catalyst (see figure 7.25). A Pt heater on top of the reaction zone was necessary to start the reaction. The temperature of ignition was around 200°C and the final product was mainly the desired NO. An advantage of small reaction chambers like these is the high safety standard compared with macroscopic devices, as the gases are immediately diluted drastically in air if the chamber is destroyed, so that no explosion is possible.

A microreactor for liquids has been presented by Desai [132]. As in many microfluidic devices, a silicon ground plane with channels and through holes has been sealed with a Pyrex cover plate. The design of this mixer allows

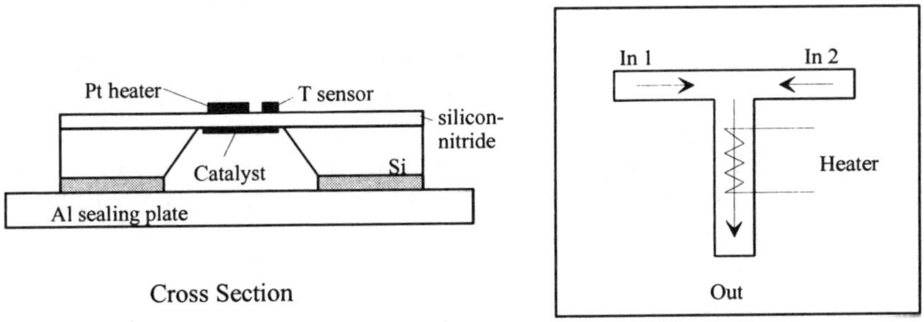

Figure 7.25: Chemical microreactor for the Pt-catalysed NH_3 oxidation [241]

the combination of 3 different chemicals via "T" junctions. The mixer has been calibrated with a fast reaction at the first intersection between dinitrophylacetate (DNPA) and NaOH, which produces dinitrophenol (DNP) and acetic acid. At the second mixer, the reaction was quenched with HCl, which neutralises NaOH. The UV absorption of the final product was measured after the second mixer, as DNPA and DNP have different absorption patterns. By varying the contact times between DNPA and NaOH and by varying the velocity of the fluids while making UV measurements, the time constant of this pseudo 1st order reaction could be determined as 0.01 s, showing a useful application for microfluidic reactors.

Another example of a microreactor is an enzyme catalysed reactor. Enzyme reactors are used in a flow-through cell for continuous monitoring of biomolecules as they provide long term stability. Most of the conventional silicon micromachined enzyme reactors have the enzyme immobilised on silicon V-grooved channels [242]. Owing to the small surface area of the enzyme, however, long term stability of this reactor is a problem. This has been overcome by improving the enzyme coupling with a porous surface of the channels to increase the surface area [243]. The design consists of 32 small channels of 10 mm length, 50 μm width and 250 μm depth. The trenches are etched with KOH into (110) silicon to obtain vertical walls. The porous layer was achieved by anodising the structures in a solution of 1 part HF (48%) and 1 part ethanol (96%) and a current through the electrochemical cell of 50 mA cm^{-2}. Then the enzyme glucose oxidase was coupled to the porous surfaces. For the oxidation of glucose with glucose oxidase immobilised on the channel walls, an efficiency increase of around a factor of 100 is gained with the porous silicon technique.

7.4 Chemical Analysis with MEMS

Chemical analysis with MEMS consists of flow injection analysis as well as DNA analysis. Flow injection analysis can be further subdivided into liquid/gas chromatography (LC, GC) or capillary electrophoresis (CE). The following sections will introduce the chemical analytical techniques used in microfluidics and review the latest devices originated from the MEMS community.

7.4.1 Introduction

The term chromatography refers to separation methods in which the components are distributed between a stationary phase and a mobile phase (eluent). The mobile phase consists of the sample to be examined and a buffer solution. Separation occurs because the components of a sample have different affinity between the stationary and the mobile phase (figure 7.26.a). Therefore the components move at different speeds along a column or other structures. The stationary phase can be the coating of the column or coated granules within the column. If a locally concentrated piece (plug) of sample is allowed to move through a column transported by the buffer then the components arrive at the exit of the column at different times. With the help of a non-selective sensing element, the individual components can be detected. The output of this sensing element over time is called a chromatogram, where each peak represents a separate sample component (figure 7.26.b). The area under the peak gives information about the relative amount of the component.

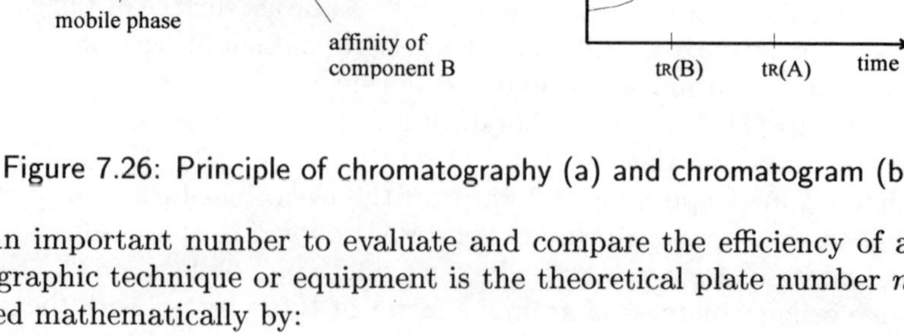

Figure 7.26: Principle of chromatography (a) and chromatogram (b)

An important number to evaluate and compare the efficiency of a chromatographic technique or equipment is the theoretical plate number n. It is defined mathematically by:

$$n = 16 \cdot \frac{t_R}{t_W} \tag{7.12}$$

where t_R is the retention time, i.e. the time required to see the maximum in concentration of a component at the exit of a chromatographic column. If

F_V is the volume flow rate of the eluent, then the product of t_R and F_V is called the retention volume V_R. The time t_W is the width of four standard deviations of the component of interest on the chromatogram (figure 7.26.b)

Another factor for comparison is the peak resolution R_S for two substances A and B, which is defined by:

$$R_S = \frac{2\,(t_R(B) - t_R(A))}{t_W(A) + t_W(B)} \tag{7.13}$$

Chromatography can be divided according to the phases into gas chromatography and liquid chromatography. Additionally, ion-exchange chromatography, paper chromatography and thin-layer chromatography are in use in analytical chemistry but have not been realised on silicon to date.

Gas Chromatography (GC)

GC is a type of chromatography in which the mobile phase is a gas, such as nitrogen or helium, and the stationary phase is a liquid or a solid. In MEMS, the stationary phase is often the wall of an etched cavity. The impact of gas chromatography on analytical chemistry is mainly found in the analysis of organic samples. A gas chromatograph (figure 7.27) consists of a flow controller through which the flow of the carrier gas (eluent) is kept constant. Then, an injection port is necessary to inject a small, calibrated volume into the carrier gas. The injection is supposed to be rapid so that the sample is swept along as a discrete plug. In order to allow the analysis of many organic compounds, the injector as well as the column and detector are often heated to temperatures up to 300 °C to keep the compounds in a vapour phase. The next part is the column which can be "packed", i.e. filled with a granular solid support coated with a thin layer of the stationary phase, or an "open tubular" column, i.e. a long tube of very small inner diameter ($< 25\ \mu$m) with the channel walls coated with the stationary phase. At the end of the column, a detector is placed which is required to provide a signal proportional to the concentration of the compound. Also, the detector needs to be insensitive to the carrier gas and not very selective to the individual components to be analysed.

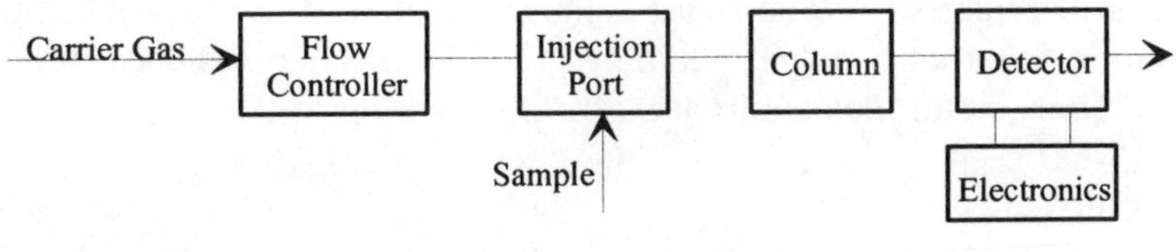

Figure 7.27: Block diagram of a gas chromatograph

Liquid Chromatography (LC)

LC includes both Liquid-Liquid Chromatography (LLC), where an aqueous stationary phase as well as an organic mobile phase are used, and Liquid-Solid Chromatography (LSC) with a solid stationary phase and a liquid as the mobile phase. LLC was invented to separate amino acids with a granular silica gel holding water as a stationary phase and using chloroform-butyl alcohol as the carrier liquid. Many other organic compounds have since then been separated on silica gel with different stationary and moving phases. In LSC, the stationary phase is a small solid particle of a material such as alumina or silica and the mobile phase is often a pure organic liquid. It is therefore most often used in "packed" columns. Modern LC, or High-Performance Liquid Chromatography (HPLC) consists of the same elements as GC: a pump, a sample injector, a column and a detector. Depending on the stationary phase, a packed or an open tubular (inner diameter $5 - 10\ \mu$m) column is used. Detection for HPLC is most often done with an ultraviolet detector at a fixed wavelength ($\sim$ 280 nm) where all aromatic organic compounds absorb intensely.

Capillary Electrophoresis

The second separation technique used in miniature chemical analysis is capillary electrophoresis (CE). Due to different electrical double layers between fluid and molecule or due to permanent charges, the velocity of molecules and/or ions from the sample differ if an electric field is applied in the fluid. As the electric field moves the whole liquid by electroosmosis in one direction, a transient separation of different molecules/ions is realised at the end of the column. Macroscopic capillaries are typically of 50 μm inner diameter and 0.5 to 1 m length. The applied potential is 20 to 30 kV. A small volume of sample (10 nl) is injected at the positive end of the capillary and the separated components are detected near the negative end of the capillary. CE detection is similar to detectors in HPLC, and includes absorbance, fluorescence, electrochemical, and mass spectrometry. The capillary can also be filled with a gel, which eliminates the electroosmotic flow. Separation is accomplished as in conventional electrophoresis but the gel-filled capillary allows higher resolution, greater sensitivity, and on-line detection. One of the most popular and sensitive methods for separation of proteins is electrophoresis in a column filled with polyacrylamide gel. The method depends upon both electrical charge and molecular size.

DNA Analysis

Micromachining is also finding an application in the field of the analysis of deoxyribose nucleic acid (DNA). DNA is a double-helix molecule consisting of a huge number of building blocks. The building blocks are formed out of 4 different components: Adenine (A), Thymidine (T), Cytosine (C) and

Guanine (G). The arrangement of the components of this 4-letter alphabet generates a gene sequence. If the sequence is for example 1000 letters-long, the sequence would be said to be 1 kilobase (1000 bases). DNA is usually double stranded, and the two strands pair with one another in a very precise way: A pairs only with T and C only with G. Thus the strand AAATTT...CGGGAC pairs with TTTAAA...GCCCTG.

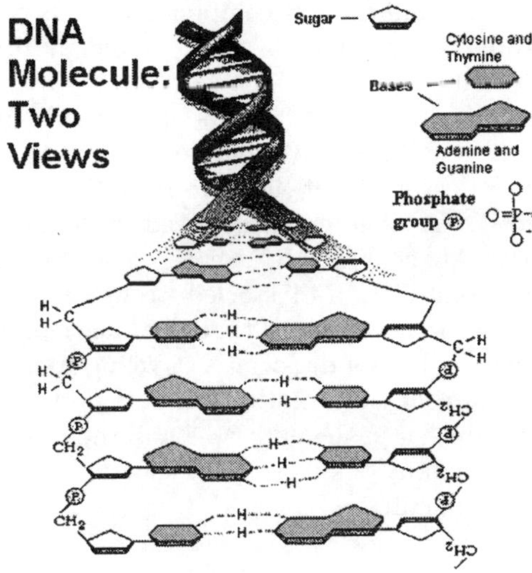

Figure 7.28: Schematic of a DNA double-helix

The polymerase chain reaction (PCR) is a technique which is used to amplify the number of copies of a specific region of DNA, in order to produce enough DNA to be adequately tested. This technique can be used to identify, with a very high probability, disease-causing viruses and/or bacteria, a deceased person, or a criminal suspect. In order to use PCR, one must already know the exact sequences which flank (lie on either side of) both ends of a given region of interest in DNA (may be a gene or any sequence). The first step of PCR is to heat the unknown DNA to around 95°C, which causes the paired strands to separate. Separately from this, primers need to be synthesised consisting of about 20 letters, which have the components of one of the flanks of the region of interest. This is done by using each of the four components. For the strand AAATTT...CGGGAC, the primer would consist of TTTAAA and/or GCCCTG. Then a large excess of primers relative to the amount of DNA being amplified is added, and the reaction mixture is cooled to allow double-strands to form again. Because of the large excess of primers, the two strands will always bind to the primers, instead of with each other.

The primer is combined with a mixture of all 4 individual components (deoxyribonucleotides) and an enzyme. The enzyme can "read" the opposing strand's sequence and extend the primer's sequence by combining the components together in the order in which they pair across from one another (Adenine to Thymidine and Cytosine to Guanine). This particular enzyme is called a DNA polymerase. Now, the enzyme synthesises new DNA strands working like a zipper. This completes the cycle in which the DNA amount has been doubled. The whole cycle is then repeated by heating the DNA again and adding more primers and more 4-component mixture. For a repetition of 20 times, approximately one million copies would be produced from one original DNA double-helix.

A second method for the amplification step in DNA analysis is the Cycling Probe Technique (CPT). This technique, as depicted in figure 7.29, is described as follows: an excess of labelled probe (DNA-RNA-DNA) which is specific to the target DNA of interest is mixed with the target DNA and hybridised (coupled). An enzyme (RNase H) is added which cleaves the DNA-RNA duplexes (combination of labeled probe and target DNA). The temperature of the reaction is chosen so that the longer probe anneals to the target and the shorter fragments dissociate. Cycling is now possible as the freed target DNA will anneal with another labelled probe. During the cycle, the amount of intact, labelled probe decreases and the amount of probe fragments increases. The amount of target DNA can thus be determined from the number of probe fragments.

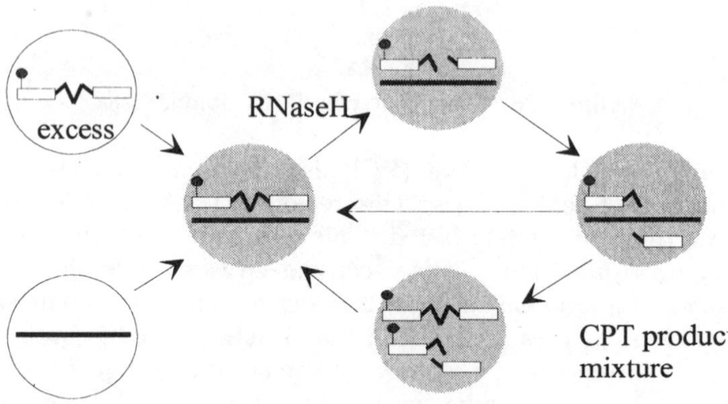

Figure 7.29: Schematic of Cycling Probe Technology [244]

Both mechanisms can be used for analysis of the target DNA. As will be shown later, silicon micromachined reaction chambers are ideal for DNA amplification and analysis because of the possibility of fast temperature cycling.

7.4.2 Flow Injection Analysis

7.4.2.1 Liquid and Gas Chromatography

The first micromachined flow injection analysis (FIA) system was a gas chromatograph developed by Terry [120]. It consists of a 1.5 m-long capillary column, an active valve with magnetic actuation and a detector element, all fabricated on a 2 inch silicon wafer. A diagram of the device is depicted in figure 7.30.

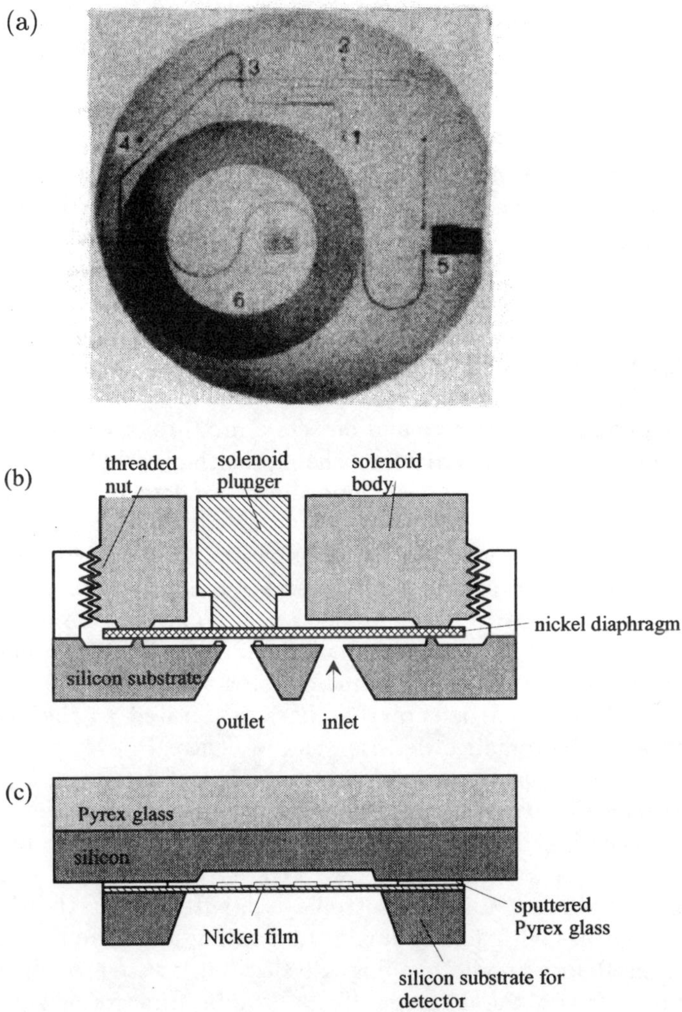

Figure 7.30: Silicon micromachined gas chromatograph [120]: a) top view, b) active valve, c) thermal conductivity sensor

The capillary with a cross section of 200 μm × 40 μm is isotropically etched into silicon and a glass wafer bonded on top to seal the structure. Through-holes are micromachined into the silicon wafer to form the connection to the capillary. The gas input is controlled by an active valve where through-holes are etched first isotropically and then anisotropically into the silicon, and a nickel membrane actuated with an electrical solenoid is bonded on top. Gas detection is done by measuring the thermal conductivity of the gas with a thermally isolated metal resistor. The resistor is biased at a fixed current level and the electrical resistance of the resistor is measured.

At first, the etched channel is primed with a gas chromatography liner to allow separation of different molecules in the gas. The system is now flushed with an inert carrier gas through inlet 2 (figure 7.30) with a pressure of 207 kPa. Then 5 nl of the sample gas is introduced into the carrier gas through inlet 1 while the valve (3) is open. After this, the valve is closed again and the sample is pushed through the column (6) by the carrier gas. Inside the column, different molecules show different travelling speeds and are thus separated over time. The sensor at the exit measures the changes in thermal conductivity of the carrier gas for the different species in the sample. In order to achieve good separation, the sample volume needs to be much less than the column volume and therefore it makes sense to integrate both the valve and the sensor on the device to avoid large dead volumes. It has been shown that this device successfully separates and detects a mixture of nitrogen, pentane, dichloromethane, chloroform, trichloroethane, thrichloroethylene and toluene. With a suitable calibration this technique allows the detection of gases within 10 ppm accuracy [120].

The next system was reported 11 years later and uses similar technology to allow chromatography of fluids [245]. The size of the device has been reduced drastically to 6 × 6 mm with a column of 6 μm × 2 μm × 15 cm with external pressure application. Unlike the previous device, the liquid chromatograph does not include any valves on the same chip for controlling the sample and carrier flows. However, a detector has been integrated at the end of the column. This conductometric detector uses a centre Pt electrode and two outer Pt electrodes in order to minimise the effect of a flowing liquid. The electrode surface area was chosen to be 100 – 200 μm^2 for channel widths in the range between 5 and 10 μm. Optimum operation for the conductometric measurement is at a frequency of 4 – 5 kHz at 1 V supply voltage. The column has been fabricated by an isotropic etch with HF/HNO_3 into silicon, covering the trenches with silicon dioxide and sealing the channels with a top plate of Pyrex. Before anodic bonding, Pt electrodes were sputter deposited on the Pyrex wafer for the detector. Tests have been performed to evaluate the conductometric sensor showing an impedance change between 0.25 and 1.25 MΩ for a variation of NaCl concentration in the range of 0.05 to 1 M. No measurements of the separation have been given.

Another device for liquid chromatography has been presented by Ocvirk

[246]. The device, which has a packed LC column, consists of a 100 μm-deep and 20 mm-long separation channel sealed with glass on the top and 85 channels of 2 μm depth at the exit. The latter allow the filling of the channel with 5 μm particles (eg. C_8-alkyl bonded phase) as the static phase for the separation. Due to the small channels, no particle greater than 2 μm can be transported beyond the separation chamber towards the detector. In this case optical fluorescence was used as the detector and the fluid was driven through the system with an external pump. The device was tested with fluorescein and acridine orange as the samples and methanol as the mobile phase. For a separation time of 11 s a theoretical plate number of 30 has been calculated. This number can be improved to 200 for a separation time of 3 min.

Recently a gas chromatograph based on the design of Manz [245] has been presented. An isotropic etch ($HF : HNO_3 : CH_3COOH = 0.5 : 89.5 : 10$) has been used [247] to produce channels of 10 μm depth. Polychlorotrifluoroethylene (PCTFE) has been sputter deposited in the channel to act as the stationary phase of GC. Tests have been conducted with the micromachined GC using a commercial mass spectrometer at the output for detection. Good performance for methane detection has been obtained in a time as short as 0.6 seconds. This has been compared with a conventional GC, which gives a signal with lower resolution after 20.4 s for the separation.

7.4.2.2 Capillary Electrophoresis

The first studies for electrophoretic separation were on an electroosmotically pumped system [114,186] for both sample and buffer. The idea of such systems was to integrate pumping of both the sample and the carrier liquid onto the same chip as the separation channels. This has been achieved by using an intersection of two capillaries, where the fluid is driven through with the electroosmotic effect of an electric field (cf. figure 7.31). Electroosmotic flow rates of 1– 2 mm s^{-1} have been realised with fields in the range of 100 and 160 V cm^{-1} [114]. The capillaries are micromachined by silicon wet etching and sealed by anodic bonding [114, 186]. To avoid short circuits through the silicon, the channels are covered with an insulating layer of silicon dioxide. However, for better performance and the use of higher voltages, silicon is not suitable and a change to glass chips is required. The idea behind this layout is to fill both channels with a buffer solution. Then the sample solution is transported from sample inlet to sample outlet (electric field between +(2) and -(2)) while the buffer channel is not activated. After switching off the flow of the sample, the buffer is transported from buffer inlet to the outlet (electric field from +(1) to -(1)). The flow movement pushes along the sample volume or plug which is defined by the intersection of the two channels. To avoid any excess sample from bleeding into the buffer, both sample inlet and sample outlet are switched to voltage potentials lower than the buffer inlet but higher than the general outlet.

The electrophoretic separation technique has been added to the electroos-

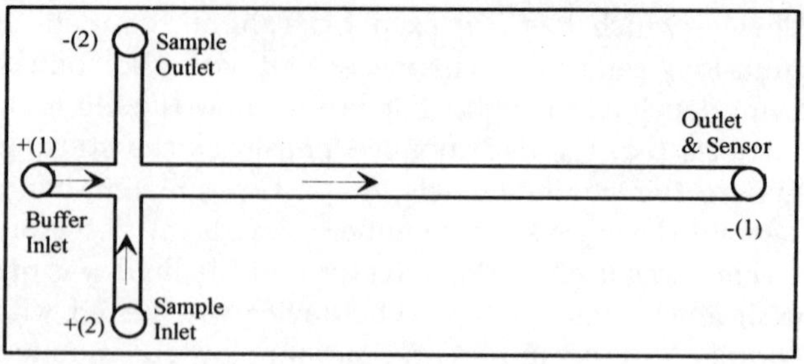

Figure 7.31: Electroosmotically driven separation channels with sample-buffer intersection [114]

motically driven separation channels by Manz and Harrison [248, 249]. The filling procedure and the layout of the channels are almost identical to the set-up described above [248]. However, due to the low breakdown strength of SiO_2 or Si_3N_4 on silicon ($7.1 - 8 \cdot 10^6$ V cm^{-1} for a voltage between a Pt electrode in aqueous solution of pH 7 and n-type silicon [114]), the chip was completely fabricated from glass. An HF solution has been used to etch channels of 12 μm depth and 50 μm width into one glass plate and another plate has been thermally bonded at 620°C to the first in order to seal the capillaries. For the experiments, six fluorescine isothiocyanate (FITC) labelled amino acids: arginine (Arg), glutamine (Gln), phenylalanine (Phe), asparagine (Asn), serine (Ser) and glucine (Gly) have been injected as a mixture into the sample inlet and the pressure of all inlets and outlets has been kept constant. After 5 and 24 mm, corresponding to analysis times of 2.5 and 14 s, the separation of the amino acids has been determined by laser induced fluorescence detection. The electric field strength for the separation was in both cases 1060 V cm^{-1}. The separation was good after 5 mm with a plate number of 5800 and became perfect after 24 mm with plate numbers between 45000 and 75000. The limit of detection has been determined as ±200 nM in terms of sample concentration. The major advantage of micro FIA is the use of small sample volumes with as low as 100 pl in this experiment and the fast analysis times which are 50 – 100 times shorter than for conventional capillary electrophoresis.

A similar device has been used by Jacobson [169] who machined channels of 5.6 μm height and 66 μm width into glass and then sealed it with a cover glass plate by direct bonding [250]. For separation, the column was treated with 1M NaOH, rinsed with water and dried at 125°C for 24 h in a helium atmosphere. Then the channels were rinsed with a 25% solution of chlorodimethyl-octadecylsilane (ODS) in toluene for 18 h at 125°C to react with the walls. Finally the channel was rinsed with toluene and acetonitrile to remove unreacted ODS. The separations could be monitored at the outlet using fluorescence. A mixture of coumarin 440 (C440), C450 and C460 at 10 and

1 μM concentration was used as the sample and a sodium tetraborate solution (10mM) with acetonitrile was used as the buffer. For the electrophoretic separation experiments, the electric field strength was up to 162 V cm^{-1} giving fluid velocities of around 0.78 mm s^{-1}. Separation was reasonable for a velocity of 0.65 mm s^{-1} with plate numbers between 1290 and 11700.

The next step to improve the separation efficiency is by synchronised cyclic capillary electrophoresis [251]. This technique allows for the elimination of very fast and very slow moving species and achieves a very detailed separation of the remaining components. A diagram of the device behaviour is depicted in figure 7.32. The principle relies on the rotation of the electric potential by 90° relative to a 4-fold axial symmetry and is described as follows. First, the sample mixture (1,2,3) is loaded into the junction between buffer and sample capillary (figure 7.32.a). Then a voltage is applied between top left and bottom right to move and separate the species (figure 7.32.b). Fast species (3) are then separated out of the sample-buffer solution (figure 7.32.c) and the voltage is rotated by a further 90 ° (figure 7.32.d). Another voltage rotation is used to move species 1 and 2 in parallel down their channels and a last rotation step of the voltage (figure 7.32.f) removes sample 1 as well. Now the components of species 2 can be separated further by several cycles of the system. After 10 cycles, 1.3 million theoretical plates are observed for 2 kV supply voltage.

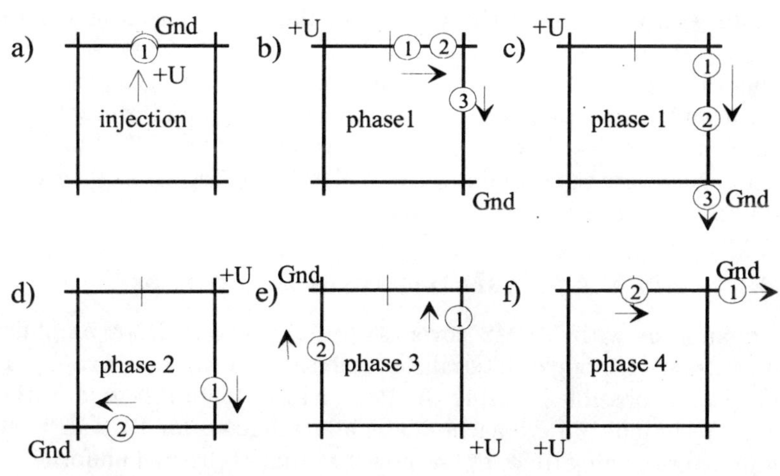

Figure 7.32: Principle of the synchronised cyclic capillary electrophoresis [251]

A further improvement to the device by Effenhauser [248] was to incorporate mixing or reactions onto the chip. This is especially interesting for labelling the analytes before or after separation to allow simple detection with fluorescence of eg. non-fluorescing amino acids [252]. Figure 7.33 shows the outline of a chip which allows reaction of the analyte before separation.

Sample and labelling agent are initially mixed in the mixing area by applying an equal potential at inlets 1 and 2 with outlet 3 connected to ground. After some time, this voltage is switched off and a voltage difference between inlet 4 and outlet 5 is created moving buffer and a small amount of reacted analyte towards exit 5 and separating it by electrophoresis. This has been performed with *o*-phthaldialdehyde (OPA) as the labelling agent and a mixture of lysine and valine (amino acids), where good separation has been achieved [252].

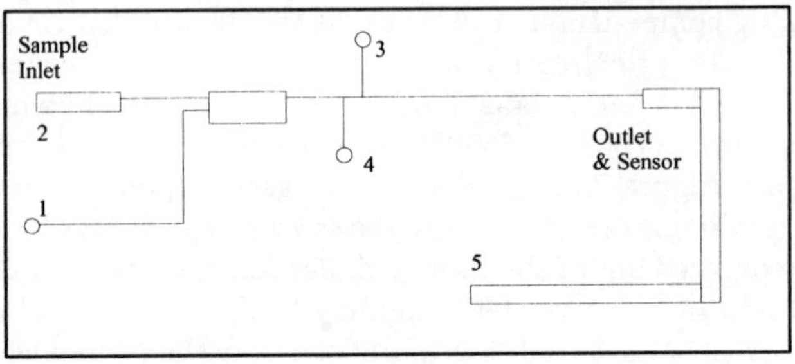

Figure 7.33: Mixing chamber and electrophoretic separation channels [252]

Sometimes the introduction of the labelling agent before the separation introduces a widening of the peaks and thus a reduction in separation efficiency. Introducing the labelling agent only shortly before the detector greatly reduces this problem [252]. For OPA as the labelling agent and a mixture of several amino acids (lysine, valine, phenylalanine, and glutamic acid) good detection was possible with 85000 theoretical plates [252]. In this case the amino acids were separated in their undetectable form and then reacted with OPA to create the fluorescent product.

7.4.3 DNA Analysis based on MEMS

DNA analysis with MEMS covers two major areas: DNA amplification for the creation of enough material for analysis, and chromatography. DNA amplification is possible by using the Polymerase Chain Reaction or the Cycling Probe Technique. Micromachining is advantageous for PCR as it allows fast temperature cycling times at low power as well as thermal uniformity throughout the reactor. One of the first portable reaction chambers for PCR was presented by Northrup (figure 7.34 [253]). The fabrication sequence for this reactor starts with the deposition of low-stress silicon nitride. This layer is used as an etch mask on one side and as an etch stop layer on the other side, leaving a thin silicon nitride membrane after the anisotropic etch. Polysilicon is then deposited onto the membrane and highly doped to form the heater. Finally metal contacts are deposited to allow connection to the heater.

The system can produce heating rates of over 5°C per second with 15 V

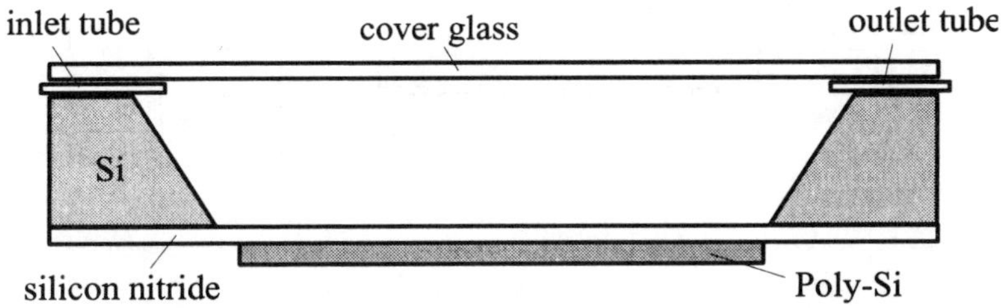

Figure 7.34: Cross section of a micro PCR device [253]

input, and cooling is possible with rates as high as -2.5°C per second. The device has been tested by performing β-globin PCR with 23 to 31 cycles. The amplification has been compared with a conventional system by running 5μl aliquots from each system on an agarose electrophoretic gel. Identical results were obtained, but the miniaturised PCR system worked over 4 times faster.

A micromachined chip for the Cycling Probe Technique has been presented by Tang [244]. CPT has been combined with a capillary electrophoretic separation method and fluorescence detection and shows that detection is possible in the nM range and shows also the possibility of detector integration. An epi-luminescence detector instead of fluorescence has been proposed also and could improve the detection range further as preliminary results have been performed with sub-nM detection.

A detection system based on a capillary electrophoresis device for DNA analysis has been presented by Webster [254]. The layout of the device is depicted in figure 7.35.

The fabrication consists of the construction of a high sensitivity diode detector in the silicon substrate and a surface mounted channel in polyimide using 20 μm-thick positive resist as a sacrificial layer. Finally, the channel is filled with a 1% agarose gel for the separation. For the operation, the DNA sample is loaded onto the inlet reservoir and then an electric field of 12.8 V cm^{-1} is applied along the channel for driving and separating the DNA molecules. A separation by 500 μm of a 1800 basepair fraction from a 5500 basepair fraction of DNA has been successfully made within 5 min using a 4.6 mm-long channel.

7.5 Biochemical Sensors for Microfluidics

We now describe sensors which can be used for chemical or biochemical substances. These sensors can be integrated into systems like the analysis systems which have been described in section 7.4.2. Another example is an insulin dispensing micropump that could be controlled by constantly mon-

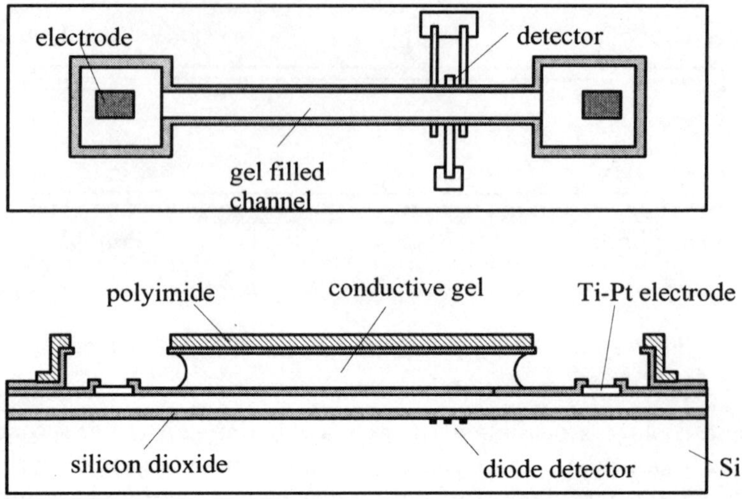

Figure 7.35: Layout of a capillary electrophoresis separation system for DNA analysis [254]

itoring the glucose level of the human blood. In order to design, fabricate or use such systems, some understanding of these sensors is required. More detailed knowledge on chemical and biochemical sensors can be found in the books published by Middelhoek [50], Schaumburg [17] and Goepel [255].

7.5.1 Semiconductor Devices used for Chemical Sensors

The detection principles of semiconductor diodes and FETs are now described. These devices are categorised by their detection principle.

7.5.1.1 Chemdiodes

This sensor type is based on heterogeneous structures like pn-junctions, metal-insulator-silicon structures and Schottky diodes. All these devices change their electrical characteristics when they are subjected to chemical measurands. Atoms or molecules surrounding a silicon structure can trap electrons or holes at the surface. This leads to a change in the electrical properties of the surface and thus allows the sensitive detection of chemicals.

In order to create an ion controlled diode [50, 256] a double layer of SiO_2 and an ion-selective membrane is attached on top of a horizontal pn-junction. A negative potential needs to be applied to a reference electrode in the solution. This creates an inverted region of holes at the surface of the formerly n-doped region. When an ion from the solution arrives at the SiO_2 - ion selective membrane interface, a change of the thickness of the depletion layer occurs. If the pn-junction is zero or reverse biased, the differential capacitance increases when the depletion layer increases. From this the ion concentration can be determined. By choosing the correct ion selective membrane, hydrogen (pH) or K^+ ions for example are measurable (figure 7.36).

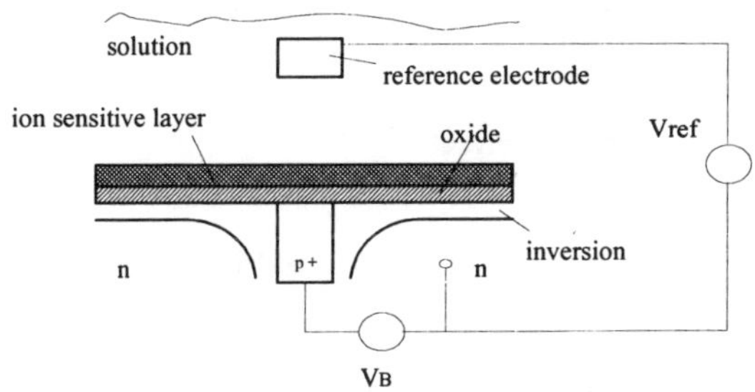

Figure 7.36: Schematic of an ion controlled diode

The same interface can be created with a Schottky junction, i.e. the junction of a metal on top of a semiconductor. At the interface, electrons of a n-type semiconductor flow into the metal until the Fermi level of both is equal. This bends the energy bands at the surface of the semiconductor in the same way as a pn-junction. The I-V characteristic of a Schottky diode is very dependent on the work function of the metal. If the work function of the metal is in turn dependent on the presence of gas atoms on its surface, a gas sensitive element is obtained. A hydrogen sensor is for example realised with Pd on CdS, ZnO or TiO_2. The sensitivity to hydrogen is explained as the Pd catalyses the dissociation of H_2 into hydrogen atoms. These are in turns transported to the Pd-semiconductor interface where they are adsorbed as dipoles. The adsorbed dipoles change the barrier height of the Schottky diode and hence the I(V) characteristic, which can be measured.

7.5.1.2 CHEMFET

Chemical Field Effect Transistors are based on MOSFET transistors and are more common for chemical and biological sensing than diodes.

An n-channel MOSFET transistor consists of heavily doped source and drain areas of identical doping on a p-type substrate as shown in figure 7.37. The channel between source and drain is covered with an oxide and a metal or polysilicon gate. With a positive bias of the metal relative to the substrate, the surface of the semiconductor is inverted and a current flows between source and drain. This drain current I_D shows a sublinear behaviour with respect to the voltage U_{DS} between drain and source at a constant gate voltage U_{GS}:

$$I_D = \frac{\mu_n C_{ox}}{2} \frac{W}{L} \left(2\left(U_{GS} - U_t\right) U_{DS} - U_{DS}^2\right) \tag{7.14}$$

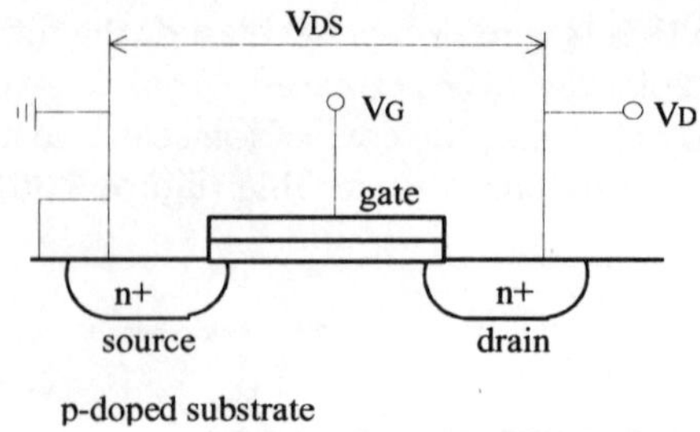

Figure 7.37: Schematic of an n-channel enhancement transistor

where $C_{ox} = \epsilon_{ox} / t_{ox}$ is the capacitance per unit area of the gate oxide, W and L are the width and length of the channel, μ_n is the average electron mobility in the channel, which is generally lower than the bulk mobility, and U_t is the threshold voltage at which the channel begins to form. The threshold voltage can be expressed as follows:

$$U_t = \phi_{ms} + 2\phi_f + \frac{Q_b}{C_{ox}} - \frac{Q_{it} + Q_f}{C_{ox}} \tag{7.15}$$

where ϕ_{ms} is the metal-silicon work function difference, ϕ_f is the Fermi level of the substrate referred to undoped material, Q_b is the depletion layer charge, Q_{it} is the charge of the interface states and Q_f is the fixed oxide charge. For $U_{DS} > U_{GS} - U_T$, I_D saturates at $I_D(\text{sat}) = \frac{\mu_n C_{ox}}{2} \frac{W}{L} (U_{GS} - U_t)^2$.

In order to allow gas detection, the gate of the MOSFET is removed completely and the gate oxide is required to be very thin of the order of a few nm. In such an ADFET, gas molecules with a permanent dipole (H_2O, HCl, CO etc.) are adsorbed at the oxide surface and affect the drain current as the fringing field of the adsorbed molecules influences the source-drain channel. These devices do however suffer from electrical noise due to their open gate construction.

An improved system of hydrogen detection is obtained when the gate of the MOSFET is formed of palladium. Hydrogen dissociates at the Pd surface and the individual H atoms diffuse through the Pd layer towards the Pd-SiO_2 interface. These atoms increase the work function ϕ_{ms} of the palladium layer and thus change the threshold voltage U_t. The change in the threshold voltage with respect to the hydrogen pressure p_{H2} can be described analytically by [50]:

$$\Delta U_t = \frac{\Delta U_{max}}{1 + k \cdot p_{H2}^{-0.5}} \tag{7.16}$$

where U_{max} is the maximum observable voltage shift and k is a constant. With this device, hydrogen concentrations of less than 1 ppm can be detected in the absence of oxygen [50].

Ions can be detected when the gate is replaced with an ionic solution and a reference electrode. A FET modified along these lines is called an Ion Selective FET (ISFET, figure 7.38). The working principle of the ISFET is based on the influence of the electrochemical potential Φ from ions at the gate oxide. This potential can be calculated as [257]:

$$\Phi = K + \frac{RT}{zF}\ln a_x \tag{7.17}$$

where K is a constant, R the gas constant, T the temperature, z the charge of the ion, F the Faraday constant and a_x the activity of the ion. In very diluted solutions, the activity a_x is identical to the concentration in the solution. The potential Φ acts like a gate voltage U_{GS} and allows the measurement of the ionic concentration or the pH of the solution.

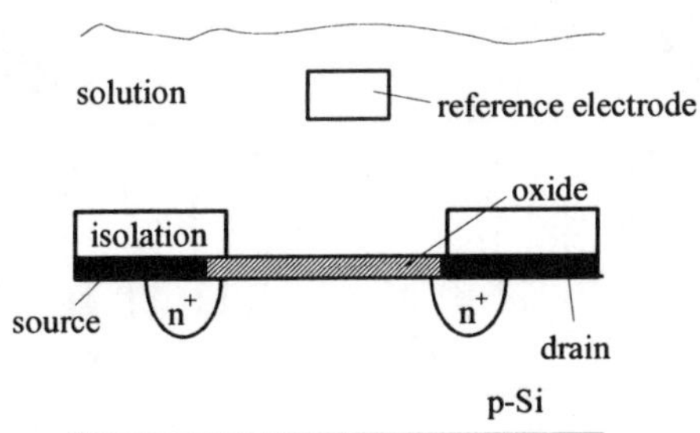

Figure 7.38: Ion Selective Field Effect Transistor

An improved version of the ISFET is the CHEMFET, where an additional layer is deposited on the gate oxide. This layer gives increased sensitivity for ions other than H^+ by permitting only specific ions through the membrane or selective reactions with the chemicals to be detected. The layer is generally spun-on or dip coated. These sensors are mainly employed in biosensors, which are a special class of chemical sensors.

One of the many examples of ISFETs has been published by Kwon [258]. The sensitivity of this sensor was around 58 mV over a pH range of 2 – 12 and a low long-term drift of 0.04 pH/day. The detection limit of CHEMFETs is usually in the ppm range as shown by Domanski [259] for hydrogen or ammonia in humid air using palladium and polyanilin as sensing layers.

7.5.2 Metal Oxide Gas Sensors

Metal oxides are semiconductors which in some cases display appreciable conductivity at elevated temperatures (200 – 500°C). The main reason for the conductivity is the presence of atomic point defects in the crystal, which can act as donors or acceptors. As an example, consider the generation of charge carriers in metal-monoxides:

One possible point defect is a metal atom M with two valence electrons at an interstitial site. The positive net charge of the metal atom deforms the periodic potential energy and the energy bands (see figure 7.39.a). The energy levels of the two valence electrons can be determined by solving the Schroedinger equation for an isolated hydrogen-like atom. The calculated levels are only slightly below the conduction band and therefore behave as donors of electrons into the conduction band [260]. The following chemical equations are therefore valid for a metal atom with two valences at an interstitial site for elevated temperatures:

$$\begin{aligned} M_i &\longrightarrow M_i^+ + e^- \text{ for temperature } \mathrm{T}_1 \\ M_i^+ &\longrightarrow M_i^{2+} + e^- \text{ for temperature } \mathrm{T}_2 > \mathrm{T}_1 \end{aligned} \tag{7.18}$$

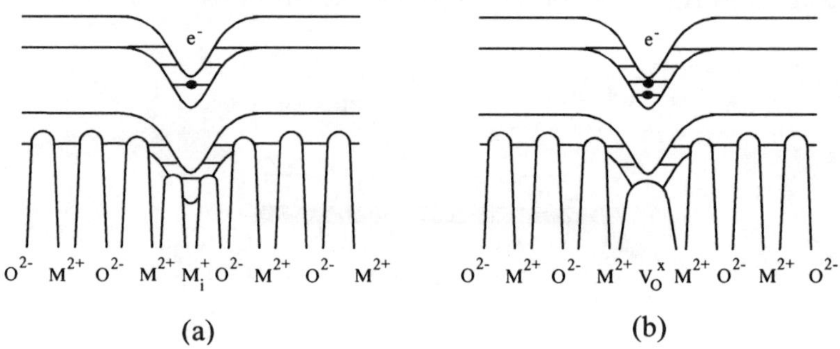

Figure 7.39: Deformation of the energy bands for metal atoms on interstitial sites (a) and oxygen voids (b) [260]

A second point defect which can cause the generation of charges is an oxygen void V_O. The oxygen void is charged positively as the oxygen atoms in the crystal normally possess a twofold negative charge. This also bends the periodic energy bands of the crystal (see figure 7.39.b), trapping valence electrons of neighbouring metal atoms. The energy levels of the trapped electrons are slightly below the conduction band and so act as donor electrons. At elevated temperatures one electron is donated from the oxygen void and the second electron is lifted into the conduction band for even higher temperatures:

$$\begin{aligned} V_O &\longrightarrow V_O^+ + e^- \text{ for } \mathrm{T}_1 \\ V_O^+ &\longrightarrow V_O^{2+} + e^- \text{ for } \mathrm{T}_2 > \mathrm{T}_1 \end{aligned} \tag{7.19}$$

The third possible point defect which can contribute to the conductivity of metal oxides is a metal void V_M. In a manner complementary to that described above, a metal void acts as an acceptor (see figure 7.40) and can give rise to one or two holes in the valence band, depending on the temperature.

$$\begin{aligned} V_M &\longrightarrow V_M^- + h^+ \text{ for temperature } \mathrm{T}_1 \\ V_M^- &\longrightarrow V_M^{2-} + h^+ \text{ for temperature } \mathrm{T}_2 > \mathrm{T}_1 \end{aligned} \tag{7.20}$$

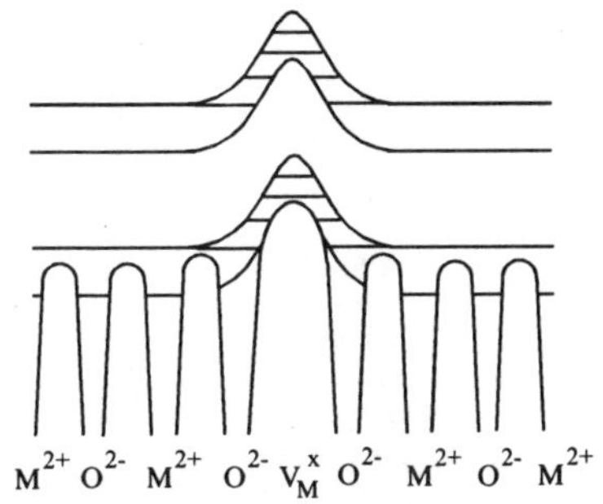

Figure 7.40: Deformation of the energy bands for metal voids [260]

Depending on the numbers of the atomic point defects the metal oxide is an n or p-type semiconductor. This can be influenced by the fabrication process but most commercial bulk metal oxides (ZnO, SnO_2) are n-type as fabricated due to a lack of oxygen.

The interaction between metal oxides and gases depends on the particular gas. Oxygen is a common component in many gases for analysis and will therefore be discussed in more detail. There is always equilibrium between the oxygen in the atmosphere and the oxygen voids in the crystal:

$$\begin{array}{c} O_2(\text{gas}) + 2V_O^+ + 2e^- \longrightarrow 2O_L \text{ for temperature } T_1 \\ O_2(\text{gas}) + 2V_O^{2+} + 4e^- \longrightarrow 2O_L \text{ for temperature } T_2 > T_1 \end{array} \tag{7.21}$$

In this way oxygen from the air (O_2(gas)) reacts at an oxygen void. This interaction takes place at different locations within the crystal and depends on the temperature. At low temperatures physisorption occurs and the oxygen forms Van-der-Waals bonds with the metal oxide (figure 7.41). At higher temperatures ionic and atomic bonds are formed at the surface and are the basis of metal oxide gas sensors. At even higher temperatures, the interaction also takes place in the volume of the metal oxide. As this interaction leads to a decrease in the number of "free" electrons in the crystal, a sensor for the partial pressure of oxygen is formed. Other oxidising gas components interact with the metal oxide crystal in the same way as oxygen. Both oxygen and oxidising components act therefore as an acceptor and change the conductivity of the crystal.

Reducing gases (RH_2) also lead to a change in the conductivity of the metal oxide by reaction with the oxygen ions at the surface of the crystal. As the oxygen ions are removed from the crystal, the valence electrons of the metal ions which are bound to the oxygen are released and injected into the conduction band:

$$RH_2 + 2O^- \longrightarrow RO + H_2O + 2e^- \tag{7.22}$$

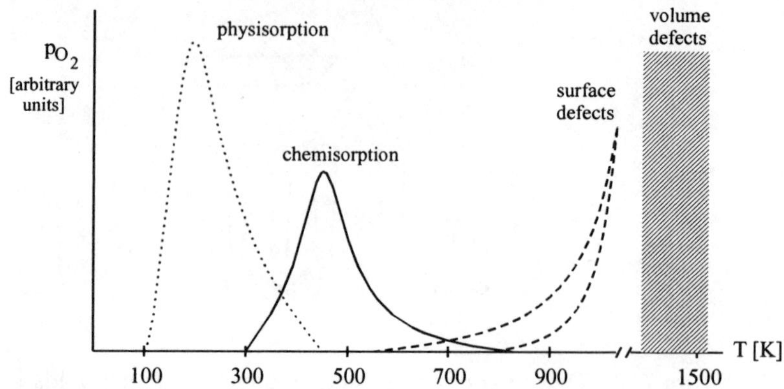

Figure 7.41: Characteristic adsorption with respect to temperature [17]

This leads to an increase of the conductivity at the surface of the metal oxide. As the maximum chemisorption (ionic and atomic bonds of the components with the metal oxide) is at different temperatures for most components (oxidising and reducing), a limited selectivity between these components can be achieved by varying the temperature. Also most reactions only alter the conductivity of the surface of the metal oxide, requiring a large surface to volume ratio for any sensor.

Research to date on metal oxide gas-sensors is very extensive especially for applications in automotive engineering. Two substrate materials are very common: alumina ceramics [261] and silicon nitride membranes [262]. Simple fabrication processes like thick-film printing or sputter deposition of feature sizes of 20 μm and larger establish a simple fabrication process for alumina substrates. Sputtering can also be employed to deposit metal oxides onto silicon nitride membranes which gives enhanced resolution compared with alumina. For example in the metal oxide sensor described by Stoev [263], a resistive heater is deposited on top of a thin dielectric membrane (figure 7.42). Above the heater, a dielectric layer and interdigited electrodes are formed. The electrodes are covered by a metal oxide layer (eg. ZnO, SnO_2 etc.). At elevated temperatures the resistance of the metal oxide between the interdigitated electrodes is measured and the amount of gas can be determined.

Recently, a ZnO based gas-sensor array has been reported by Althainz [264] with a detection limit of well below 1 ppm for most organic vapours. Although the sensor response is rather fast, the computing time with pattern recognition is about 100 s.

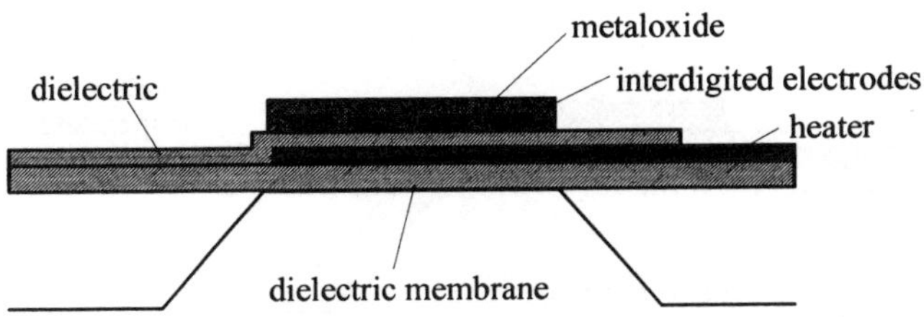

Figure 7.42: Schematic of a metal oxide gas sensor

7.5.3 Catalytic Gas-sensors

Catalytic gas-sensors [265] or pellistors detect gases in an indirect way. The operating principle is to measure the exothermic heat of a reaction at the surface of the sensor. The generated heating power of an exothermic chemical reaction, which changes the kinetic energy of the components from W_1 into W_2, can be described as follows [17]:

$$P_{heat} = \frac{\partial n}{\partial t} \cdot (W_2 - W_1) \tag{7.23}$$

where $\partial n/\partial t$ is the reaction rate. This leads to the heating of the sensor system by an amount proportional to the thermal resistance R_{th} between it and its surroundings: $\Delta T = P_{heat} \cdot R_{th}$. The concentration of chemical substances can then be determined from the rise in temperature.

The reaction is usually an oxidation of the gas of interest and is conducted at elevated temperatures using a catalyst. Both the catalyst and the specific temperature allow some selectivity of the sensor against other gas components. The reaction heat can be determined by a measurement of the temperature of the sensor surface, as this surface is heated by the reaction. Pellistors are usually operated at around 500°C due to the low reactivity of gases at room temperature. Catalysts are necessary to avoid even higher operating temperatures.

In the designs by Krebs [265] a metallic heater and a heat sensor are deposited on a thermally insulating, dielectric membrane. The dielectric layers need to be of low-stress to avoid buckling at room temperature or breaking at operational temperatures. The catalyst is deposited on top of another dielectric layer. A schematic of the device is depicted in figure 7.43.

The sensitivity of the device described by Krebs [265] for example is 13 mV per % of methane in air. A new catalytic gas-sensor has been presented recently [266] providing sensitivities to methane and butane gases of 2.2 mV and 8.9 mV for 1000 ppm, respectively. The applied power was in the range 350 to 400 mW, being much smaller than that needed for conventional pellistors.

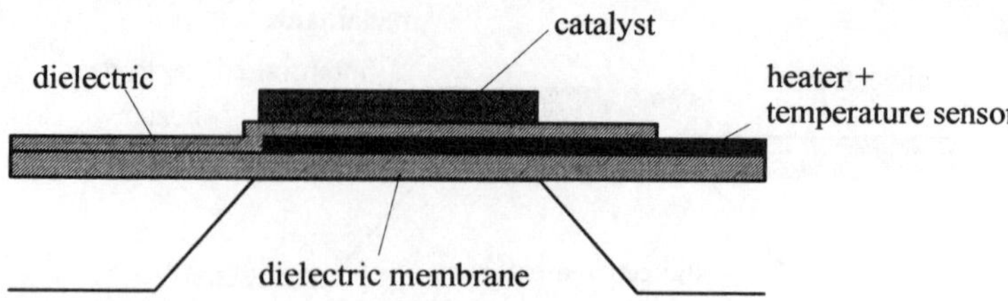

Figure 7.43: Catalytic gas sensor

7.5.4 Optical Sensing Techniques for Chemicals

The last chemical sensing technique described here is based on optical detection. This can be done by intensity absorption of an incident light beam as described by Effenhauser [248] or others [267, 268], or alternatively a monolithic Mach-Zehnder interferometer can be employed [269].

- Absorption: Silicon sensors based on optical absorption are only partly micromachined. This means that the device containing the absorbing component is micromachined and the optical detection is done with optical fibres and photodetectors.

  Monochromatic light travels through a volume which contains the analyte and a background fluid. The background fluid is required to show negligible absorption at this wavelength. Light of this wavelength is absorbed by the molecules of the analyte, which weakens the incoming light beam. This absorption is proportional to the concentration of dissolved molecules. In order to allow good sensitivity, the optical pathlength for the absorption has to be rather large.

  One example of such a sensing device is described by Effenhauser [248]. In this set-up, the 488 nm line of an argon ion laser is coupled into an optical fibre and from there with a 45 degree angle onto the glass covered flow channels. A fluorescence detector is mounted perpendicular to the silicon surface. The detector is used at the end of a separation channel and a detection volume of 15 pL allows plate numbers of up to 160000.

  A fully micromachined absorption sensor has also been presented recently [270]. Both an optical fibre and a mesa-photodiode have been integrated on the silicon wafer. The lateral optical fibre is interrupted perpendicularly by a flow channel. There the beam is coupled into the absorbing fluid before it is guided back into the second part of the waveguide leading to the photodiode. The device has been fabricated, but no measurements have been presented so far.

- Mach-Zehnder interferometer: Chemical sensors of this type usually have the optical fibre integrated on silicon. The optical fibre which guides a laser beam splits into two waveguides, one of which uses water as a cladding in the detection zone. Proteins dissolved in the water increase the refractive index and therefore change the speed of the light travelling through the waveguide. The phase shift produced when compared with the reference arm creates interference at the output port and hence detection of the chemical species. By using the coupler structure shown in figure 7.44, the detection limit can be reduced down to 0.1 ppb [269] for the detection of the herbicide simazine.

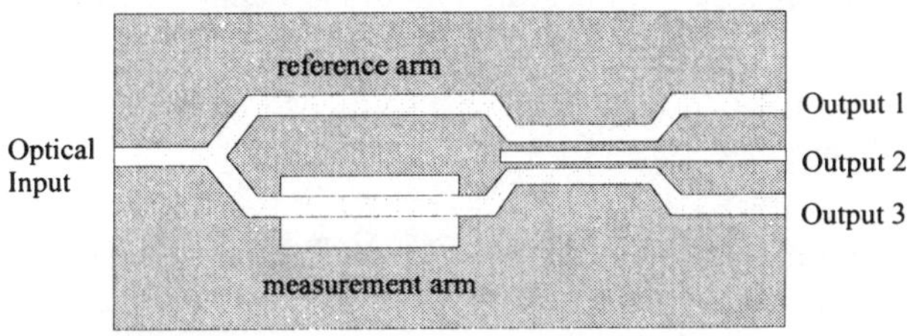

Figure 7.44: Principle of a coupler structure for a chemical sensor with optical waveguides

7.5.5 Biosensors

Biosensors are obtained by introducing a biological sensing element, such as an antibody, enzyme or cell in close contact with a physical-chemical transducer [271]. Because the molecular recognition of biological species can be made very specific, biosensors show high selectivity as compared to chemical sensors. However, in order to achieve a good sensitivity, a highly efficient coupling between biological element and transducer element is needed. Biologically sensitive materials are: enzymes, organelles, membrane components, bacterial cells or whole cells, antibody or antigen or whole slices of mammalian or plant tissue [271]. An overview of the transducing effects used for biosensors is given below:

- Potentiometric Biosensors [272]: An enzyme is placed on top of the gate oxide layer of an ISFET (figure 7.45.a). This enzyme catalyses a reaction of biological matter with ionic products. The ionic concentration can be determined by the ISFET and thus the concentration of the biological substance is measured. The detection of urease in blood is an example of one of the many applications of ISFET devices. The sensor uses the electrolysis of hydrogen peroxide, one of the by-products of the oxidation of glucose, and a Pt reference electrode [273], which gives a high sensitivity of 9 mV mM^{-1}.
- Conductometric Biosensor [272]: Chemical reactions produce or consume ionic species and thereby alter the overall conductivity of a solu-

tion. For example, interdigitated electrodes covered with an immobilised enzyme can be used to detect a species like urea in a solution. This species (urea) contacts with the enzyme (urease) and changes the conductivity of the solution. As the enzyme is permeable for ionic species, the conductivity change allows selective detection at the electrodes. Another example is a sensor for glucose based on a flow-injection conductometric manifold equipped with an immobilised glucose oxidase minireactor [274]. The dynamic range of the sensor is from the tens of μM to the order of mM and stability could be achieved for at least several months.

The oxidation of glucose with the glucose oxidase (GOD) enzyme is described by:

$$\begin{aligned} &\text{glucose} + O_2 + \text{GOD} \longrightarrow \text{gluconolactone} + H_2O_2 \\ &H_2O_2 \longrightarrow 2H^+ + O_2 + 2e^- \end{aligned} \tag{7.24}$$

and the hydrogen ions can be detected by the conductometric sensor.

- Amperometric Biosensor [272]: An electrically inactive species reacts with the help of enzymes to form electroactive products which can be detected electrochemically. This method is highly specific and allows high selectivity of the sensor. A constant potential between working and counter electrode is used, and the resultant steady state or limiting current is measured. The current is a result of an electrode reaction taking place only at the applied voltage. The current is proportional to the concentration of the electroactive species in the surrounding liquid. One possible application of this sensing type is the detection of glucose [272]. The electrodes of such a device can be covered with immobilised glucose oxidase (GOD) in an enzyme-polymer membrane for detection. The detection reaction is the same as that described above for the conductometric sensor. The device by Sethi [272] employs this technology and allows glucose detection in the mM concentration range.

- Surface-Acoustic-Wave Biosensor [272] (figure 7.45.b): Surface acoustic wave propagation properties are dependent on the nature of the surface. In a biosensor, an antibody or antigen is immobilised at the surface. When a bio-molecule couples with the antibody or antigen, the characteristics (damping and velocity) of surface-acoustic waves are changed, allowing the biological substance to be measured. An example is the determination of organophosphorus pesticide based on the principle of enzyme coupling [275]: The calibration of the sensor was done by correlating the enzyme activity with various concentrations of pesticide compound in a buffer solution. Here the detection limit of the sensor is approximately 81 ng ml^{-1}.

- Optical Biosensor [272] (figure 7.45.c): Molecular absorption on an optical fibre changes the refractive index and thus the velocity within the waveguide. If a reference beam is used, interferometry can be used for biological sensing as described above in the chemical section. An example of a device with integrated Mach-Zehnder interferometer has been presented recently [276]. The sensor design incorporates a three-waveguide coupler structure as depicted in figure 7.44 which gives advantages in terms of signal referencing and in establishing and maintaining a sensitive operating point. The detection limit for protein loading has been estimated at 5 pg mm^{-2}.

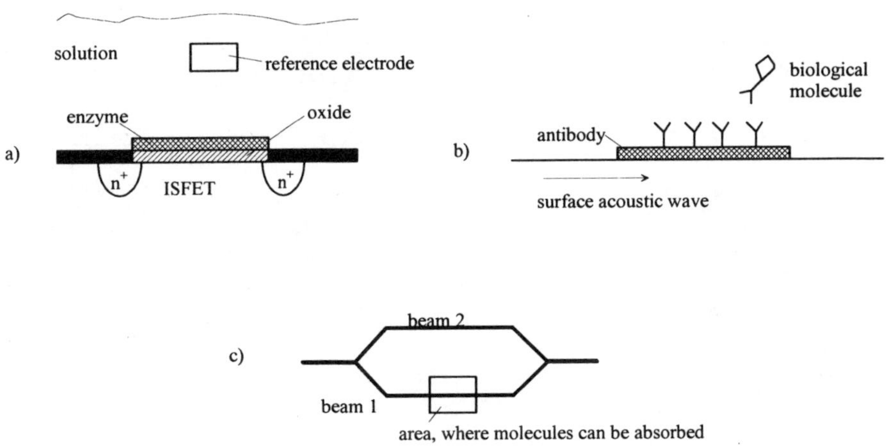

Figure 7.45: Biosensors : a) modified ISFET, b) surface-acoustic-wave sensor, c) optical sensor

All the biosensors described above are chemical sensors with a transducing element. The most common sensing element is an immobilised enzyme. In order to deposit this enzyme on top of the transducing element several approaches can be taken [272]:

- entrapment in a polymer matrix (photosensitive polyvinilpyrodoline or photostructurable membrane)
- physical adsorption at a solid interface (no long term stability guaranteed)
- crosslinking with bifunctional agents (links proteins with each other and with the support)

- covalent bindings to the support (bonding by using nucleophilic functional amino acid groups that are not concerned with the function and activity of the biorecognition itself).

CHAPTER 8

Integrated Microfluidic Systems

In this last chapter we deal with microfluidic systems. System integration is one of the most important topics in microfluidics, as the main applications lie in the medical, biological and chemical fields where it is desirable to integrate actuation and sensing elements on one device. The lesson to be learned from microelectronics is that mass production of systems is the cheapest way to produce devices. The hope therefore is that the integration of microfluidics will lead to cheap and disposable dosing systems, implantable drug delivery systems, blood gas monitoring systems, chemical analysis systems, bioreactors and genetic diagnostics systems. Examples of all of these systems will be presented here after the basic technology for integration has been described.

8.1 Microsystems Technology

By microsystems technology we mean the integration of electronics together with micromechanical devices, as is currently used in pressure sensors and accelerometers. In this section we first discuss the advantages and disadvantages of microsystem integration and then give some examples of integrated devices.

8.1.1 Advantages of Integration

Let us first consider a sensing system. A general system might contain one or more transducers that convert mechanical, optical or electrical signals from the outside world into electrical signals. If as is usually the case the signal is analogue in nature, it may be amplified, filtered, and even converted to digital form using an analogue-to-digital converter (ADC). When several sensors are connected to a computer, the individual digital signals are multiplexed and sent to the internal bus of the computer. From there, the signals are stored, processed or outputted. A sketch of such a sensing system can be found in figure 8.1.

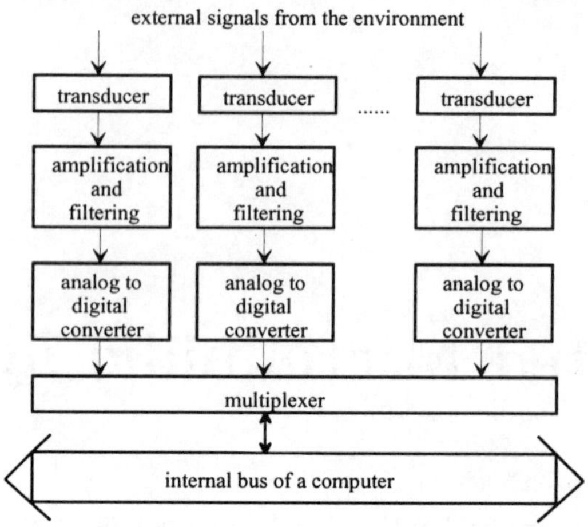

Figure 8.1: Schematic of a micromechanical, microoptical or microfluidic sensing system

Integration of electronics with microsystems can improve the performance of the system. For example, the integration of a transducing element and a sensing circuit may reduce the noise substantially. This may be because of the short distance between the sensor and the electronics, which eliminates or reduces pick-up, or may be because of the reduced interface capacitance. Such techniques are essential when the signal level is small compared to the noise.

Integration is not always appropriate if the signal to noise ratio is reasonable. It is sometimes difficult and expensive to integrate circuit and sensor technologies. Therefore this should only be done either where there is no alternative or when the process development expenses can be recovered by the sale of a marketable product.

8.1.2 Integration of Micromechanics and Microelectronics

Integration of electronics with micromechanical devices is now quite advanced in the area of pressure sensors and accelerometers. Pressure sensors play a major role in microfluidics because the pressure in fluids and gases has to be monitored in many applications.

Electronics can be integrated with any of the pressure sensors described in section 6.5.2 (capacitive [210], piezoresistive [1] and resonant [215]). As an example, we look at a piezoresistive pressure sensor built in 1.0 μm CMOS technology [1]. A cross section of the sensor can be found in figure 8.2.

The first process steps are to form the piezoresistive sensing element and

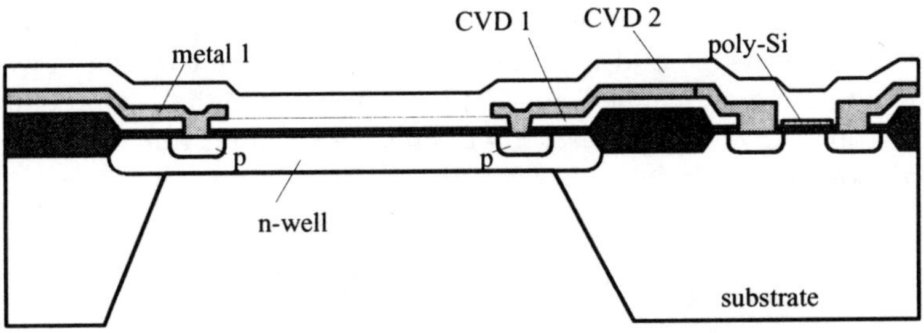

Figure 8.2: Cross section structure of the silicon membrane for pressure sensors [1]

the circuitry on the front-side of the wafer by using a standard CMOS process flow. The piezoresistors are obtained from the standard PMOS transistor source and drain doping step. Then the back side of the wafer is covered with silicon nitride and double sided lithography is used to pattern the back, aligned to the sensing elements on the front. Finally, the back is etched with TMAH (tetra methyl ammonium hydroxide), while the front side is protected by sealing it with an o-ring against a cover plate. The etch stops at the n-well of a pn-junction by using the electrochemical method. Thus, a membrane thickness of 5 μm is obtained.

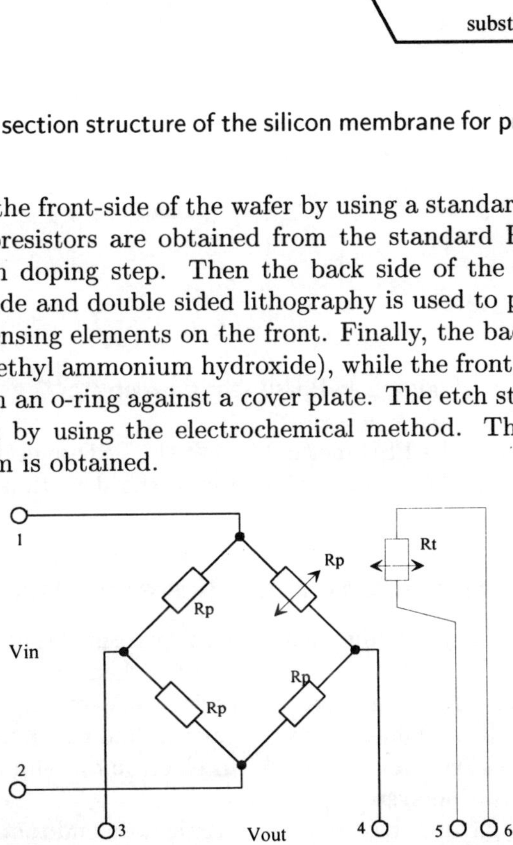

Figure 8.3: Wheatstone bridge and temperature compensation of a silicon pressure sensor

The use of TMAH for silicon etching avoids any potassium contamination of the CMOS transistors and it is found that the electronic circuits work perfectly. The circuit layout of a simple piezoresistive pressure sensor consists generally of a resistive Wheatstone bridge and an external resistor for temperature compensation (figure 8.3).

The detection of the resistance changes can be made with a simple differential amplifier circuit or with a more complicated phase sensitive detector

(PSD) system as shown in figure 8.4. With this, a signal is fed into both the Wheatstone bridge and a phase shifter. The output of the Wheatstone bridge is amplified, multiplied by the phase shifted original signal and finally passed through a low pass filter that removes medium and high frequency noise components.

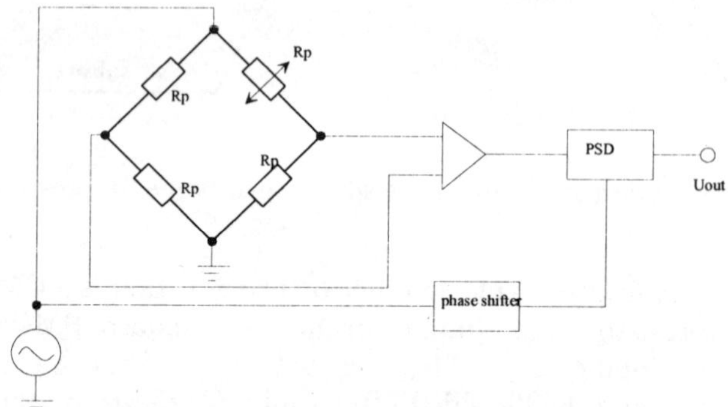

Figure 8.4: Phase sensitive detector

The advantage of the PSD method is that the PSD rejects drift, offset and amplifier 1/f noise [277], and therefore allows the detection of small output signals.

8.1.3 Integration of Microfluidics and Microelectronics

The integration of microfluidics with microelectronics is far less advanced, but simple integrated systems are available for chemical and biological systems. In this case individual chemical sensing elements are often integrated with electronics. More sophisticated systems containing more than one sensor can be used as an electronic tongue or an electronic nose (sometimes called olfactometric measuring systems).

The most complicated part of an electronic nose and tongue system is the data handling. For this, pattern-recognition or neural networks are necessary to distinguish different chemicals. In the food market for example, sensors are necessary for organic acids, alcohols, amines, sulphides and carbonyls to be able to classify good and bad quality [278].

One example of an electronic nose has been developed and described by Hartinger [279]. This system is still a hybrid system as the individual components are not integrated. The device can be split into three levels: the physical-chemical level, the electronic level and the information level. The physical-chemical level comprises the sample handling and the sensors. These are based on semiconductor gas sensors and are operated at temperatures between 200 and 300°C. Because semiconductor gas sensors have different gas sensitivity at different temperatures, an array of these sensors gives information not only on one specific gas but on a whole series of different gases. The electronic level is responsible for controlling the temperature at each sensor (±0.1°C) and for measurement of capacitive or resistive changes. The electronic level also has to convert analog signals into digital signals. These signals are then analysed by a computer which forms the third level. A set-up of 8 different metal-oxide semiconductor gas sensors has been successfully applied to distinguish between hazelnuts roasted for various times at 180°C.

Another interesting device has been presented by Gumbrecht [280]. This device includes several sensors for liquids and is integrated on a single silicon chip. The whole system was developed for analysing blood gases (pH=-log[H^+], pO_2=-log[O_2] and pCO_2=-log[CO_2]) extracorporeally. It includes 5 ISFETs with impedance converters fabricated with a 5μm aluminium-gate MOS process. The silicon chip contains a heater, a temperature sensor, an amperometric transducer, a reference electrode, an indirect potentiometric transducer, a direct potentiometric transducer, a pseudo reference electrode and an impedance converter. The complete system showed satisfactory performance in many evaluation studies [281]. However, due to limited lifetime of the sensors, market acceptance will only be achieved if the sensing system is a cheap mass product. Therefore it is important to transfer the sensor design to industrial CMOS lines.

An important application of electronic noses or tongues is in the food processing market and Natale [278] describes a case study in which an electronic nose was compared with a human nose. The sensor array of this electronic nose consisted of eight quartz resonant chemical sensors coated with metalloporphyrines. The analysis was carried out with tomato pastes of different quality (good to bad in three different categories). The electronic nose was able to successfully distinguish all categories. This shows its suitability for classifying food quality.

Another electronic tongue has been described by Savoy [282]. This sensing system is a hybrid device consisting of a silicon carrier, polystyrene microspheres and a CCD camera (see figure 8.5). The polystyrene spheres are fixed in micromachined cavities and coated with different indicator molecules that change colour or fluoresce on contact with certain species. This change in colour is detected optically with the CCD camera. The advantage of using microspheres is the simple technology of just placing the spheres into their cavity. In order to allow the light to pass through the system, the cavities are sealed with transparent membranes. A first prototype system recognised pH, Ca^{+2}, Ca^{+3} and sugar successfully.

The electronic tongue described by Savoy is based on combinatorial chemistry, where a number of different receptors (molecules that perform any physical or chemical change when they come into contact with certain chemicals) have to be found. These receptors are permanently attached to the microspheres. None of the receptors needs to react exclusively to a specific analyte

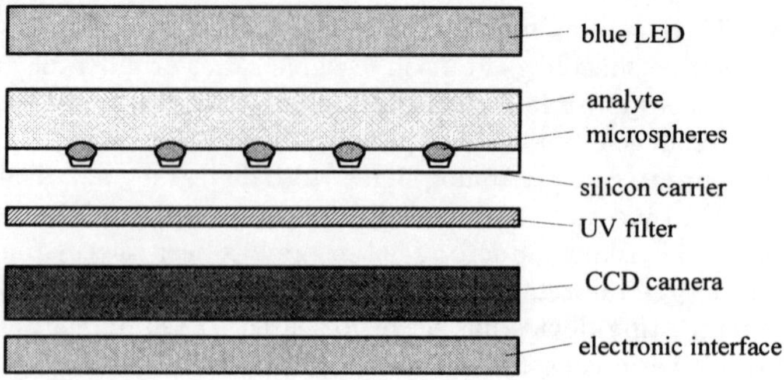

Figure 8.5: Electronic tongue with colorimetric and fluorescent detection. The whole device consists of 10x10 cavities, each measuring several hundred μm. The microspheres placed in these cavities are typically 50 - 100 μm in diameter and coated with different receptors

as pattern recognition is used for detection. For example, some of the receptors change their colour in contact with certain chemicals. Depending on the combination of receptors which change colour, it is possible to identify the correct chemical.

8.2 Connection Systems for Microfluidic Devices

One of the most important parts of microfluidic systems is the fluid-interconnection of the different devices. This can be done macroscopically with ordinary tubing or with micromachined channels. Ordinary tubing has the disadvantage of a large dead volume, so reducing some of the advantages of microsystems. Therefore, it is necessary to develop techniques that allow connection of different microfluidic devices on a micron scale. Some of these interconnection systems are described in this section.

8.2.1 Vertical System Integration

The first way of integrating microfluidic devices is by vertical integration [137]. Here the different components are mounted on top of each other as shown in figure 8.6. The whole system consists of pumps and a chemical sensor and forms a simple chemical analysis system. A leakfree interconnection is obtained by the use of small 0-rings or deposited sealing rings between the different modules. Finally, the whole stack is clamped together. A great advantage of this approach is its simplicity. However, as with most other approaches, the presence of gas bubbles or particles creates major problems.

The system consists of 5 different layers which are made of silicon or Pyrex. The module size for each layer is 22 × 22 mm^2. The through holes measure 1 × 1 mm^2 and the flow channels on the passive glass plate are 600 μm wide and 200 μm deep. The pumps are driven with a surface mounted piezoelectric actuator that is glued on top of the membrane. The detectors used in the first prototype were ISFETs [283] and the system has successfully detected changes in the pH value of different calibration solutions.

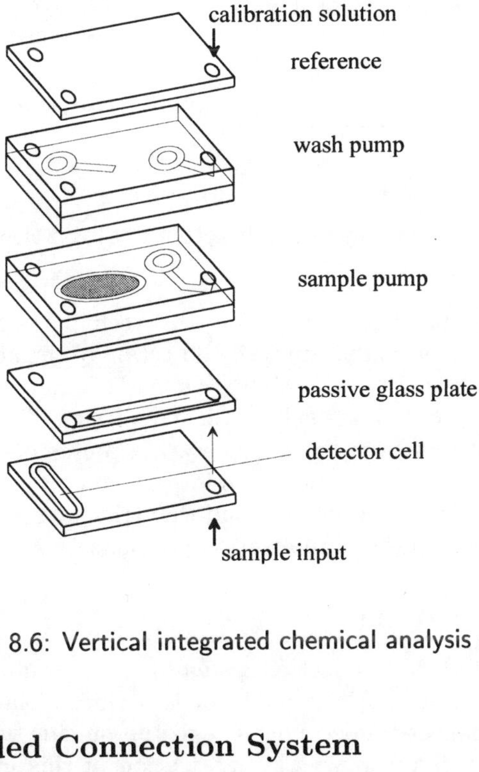

Figure 8.6: Vertical integrated chemical analysis system

8.2.2 Moulded Connection System

The next system that has been described in the literature is a so called mixed circuit board (MCB) made of plastics (figure 8.7 [284]). This system works like a printed circuit board (PCB). It consists of a PCB top layer, a channel layer and a PCB bottom plate. The top layer not only holds the individual microfluidic devices but can also contain the electronic circuit, supplying the different devices. The demonstrator consisted of an epoxy printed circuit board attached to a transparent polycarbonate substrate. The individual components were glued onto the PCB.

A major advantage of horizontal integration is the possibility to use high volume 'pick and place'-production processes. Although bubbles still can occur in such a system, it has the advantage that one can trace particles or bubbles because the back plane is transparent.

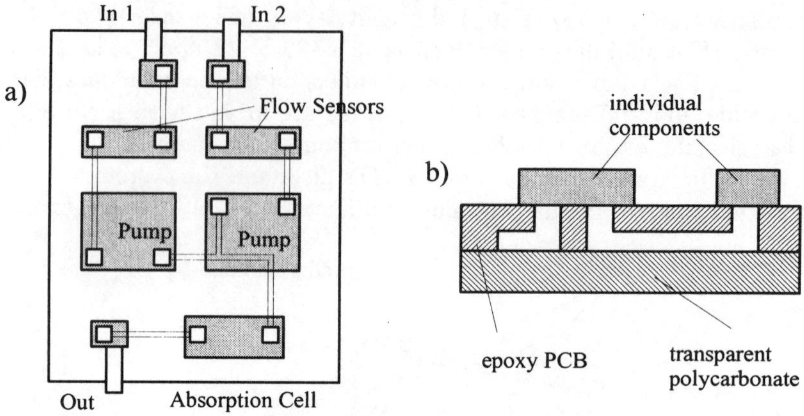

Figure 8.7: Plastic PCB for microfluidic systems: a) top view, b) cross section

A demonstrator has been built using this technology for the purpose of measuring chemical reaction products by detection of the absorption of an incident light beam [285]. Here the MCB was made from epoxy PCB attached to a transparent polycarbonate substrate. The design allowed the mixing of sample and reagent liquids in the appropriate amounts on-board, and detection was done by optical absorption, also integrated on the MCB. The system worked with dyes of different colours, and the light intensity passing through the absorption unit identified the reaction products.

8.2.3 Silicon Tubing

Another approach for interconnection systems has been published by Gonzalez [286]. This "Lego(TM)-like" assembly of two silicon plates provides both breadboard interconnects and connectors to the outside world. The first of these is depicted in figure 8.8. The fabrication of this interconnect starts by etching the fluidic via. Then, the fluidic channels are etched with KOH. Both silicon wafers are now anodically bonded to Pyrex substrates in order to cap the flow channels. Next a UV curable polysiloxane is spun on the lower silicon-glass stack, exposed in the area of the compression seal (0-ring) and developed. Finally, a wafer saw is used to cut the interlocking fins into the surface of the silicon wafers. The groove period needs to be twice the width of the saw cut so that upper and lower silicon chips interlock. The locking force is mainly determined by the friction of the two surfaces in contact and the spring constant of the fins themselves.

In order to connect a microfluidic system to the outside world, a silicon interface has been described also [286]. The interface consists of two silicon wafers that are fusion bonded together (see figure 8.9). Prior to bonding,

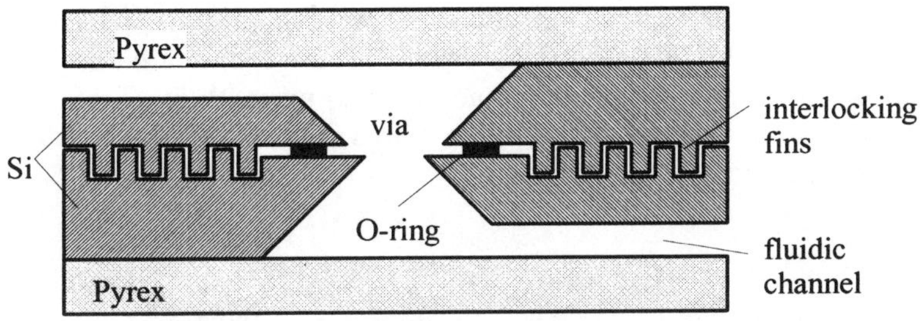

Figure 8.8: Cross sectional view of a fluidic interconnect with a compression seal

fluidic channels can be etched. These fluidic channels can also be used in conjunction with the interconnection system described previously. In this case a via has to be etched at the same time as the tube connector and the interlocking fins have to be sawn on top of the upper wafer. A whole microfluidic breadboard has been achieved on a silicon wafer by combining these two technologies. The interconnect has also been tested for leakage. No leaks were visible and pressures up to 137 kPa were possible without damage.

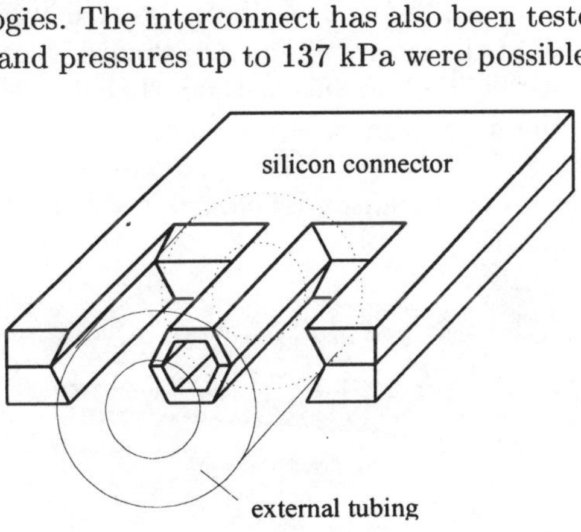

Figure 8.9: Cross sectional view of a tube connector which allows the connection to the outside world

8.2.4 Pyrex-Silicon Microfluidic Circuit Board

A system based on a microfluidic circuit board consisting of Pyrex and silicon has been described by the present authors [287]. The system features a micromachined chemical reaction system and is based on the integration of two pumps, two flowsensors and a micromixer. The interconnection of the discrete microfluidic devices is realised with a microfluidic circuitboard (figure 8.10). The devices for the reaction system are simply mounted on top of the microfluidic circuitboard using the anodic bonding method. This allows the creation of a whole system on a single chip without the use of adhesive or macroscopic interconnections.

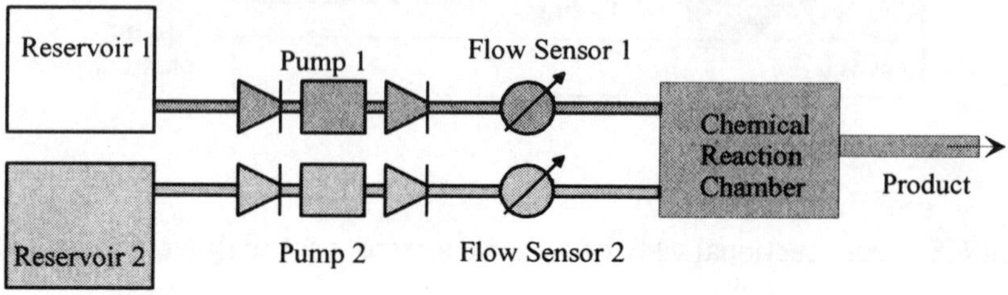

Figure 8.10: Principle of a chemical reaction system on demand

The principle of the microfluidic circuitboard is shown in figure 8.11. The circuitboard consists of a Pyrex wafer in which channels and connection holes are etched and an anodically bonded silicon bottom wafer is used to seal the channels. On top of the microfluidic circuitboard discrete devices can be mounted by using the anodic bonding method. The discrete devices are thus connected without the need of macroscopic pipework. With the microfluidic circuitboard it is possible to create whole systems on a single chip. Different microfluidic systems may be realised by changing the mask layout.

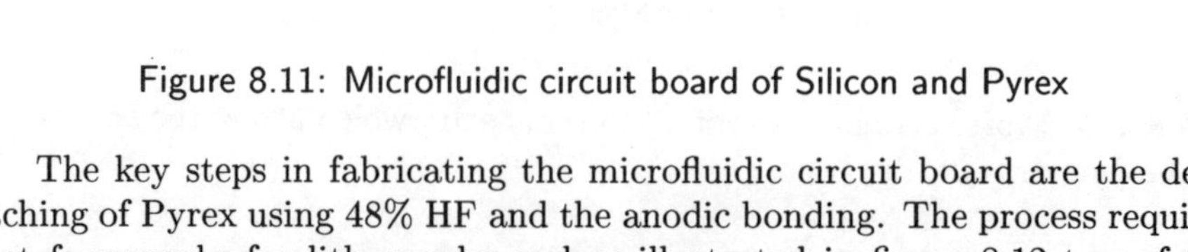

Figure 8.11: Microfluidic circuit board of Silicon and Pyrex

The key steps in fabricating the microfluidic circuit board are the deep etching of Pyrex using 48% HF and the anodic bonding. The process requires just four masks for lithography and as illustrated in figure 8.12 two of the steps require only coarse alignment.

The micromachining of the Pyrex wafer can be divided into two major parts. The first part is creating the preconditions for double sided alignment. Cr/Au-layers (40nm and 400 nm) are evaporated on both sides of the Pyrex wafer and patterned using photolithography so that front side alignment marks can be etched. Then the wafer is exposed to 48% HF for 90 seconds to create the 12 μm-deep alignment marks. The Cr/Au-layers are then stripped off and thicker second layers of Cr/Au (40nm and 600 nm) are evaporated onto the Pyrex. The back side Cr/Au layer is now opened with two big square holes to expose the alignment marks on the front (figure 8.12.a).

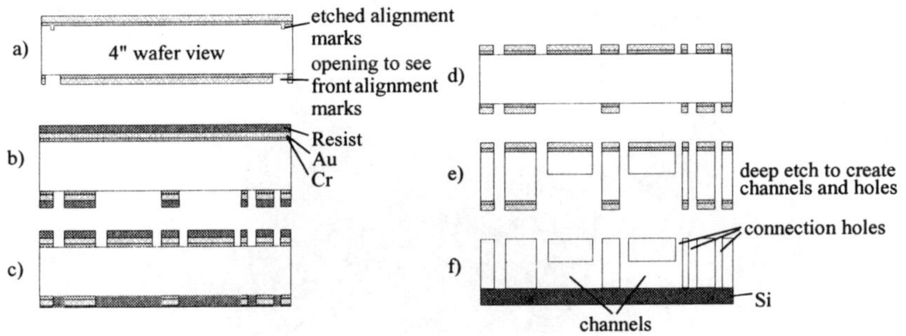

Figure 8.12: Fabrication steps for a microfluidic circuit board

The same layers are now used for the second part of the fabrication process, which is the creation of the channels and holes into the microfluidic circuit board. The back contains the structures for the channels and holes. On the front side are the structures for the connection holes to the different microfluidic devices, which are placed on top of the circuit board in a final step. The back is aligned through the previously opened windows in the Cr/Au-layer to the etched marks on the front. Then resist, gold and chrome are patterned while the front Cr/Au-layer is protected with resist (figure 8.12.b). Then the front mask is aligned to the etched marks on the front. The back is protected during the front side etch of the gold and chrome (figure 8.12.c). After that the resist is stripped.

Now, after both sides of the Cr/Au-layer on the Pyrex are structured (figure 8.12.d), the wafer is exposed to 48% HF at room temperature for 32 minutes to etch from both sides equally into the 500 μm-thick Pyrex (figure 8.12.e). The Pyrex machining is finished by stripping the gold and chrome from both sides. To complete the microfluidic circuit board the channels are sealed by bonding a silicon base wafer at 400 V and 390°C for 15 minutes (figure 8.12.f).

Finally, the individual devices are anodically bonded onto the top of the Pyrex as seen in figures 8.11 and 8.13. The size of the microfluidic circuitboard as shown in figure 8.10 is 3.5 cm × 3.0 cm × 0.1 cm.

Initial tests of the reaction system were made by mixing two inks, red and green. Before use the inks were diluted with ethanol (5 parts ink, 5 parts ethanol). Pressure was applied by controlling the heights of the liquid columns at the inlets, but should also be possible with the micropumps. Neither the fluidic interconnects of the microfluidic circuitboard, nor the anodically bonded joints of the discrete devices to the Pyrex plate, showed any leakage. This technology for a microfluidic circuitboard is therefore viable and applicable to mass fabrication of chemical reaction systems (figure 8.13).

Figure 8.13: Photograph of a fabricated chemical reaction system in comparison to a British penny

8.3 Applications of Integrated Microfluidic Devices

The most interesting topic in microfluidic systems is the integration of individual devices. In general, microfluidic systems have great potential to be used in fields like chemistry, environmental monitoring or medicine. This section will therefore cover many different systems from the application point of view.

8.3.1 Micro Dosing Systems and Drug Delivery

The first microfluidic systems to be described were micromechanical dosing systems [35,288]. A dosing system consists of a micropump and an associated flowsensor, and can be used for example in chemical analysis or drug delivery. One of the first systems is based on a piezoelectrically actuated pump and the flowsensor is of the differential pressure type [130]. Figure 8.14 shows the basic layout of this device.

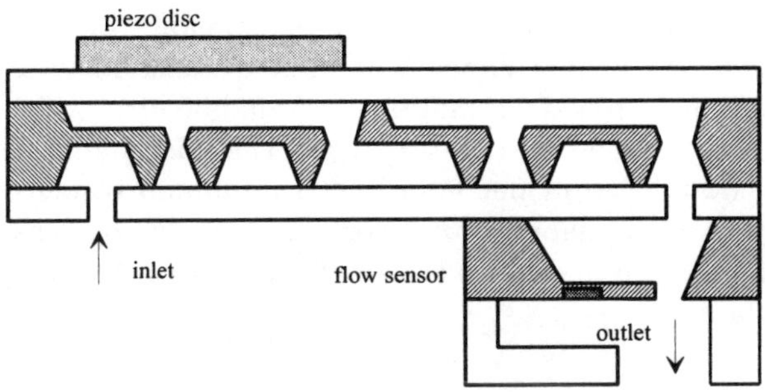

Figure 8.14: Dosing system with piezoelectric pump and flowsensor [288]

This system maintains constant output flow rates of 10 to 100 μl min^{-1} below 3500 Pa. The response time for the system was quite fast, of the order of 200 msec. The fabrication of the device utilises anodic bonding of silicon and glass. Holes are drilled into the main glass board and then the pump and flowsensor are bonded anodically from both sides.

The second system is based on thermopneumatic actuation of the pump and a thermal flowsensor [35]. This combination allows higher integration as both devices operate on thermal effects. The fabrication is based on etching of valves, membranes and bridges and requires only one silicon wafer. Glass wafers are bonded onto both sides of the silicon wafer. The system is depicted in figure 8.15.

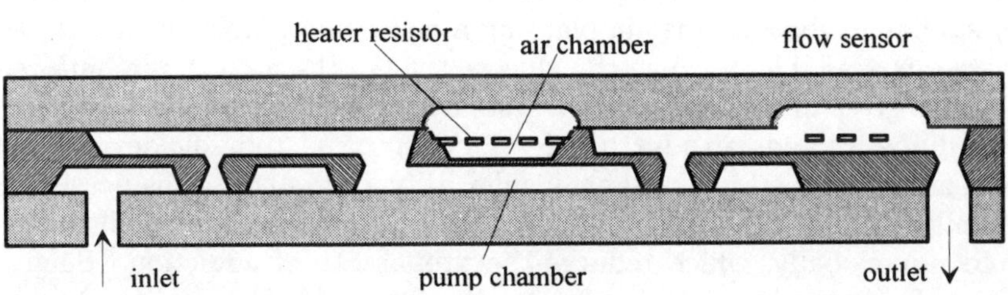

Figure 8.15: Dosing system with thermopneumatic pump and flowsensor [35]

These dosing systems are the first step towards the realisation of drug delivery systems as needed in medicine. Drug delivery could be used for example for insulin, painkillers like morphine or anaesthetics. The purpose of an insulin pump is to mimic the action of the pancreas by continuously giving the body a supply of insulin to maintain blood glucose levels within a normal range. For this the pump has to work 24 hours a day.

At present insulin as well as painkillers are given in high doses to last for several hours. This results in initially high concentrations well above the required level being used, with consequent wastage of insulin or possible addiction in the case of morphine. The pressure necessary for a dispensing system for the human blood circulation is around 30 kPa, which is achievable with today's micropumps. At the same time, the micropump has to pump at rates between 0 and 250 μl per hour and a minimum dispensing increment of 1 μl. A system for diabetics will consist of a glucose sensor, an insulin reservoir, a micropump and a flowsensor (figure 8.16).

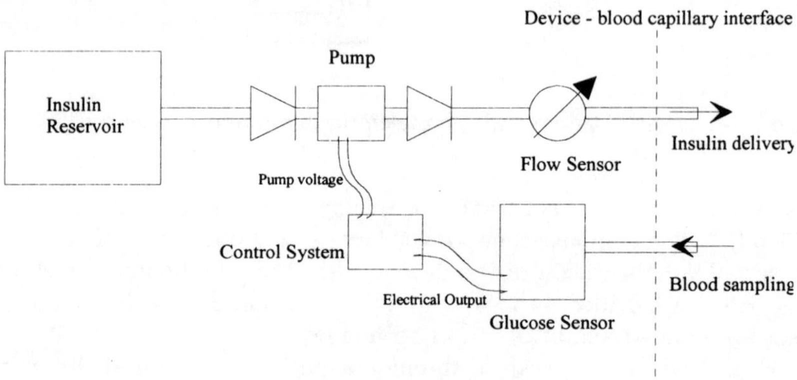

Figure 8.16: Schematic diagram of an insulin delivery system

Macroscopic insulin pumps have already been implanted into the human body [289]. These pumps are implanted in a person's abdominal wall, using surgery similar to that used for implanting a pacemaker. However, no closed loop system is available in which the glucose level of the patient is monitored and insulin is given on demand. Such a system could much improve the quality of life for diabetics and, with further development of microfluidic devices and glucose sensors, such an implantable insulin delivery system seems possible.

A similar system could be used for painkillers, to inject the medicine locally and not globally, which reduces the probability of addiction. Such a system has been developed and tested at the Maternity Hospital in Dublin, Ireland [290]. Unlike the devices described above, it uses a micropump that is electrochemically driven and contains no moving parts. An electrolytic reaction generates a gas that compresses a reservoir containing medication. With the ongoing worldwide research activities in microfluidics, systems which include the pump and sensing elements might be anticipated in the near future. Again, the quality of life for people in need of painkillers or other drugs could be greatly improved.

8.3.2 Chemical Microplants

A technology suitable for the mass production of micromachined chemical reaction systems has already been described in section 8.2.4. Such a system could handle any chemical reaction that has to be carried out immediately before the product is used. There are many potential (bio-)chemical applications for a microreactor system. A short summary of some of these applications is given below [287]:

- Measurement of reaction rate over the length of a capillary (stopped flow): determination of the parameters of a reaction (order etc).
  In the stopped-flow technique, reactant solutions are delivered from drive syringes to a mixing chamber and then to an observation cell. The flow of mixed reactants is stopped suddenly and the course of the reaction is monitored by a suitable method. The most commonly employed method is absorption spectrophotometry. The advantage of this technique over conventional spectrophotometry arises partly from the rapid mixing of reactant solutions which enables fast reactions to be followed. In addition, because the cell is fixed, very precise measurements of absorption changes (as low as 0.001) can be made, and errors due to slight changes in the position of the cell following its removal and replacement, which occur in conventional spectrophotometry, are completely eliminated. Precise measurements of initial rates of reaction can be made. Stopped-flow methods of analysis are rapid, sensitive and selective and have the potential for automation.

- Resolution of a mixture of different chemicals using immobilised enzymes (eg amino acid resolutions).

- Reactions with particle beads within nl wells, where the particles (silica or polystyrene) are 3 μm in diameter and coated with reactive groups like amino groups. The smaller the particles, the more tests are possible simultaneously. Another possibility is DNA synthesis, which uses the same principle, as well as continuous flow synthesis where very small scales are needed. A possibility is also to immobilise amino groups on the silicon surface of a mixer (section 7.3). The chemicals are then flow-injected into the device, where they react with the immobilised reactive group. This is very interesting for drug discovery, as such reactions have to be executed with millions of different chemicals (screening - see also section 8.3.4).

- Enzyme + Substrate reactions: Very common liquid enzymes are proteases which are used for numerous different fluidic substrates. Often the reaction changes the colour of the incoming liquids, which could be observed through a Pyrex window in the silicon chip. However, it is also possible to immobilise the enzyme on the surface of the micromixer and introduce a substrate and/or an inhibitor.

- Reduction of the size of reaction vessels: Possible catalysis and/or high temperatures and/or high pressure application. The big advantage of these systems is the small reactor volume, giving homogeneous reactions. If heating is required, the heat distribution is very even and the products are continuously flushed out. The small volumes involved lead to less hazards if one reactor is destroyed in a parallel array of devices (for one reactor: fluid streams are in the $\mu l\,\mathrm{min}^{-1}$ range). Examples of such dangerous reactions are: production of cyclohexyl isocyanate by the homogeneous phosgenation of cyclohexylamine at high temperatures; methyl isocyanate at high temperature; heterogeneous catalytic oxidation of monomethylformamide; production of hydrogen cyanide by the catalytic reaction of methane and ammonia; production of phosgene by the catalytic reaction of carbon monoxide and chlorine gas.

- High temperature (easily up to 150°C, which is the upper limit for most reactions in organic chemistry) and light (energy) induced reactions are possible. The advantage is that no solvents or reaction chemicals evaporate, due to the enclosed system. Examples of these reactions are Diels-Alder reactions or cyclo-additions. Some of the cyclo-additions are forbidden thermally but can be initiated by the use of light. A big advantage is the small thickness of the fluid layers, as absorption is rather small for 5μm-thick fluid streams as compared to large reaction chambers.

- Many reactions in organic chemistry are carried out at the point of use. This is especially interesting for very reactive reactions, where the chemicals A and B are mixed initially and the product is then used to introduce and react with chemical C. An example is:

$$\begin{gathered} R - COOH + \frac{1}{2}Cl_2 \longrightarrow [R - CO - Cl] \\ [R - CO - Cl] + R' - NH_2 \longrightarrow R - CO - NH - R' + HCl \end{gathered} \tag{8.1}$$

To summarise, the advantages of a microreactor system compared with conventional chemical reactors are:

- very clean reactions are possible because the system is small and therefore very homogeneous.

- the heat distribution is very even and the products are continuously flushed out.

- safe reactions are guaranteed, as the rupture of one microreactor releases only small quantities of chemicals.

8.3.3 Medical, Chemical and Environmental Monitoring

Another field where microfluidic systems could have a major impact is the monitoring of medical and environmental fluids and gases. One of the main advantages of microfluidics is the small sample size, so avoiding waste and allowing analyses which are sometimes not possible if the sample is too small for a macroscopic tool (genetic analysis). Due to the small size, the energy consumption is also low, thus making possible transportable analysis tools with the capability of large laboratory equipment.

One example in medicine is a sensor chip for monitoring hydrogen, oxygen, carbon dioxide and temperature which has been described by Arquint [291]. The sensor chip provides a flow-through channel defined by a 600μm-thick polymeric ring (see figure 8.17). It consists of a reference electrode, 4 amperometric pO_2 sensors, two ISFET-based pCO_2 sensors, one pH-ISFET sensor and a temperature sensor. The sensor cells, apart from the pH sensor which is not covered, are solid-state transducing elements covered with a hydrogel layer.

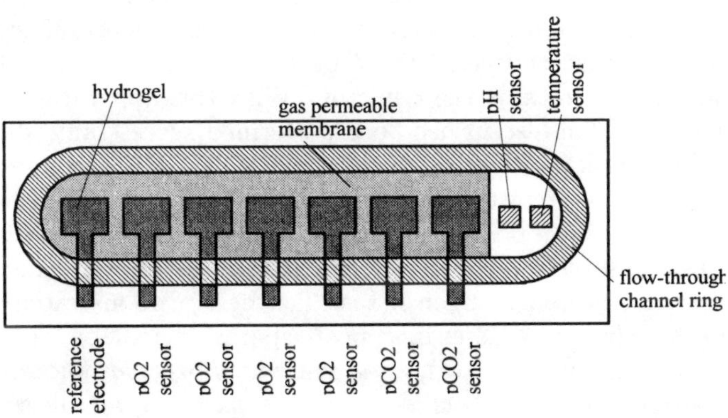

Figure 8.17: Blood gas sensor chip for pO_2, pCO_2 and pH [291]

The response of all the sensors is quite linear with respect to the individual blood gas components. The sensitivity of the pO_2 sensor is 0.36 nA / mmHg pO_2, the sensitivity of the pCO_2 sensor is 39 mV / decade pCO_2 and the sensitivity of the hydrogen sensor is 51mV / pH. The devices can be continuously operated for over two months with good stability. The system has been tested successfully with aqueous solutions, transfusion blood and whole blood.

The next example for clinical diagnostics and/or chemical monitoring is based on electrophoretic separation [292, 293]. The use of electroosmotic pumping, for example, allows pre-separation mixing of fluorescent compounds on chip, mixing of reagent solutions or post-separation fluorescent labelling on chip. An example is given in figure 8.18, where two labelled amino acids (Arg-

FITC and Tyr-FITC) can be mixed with fluorescein and then separated [293].

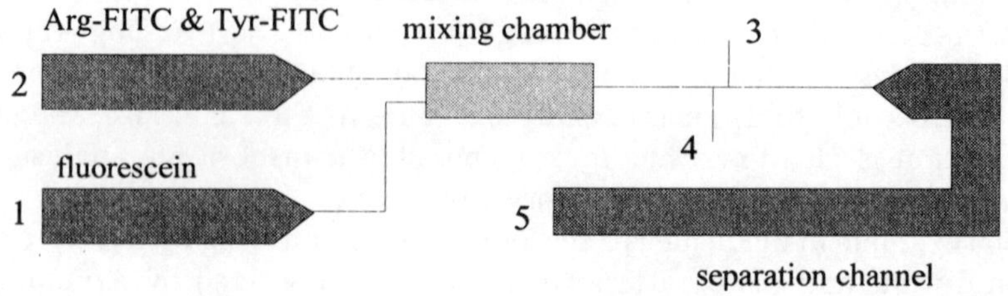

Figure 8.18: Separation of amino acids with electrophoresis [293]

Samples are introduced through reservoirs 1 and 2, and the other reservoirs are filled with a buffer. At first, a voltage between 1 and 3 as well as 2 and 3 drives the liquids into the mixing chamber where diffusive mixing occurs. Then a voltage between 4 and 5 pushes the mixture through the separation channel and towards the detector. With this apparatus, separation of amino acids and fluorescein has been performed successfully, showing the possibility of fast reactions with biological molecules and their detection.

Using a similar system to that described above, a more complex serum protein analysis has been performed on-chip [292]. In the clinical diagnosis of various states of disease it is important to separate and analyse quantitatively the human serum proteins. Such a task has been demonstrated by Chien [292], where labelling was done in a post-separation reactor with TNS (2-toluidinonaphthalene-6-sulfonate), a reagent which becomes fluorescent upon binding to various proteins. Due to the post-separation, any degradation of the separation quality is avoided. The detection is done optically on the same chip using laser induced fluorescence.

The same apparatus also allows reactions to take place before the separation, as described before. This has been demonstrated by performing a reaction of immunoreagents on-chip, followed by on-chip separation, which represents an increased level of automation [292].

A very interesting tool from the system point of view is a chemical analysis system which has been developed as a microfluidic demonstrator [136]. It is based on a silicon-on-Pyrex backplane like the microfluidic circuit board described earlier. On this board passive check valves, pumps, flowsensors and conductivity sensors are mounted with conventional adhesive. The layout of the whole system is depicted in figure 8.19.

During any analysis, pumping unit PU 1 performs the sample uptake, PU 2 provides for controlled cleaning of the system and PU 3 is for sensor calibration. The thermal flowsensors are to control the flow rate through the

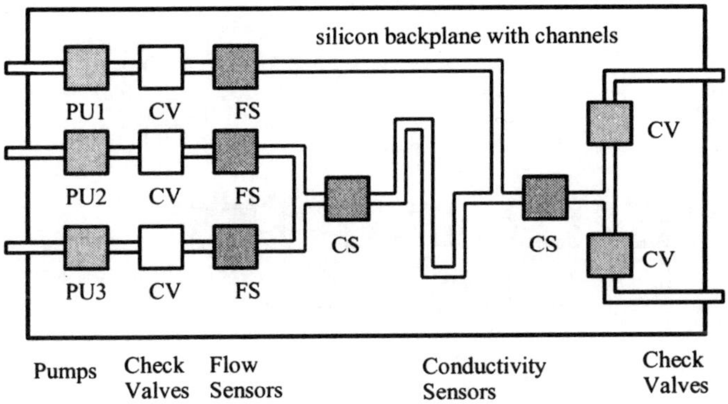

Figure 8.19: Chemical microanalysis system - basic setup [136]

different pumps and operate in pulsed mode. The frequency of the flowsensor had to be adjusted at the beginning of the operation. At low pulse frequencies the flowsensor could not follow the output of the micropump and at high frequencies the heat generated in the heating part of the flowsensor is not large enough. 10Hz was the frequency required for optimum operation. With the combination of micropump and flowsensor a constant delivery is possible. The conductivity sensors measure the resistivity of the solution via platinum electrodes at different points.

Priming of the whole system was done with CO_2 which reduces the possibility of air bubbles. The whole system was tested with $CaCl_2$ of various conductivities. Good separation was possible with 2, 3, 5 and 10 mM solutions. No interference from the micropump driving signal was observed, as the pump diaphragm is connected to electrical ground. However, air bubbles created major problems for the system. This can only be resolved when micropumps are able to generate larger pressures to drive bubbles through the system. Then such a system can show good performance in chemical analysis.

A similar device for environmental or medical monitoring has been proposed and partly realised by the same group [294]. The whole system is based on the same silicon backplane as before. This demonstrator is a mobile instrument for on-site water analysis, so nitrate, pH-value, phenolic compounds and oxygen have been chosen for analysis. Although the demonstrator used external pumps the final system will have fully integrated micromachined pumps. The sensing is done in the first instance with amperometric sensors and an optical cell, but a switch to ISFETs will also be possible. The flow scheme of the whole system is depicted in figure 8.20.

Another hybrid device for clinical monitoring has been presented recently [295] (figure 8.21). The basic idea for this system is that assays of transaminases present in the serum give information on the course of liver diseases or myocardial infarction. For example, elevated glutamic oxaloacetic transami-

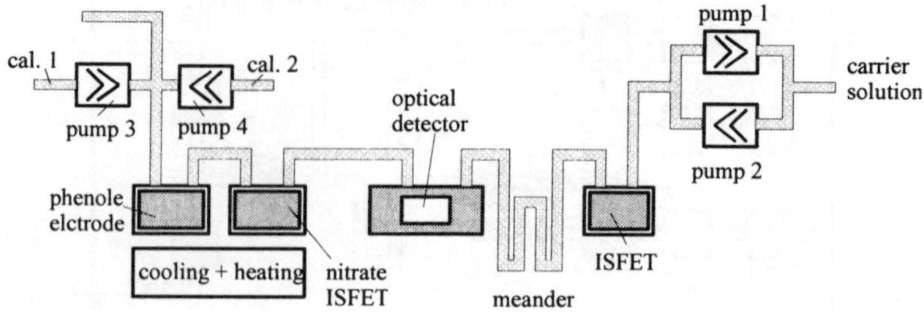

Figure 8.20: Flow scheme for the detection of phenole and nitrate with a microfluidic system [294]

nase (GOT) indicates severe damage to the cells of heart and liver, and raised plasma activity of glutamic pyruvic transaminase (GPT) indicates severe liver diseases like viral hepatitis or toxic liver necrosis. A transportable device for the assay of GPT and GOT could improve the point-of-care testing, especially in the case of emergency medicine.

The system for the assay of the transaminases GPT and GOT consists of a printed circuit board (PCB) which performs both electrical and fluidic functions. It comprises a mixing coil of 170 mm length and the connection to a biosensor array. The multi-biosensor consists of two glutamate sensors, two inactive sensors for reduction of electronic noise and a silver/silverchloride reference electrode. The technology for the PCB uses dry resist layers for the generation of channels [296]. A schematic view of the whole system is given in figure 8.21.

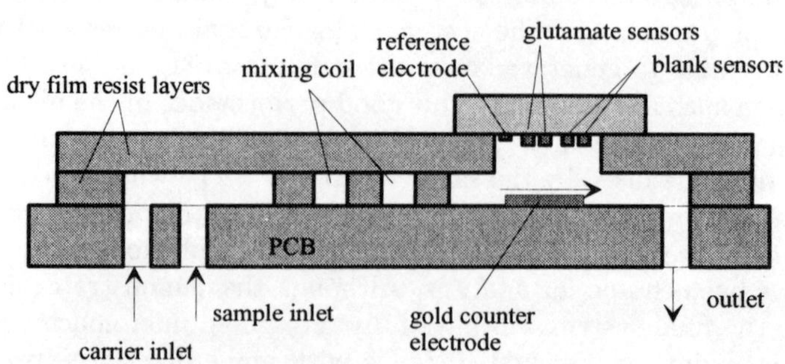

Figure 8.21: Schematic cross section of a miniaturised liquid handling system for rapid liver enzyme assay [295]

The device is operated as follows: a sample and a substrate solution are pumped at constant flow rates with two external pumps into the system. In the mixing coil the sample (GOT or GPT) catalyses a reaction of reagents which are components of the substrate solution. The resulting components (in both cases glutamate and another substance) can then be detected with the biosensor array. For the detection, the flow is stopped and the mixture needs to stay in contact with the gluatamate sensors for approximately 100 s. The device has several advantages compared to conventional spectro-photometric techniques. On the one hand it is faster than the analysis with conventional devices, and on the other hand it reduces the required sample volume.

8.3.4 Drug Discovery and Genetic Diagnostics

Drug discovery is one of the most interesting possible applications of microfluidics. Originally, drugs were discovered by analysing natural products to find the active molecular species. These natural products were already known to be drugs. The active species were then extracted from the natural products, purified, their structure analysed and then reproduced by chemical synthesis.

During the last few decades, chemical intuition or synthesis has been applied to discover new drugs. Once a chemical is found to be active against a relevant target, it rarely possesses all of the properties required to become a drug. These chemicals then have to be tested on animals and human beings to determine their interaction with other biological substances and their side effects. In order to discover drugs without these side effects, chemical analogues of the active compound must be synthesised and retested until the desired characteristics is found. However, this is a very time consuming method for drug discovery. For this reason combinatorial chemistry (cross reacting reagents) and random screening (randomly testing reaction products from combinatorial chemistry for a certain effect) of chemical libraries is the latest trend. Here, different starting products are combined and react to give a new product. For example, if 100 products A are reacted with 100 products B, 10000 products AB are obtained which can be put through quality control and bioassays (figure 8.22). With conventional chemical equipment, combinatorial chemistry is quite fast and the analysis is becoming also reasonably fast. However, there is no automated link between the combinatorial chemistry and the screening. Using human labour for connecting the two processes increases the time for any analysis drastically. This is where micromachining can make a contribution because both combinatorial chemistry and screening can be carried out on one single chip, i.e. a chemical microprocessor.

For such a chemical microprocessor, several elements are necessary: a boolean operation, where two products can be reacted and analysed; a serial to parallel converter etc.

Such a serial to parallel converter has been presented by Manz [297]. It is based on electroosmotic flow in 17 separation channels (figure 8.23). By using only 4 electrodes, 17 samples that are located one after the other in a feeding channel can be directed into 17 different channels, where they can

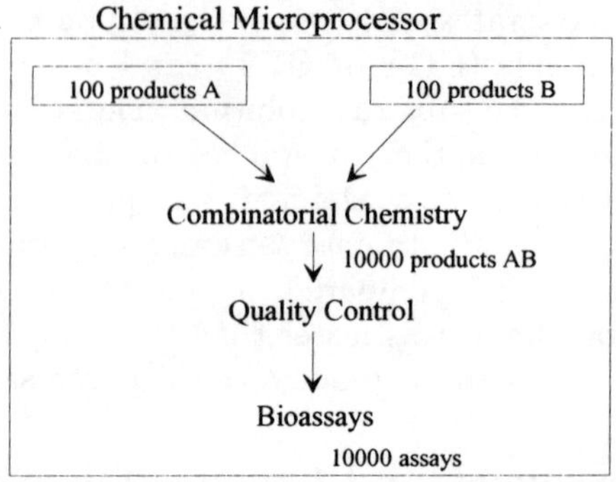

Figure 8.22: Concept for drug discovery and screening [297]

be analysed optically in parallel. Although this is only one component of the chemical microprocessor, it is likely that we will see such a microprocessor quite soon.

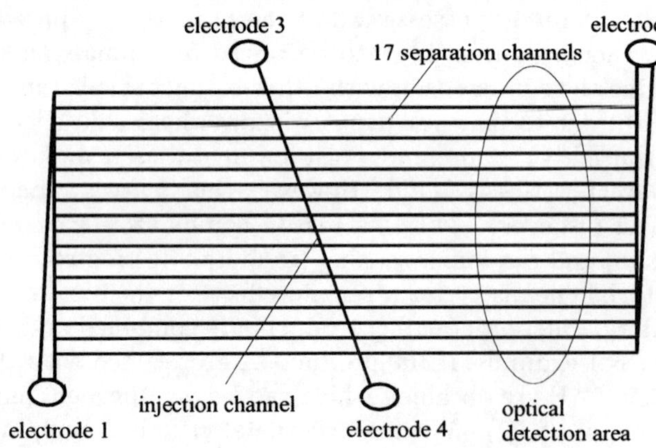

Figure 8.23: Schematic drawing of a 17 channel device for use with 4 external electrodes [297]

As described already in section 8.3.2, screening of drugs is also possible with small reaction cells. Orchid Biocomputer Inc. [298] has already described a device with 144 wells or reaction cells on one chip and is planning to scale up to 10000 wells on one chip (see figure 8.24). These reaction cells are capped but connected to the outside world by fluidic channels, avoiding contamination which might occur in open cells. Two different chemicals are fed into each reaction cell, where the reaction takes place. Product detection and identification are accomplished using optical spectroscopy. The chip has been fabricated using silicon, glass or plastics to allow as many chemicals to be used as possible. Compared to conventional wells of approximately 100 μl volume, the volume of the microfabricated well is only in the nl range so the costs for expensive chemicals are reduced dramatically. This bio-chip can be used for small molecule synthesis as well as genomic screening (both in the sector of combinatorial chemistry).

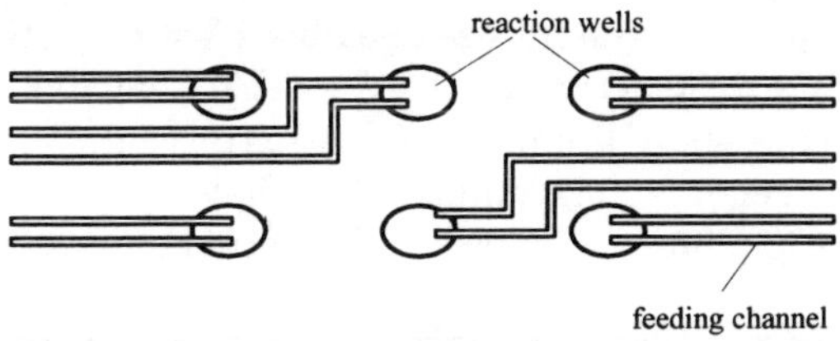

Figure 8.24: Schematic of a 6-well synthesis chip for drug discovery [298]

A similar system for drug discovery on a silicon chip has been described by Bousse [299]. Here, a whole-cell biosensor has been developed that performs functional rather than analytical analysis. In this case the output of the sensor is the effect of a stimulus on a living system. This is especially useful in assays of drugs to determine if they are toxic or environmentally harmful. By the use of micromachining, this chip allows high throughput, which makes it very suitable for use as the final part of the chemical microprocessor previously described.

Apart from combinatorial chemistry, genetic diagnostics is also an excellent application for microfluidics in modern analytical biochemistry. This application is concerned with the genetic information stored in DNA molecules (see section 7.4). The information can be extracted by a series of enzymatic and other reactions well known in molecular biology. Genetic diagnostics can be used in farming, agriculture and most of all in the human genome project. About 400 diseases can now be diagnosed by this technique [300], and man has approximately 100000 genes that could be analysed for defects. Therefore, this sector could provide an enormous market for microfluidics because time and quantity for the analysis are greatly reduced with micromachined devices.

The whole process of information retrieval requires thermal cycling for DNA amplification, mixing with reagents, labelling and fragmentation analysis, all processes that can be performed by microfluidic devices. DNA amplification is done with a PCR reaction, and the detection is done by labelling the genes with fluorophore dyes that emit light when bound to the DNA and are extremely sensitive to the detection of individual molecules [300]. Separation of DNA may also be required during the analysis to allow the detection of even smaller samples. As DNA fragments are negatively charged in solution, electrophoresis can be applied to move the DNA fragments (see section 7.2). Electrophoresis can also be applied to perform separation, as the mobility of DNA fragments in gels (consisting of entangled polymer matrices) is roughly inversely proportional to the logarithm of the fragment size [300].

Single devices that can perform parts of the genetic analysis have already been described in the previous chapters. Some very interesting systems, however, have been presented recently by Anderson [301], Woolley [302] and Burns's [303]. Anderson's system uses flexible plastics as the basic material and micromachining to structure it. The system is capable of performing an entire assay as it consists of a unit for DNA extraction, PCR, fragmentation of the DNA and other handling units. The whole system is operated under computer control using internal valves and external pumps to steer the fluids.

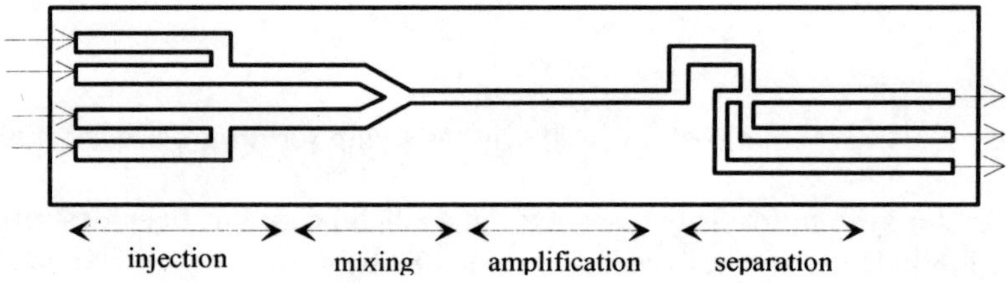

Figure 8.25: Integrated glass device for DNA amplification and detection [303]

The system described by Woolley consists of a vertical PCR microwell coupled with a planar glass chip capillary electrophoresis (CE) device to perform both amplification and separation in less than 45 min.

Burns' system incorporates simple injectors, mixers, a heating chamber and separation channels with electrodes (figure 8.25) on a single silicon chip of 3cm length. Detection is possible with integrated diode detectors, covered with a thick optical filter that blocks any UV excitation necessary to stimulate fluorescence in the DNAs. The channels are etched in a Pyrex wafer that is bonded to the silicon chip. The device provides transport, amplification, separation and detection of DNA on a single chip.

Several similar genetic analysis systems have been presented (see [300]). Instead of glass, however, these systems use flexible plastics for the fluidic channels. The plastic layers are bonded easily to substrates which are not perfectly flat and exhibit good biocompatibility, which is a major issue in genetic analysis. Unfortunately, plastic is more susceptible to the evaporation of buffer and gels due to the large solid-phase diffusion coefficients. This can create bubbles that block the flow, but can be avoided by coating the plastic walls with a thin glass layer.

8.4 Conclusion

As shown above, microfluidic systems can be quite sophisticated devices that can simplify drug dispensing, drug discovery, genetic analysis and other medical, biological or chemical tasks. However, one of the most problematic parts of these systems is the packaging with its connections to the outside world. Integrated systems need connectors for both electronics and fluids. The electronic part can be made on one side of the chip, and completely covered by epoxy as used in microelectronic chip packages, and the open fluid connection to macroscopic devices can be made on the opposite side. However, leak-free and bubble-free connections are required to allow small volumes of the fluid to be examined. Here much work has still to be done by scientists and industry to guarantee the success of microfluidics.

Abbreviations and Acronyms

ADC	Analogue to Digital Converter
BPSG	Boro-Phospho-Silicate Glass
CFD	Computational Fluid Dynamics
CE	Capillary Electrophoresis
CMOS	Complementary Metal Oxide Semiconductor Device
CPT	Cycling Probe Technique
CVD	Chemical Vapour Deposition
DNA	Deoxyribose Nucleic Acid
EHD	Electrohydrodynamics
EDP	Ethylene Diamine Pyrocatechol
EDX	Energy Dispersive X-Ray Imaging
FEM	Finite Element Method
FET	Field Effect Transistor
FIA	Flow Injection Analysis
GC	Gas Chromatography
HMDS	Hexamethyldisilazan
IBE	Ion Beam Etching
IC	integrated circuit
ISFET	Ion Sensitive FET
KOH	Potassium Hydroxide
LC	Liquid Chromatography
LIGA	Lithography Electroplating and Moulding
LPCVD	Low Pressure Chemical Vapour Deposition
LTO	Low Temperature Oxide
MEMS	Micro Electro Mechanical Systems
MOS	Metal Oxide Semiconductor
PCB	Printed Circuit Board
PCR	Polymerase Chain Reaction
PECVD	Plasma Enhanced Chemical Vapour Deposition
PMMA	Polymethyl-Methacrylate
PSD	Phase Sensitive Detector
PSG	Phosphorus Silicate Glass
PVD	Physical Vapour Deposition
PZT	Lead Zirconium Titanate
RCA(TM)	Cleaning Solution
RF	Radio Frequency
RIE	Reactive Ion Etching
RIBE	Reactive Ion Beam Etching
RTA	Rapid Thermal Annealing
SEM	Scanning Electron Microscope
TEOS	Tetraethoxysilane
TMAH	Tetra Methyl Ammonium Hydroxide
UV	Ultraviolet light

References

[1] J. Bausells, J. Carrabina, A. Merlos, S. Bota, and J. Samitier, "Mechanical sensors integrated in a commercial CMOS technology," in *Proceedings 10th European Conference on Solid-State Sensors 1996*, (Leuven, Belgium), pp. 367 – 370, 1996.

[2] R. Zengerle, "Microfluidics," in *Proceedings Micromechanics Europe 1998*, (Ulvik, Norway), pp. 111 – 122, 1998.

[3] P. Gravesen, J. Branebjerg, and O. S. Jensen, "Microfluidics - A review," *Journal of Micromechanics and Microengineering*, vol. 3, no. 4, pp. 168 – 182, 1993.

[4] F. M. White, *Fluid Mechanics.* New York: McGraw-Hill, 1979.

[5] I. H. Shames, *Mechanics of Fluids.* New York: McGraw-Hill, 1992.

[6] T. Gerlach, *Ein neues Mikropumpen-Prinzip mit dynamisch passiven Ventilen.* PhD thesis, Technische Universität Ilmenau, 1996.

[7] W. M. Swanson, *Fluid Mechanics.* New York: Holt, Rinehardt & Winston, 1970.

[8] M. Heschel, M. Müllenborn, and S. Bouwstra, "Fabrication and characterization of truly 3-D diffuser/nozzle microstructures in silicon," *Journal of Microelectromechanical Systems*, vol. 6, no. 1, pp. 41 – 47, 1997.

[9] M. Müllenborn, H. Dirac, J. W. Petersen, and S. Bouwstra, "Fast three-dimensional laser micromachining of silicon for microsystems," *Sensors and Actuators*, vol. A 52, pp. 121 – 125, 1996.

[10] A. Olsson, P. Enoksson, G. Stemme, and E. Stemme, "A valveless planar pump isotropically etched in silicon," in *Proceedings Micromechanics Europe 1995*, (Copenhagen, Denmark), pp. 120 – 123, 1995.

[11] J. H. Jerman, "Electrically-activated, normally closed diaphragm valves," in *Proceedings Transducers 1991*, (San Francisco, USA), pp. 1045 – 1048, 1991.

[12] J. Pfahler, J. Harley, and H. Bau, "Gas and liquid flow in small channels," *ASME Proceedings*, vol. 32, pp. 49 – 60, 1991.

[13] H. Matsumoto and J. Colgate, "Preliminary investigation of micropumping based on electrical control of interface tension," in *Proceedings IEEE Micro Electro Mechanical Systems 1990*, (Napa Valley, California), pp. 105 – 110, 1990.

[14] R. Zengerle, M. Leitner, S. Kluge, and A. Richter, "Carbon dioxide priming of micro liquid systems," in *Proceedings IEEE Micro Electro Mechanical Systems 1995*, (Amsterdam, The Netherlands), pp. 340 – 343, 1995.

[15] C. Gerthsen, H. O. Kneser, and H. Vogel, *Physik*. Berlin: Springer Verlag, 1992.

[16] A. Heuberger, ed., *Mikromechanik*. Berlin: Springer Verlag, 1990.

[17] H. Schaumburg, *Sensoren*. Stuttgart: B. G. Teubner, 1992.

[18] J. Merhaut, *Theory of Electroacoustics*. London: McGraw-Hill, 1981.

[19] R. D. Blevins, *Formulas for natural frequencies and mode shapes*. New York: Van Nostrand Reinhold Company, 1979.

[20] A. P. Boresi, R. J. Schmidt, and O. M. Sidebottom, *Advanced Mechanics of Materials*. New York: John Wiley & Sons, 1993.

[21] S. Timoshenko, ed., *Theory of Plates and Shells*. New York: McGraw-Hill, 1940.

[22] M. Royer, J. O. Holmen, M. A. Wurm, O. S. Aadland, and M. Glenn, "ZnO on Si integrated acoustic sensor," *Sensors and Actuators*, vol. 4, pp. 357 – 362, 1983.

[23] E. L. Eriksen, ed., *Collected Papers of S. P. Timoshenko*. London: McGraw-Hill, 1953.

[24] P. R. Lancaster and D. Mitchell, *The Mechanics of Materials*. London: McGraw-Hill, 1967.

[25] Landold and Bornstein, *Numerical Data and Functional Relationships in Science and Technology*, vol. 3. Berlin: Springer Verlag, 1966.

[26] A. D. Johnson, J. D. Busch, C. A. Ray, and C. Sloan, "Fabrication of silicon based shape memory alloy micro actuators," *Mat. Res. Soc. Symp. Proc.*, vol. 276, pp. 151 – 160, 1992.

[27] W. L. Benard, H. Kahn, A. H. Heuer, and M. A. Huff, "A Titanium-Nickel shape memory alloy actuated micropump," in *Proceedings Transducers 1997*, (Chicago, USA), pp. 361 – 364, 1997.

[28] R. Zengerle, A. Richter, and H. Sandmaier, "A micro membrane pump with electrostatic actuation," in *Proceedings IEEE Micro Electro Mechanical Systems 1992*, (Travemünde, Germany), pp. 19 – 24, IEEE, 1992.

[29] F. C. M. van de Pol, H. T. G. van Lintel, M. Elwenspoek, and J. H. J. Fluitman, "Thermopneumatic micropump based on micro-engineering techniques," *Sensors and Actuators*, vol. A 21, no. 1 - 3, pp. 198 – 202, 1990.

[30] D. de Bhailis, C. Murray, M. Duffy, J. Alderman, G. Kelly, and S. C. O Mathuna, "Modelling and analysis of a magnetic microactuator for use as a micropump," in *Proceedings Micromechanics Europe 1998*, (Ulvik, Norway), pp. 256–259, 1998.

[31] B. Wagner and W. Benecke, "Microfabricated actuator with moving permanent magnet," in *Proceedings IEEE Micro Electro Mechanical Systems 1991*, (Nara, Japan), pp. 27 – 32, 1991.

[32] A. Dogan, S. Yoshikawa, K. Uchino, and R. E. Newnham, "The effect of geometry on the characteristics of the moonie transducer and reliability issue," in *Proc. IEEE Ultrasonics Symposium*, pp. 935 – 939, 1994.

[33] S. Timoshenko, "Analysis of bi-metal thermostats," *Journal of the Optical Society America*, vol. 11, pp. 233 – 255, 1925.

[34] J. D. Busch and A. D. Johnson, "Prototype micro-valve actuator," in *Proceedings IEEE Micro Electro Mechanical Systems 1990*, (Napa Valley, California), pp. 40 – 41, 1990.

[35] T. S. J. Lammerink, M. Elwenspoek, and J. H. J. Fluitman, "Integrated micro-liquid dosing system," in *Proceedings IEEE Micro Electro Mechanical Systems 1993*, (Fort Lauderdale, FL, USA), pp. 254 – 259, 1993.

[36] P. B. Koeneman, "Conceptual design of a micro power supply for a mems smart bearing," Master's thesis, Univ. Texas, Austin, USA, 1995.

[37] P. B. Koenemann, I. J. Busch-Vishniac, and K. L. Wood, "Feasibility of micro power supplies for MEMS," *Journal of Microelectromechanical Systems*, vol. 6, no. 4, pp. 355 – 362, 1997.

[38] n.n., *ANSYS V5.3 Manual.* New York: Ansys Technologies.

[39] M. Koch, A. G. R. Evans, and A. Brunnschweiler, "Simulation and fabrication of micromachined cantilever valves," in *Proceedings 10th European Conference on Solid-State Sensors 1996*, (Leuven, Belgium), pp. 849 – 852, 1996.

[40] J. Ulrich and R. Zengerle, "Static and dynamic flow simulation of a KOH-etched microvalve using the finite-element method," *Sensors and Actuators*, vol. A 53, no. 1 - 3, pp. 379 – 385, 1996.

[41] M. M. Athavale, H. Q. Yang, and A. J. Przekwas, "Coupled fluid-thermo-structural simulations methodology for MEMS applications," in *Proceedings Transducers 1997*, (Chicago, USA), pp. 1043 – 1046, 1997.

[42] M. Koch, A. G. R. Evans, and A. Brunnschweiler, "Coupled FEM simulation for the characterization of the fluid flow within a micromachined cantilever valve," in *Proceedings Micromechanics Europe 1995*, (Copenhagen, Denmark), pp. 160 – 163, 1995.

[43] R. Zengerle and M. Richter, "Simulation of microfluid systems," *Journal of Micromechanics and Microengineering*, vol. 4, no. 4, pp. 192 – 204, 1994.

[44] J. Tiren, L. Tenerez, and B. Hök, "A batch fabricated non-reverse valve with cantilever beam manufactured by micromachining of silicon," *Sensors and Actuators*, vol. 18, no. 3 - 4, pp. 389 – 396, 1989.

[45] S. Park, W. H. Ko, and J. M. Prahl, "A constant flow-rate microvalve actuator based on silicon and micromachining technology," in *Technical Digest IEEE solid-state sensor and actuator workshop*, (Hilton Head Island, SC, USA), pp. 136 – 139, IEEE, 1988.

[46] K. Meyberg and P. Vachenauer, *Höhere Mathematik.* Berlin: Springer Verlag, 1990.

[47] R. Zengerle, J. Ulrich, S. Kluge, M. Richter, and A. Richter, "Bidirectional silicon micropump," *Sensors and Actuators*, vol. A 50, no. 1 - 2, pp. 81 – 86, 1995.

[48] T. S. J. Lammerink, N. R. Tas, M. Elwenspoek, and J. H. J. Fluitman, "Micro-liquid flow sensor," *Sensors and Actuators*, vol. A 37 - 38, pp. 45 – 50, 1993.

[49] O. N. Tufte, P. W. Chapman, and D. Long, "Silicon diffused-element piezoresistive diaphragms," *Journal of Applied Physics*, vol. 33, pp. 3322 – 3327, 1962.

[50] S. Middelhoek and S. A. Audet, *Silicon Sensors.* London: Academic Press Ltd, 1989.

[51] S. M. Sze, *VLSI Technology.* New York: McGraw-Hill, 1988.

[52] J. W. Gardner, *Microsensors: Principles and Applications.* New York: John Wiley & Sons, 1994.

[53] K. E. Petersen, "Silicon as a mechanical material," *Proceedings IEEE*, vol. 70, no. 5, pp. 420 – 457, 1982.

[54] M. D. B. Charlton and G. J. Parker, "Nanofabrication of advanced waveguide structures incorporating a visible photonic bandgap," in *Proceedings Micromechanics Europe 1997*, (Southampton, UK), pp. 238 – 241, 1997.

[55] T. Kamins, *Polycrystalline Silicon for integrated circuit applications*. Boston: Kluwer Academic Publishers, 1988.

[56] D. Lu, Z. Li, X. Liu, I. Huang, Y. Fu, and G. Lin, "Study on the structure and properties of lead zirconate-titanate thin film," *Zhenkong Kexue yu Jishu Xuebao/Vacuum Science and Technology*, vol. 16, no. 3, pp. 206 – 211, 1996.

[57] M. Sakata, S. Wakabayashi, H. Goto, H. Totani, M. Takeuchi, and T. Yada, "Sputtered high d31 coefficient PZT thin film for micro actuators," in *Proceedings IEEE Micro Electro Mechanical Systems 1996*, (San Diego, CA, USA), pp. 263 – 266, 1996.

[58] R. Bruchhaus, D. Pitzer, R. Primig, W. Wersing, and Y. Xu, "Deposition of self-polarized PZT films by planar multi-target sputtering," *Integrated Ferroelectrics*, vol. 14, no. 1 – 4, pp. 141 – 149, 1997.

[59] W. Tjhen, T. Tamagawa, C. P. Ye, C. C. Hsueh, P. Schiller, and D. L. Polla, "Properties of piezoelectric thin films for micromechanical devices," in *Proceedings IEEE Micro Electro Mechanical Systems 1991*, (Nara, Japan), pp. 114 – 119, 1991.

[60] K. E. Petersen, "Dynamic micromechanics on silicon: Techniques and devices," *IEEE Transactions on Electron Devices*, vol. ED 25, no. 10, pp. 1241–1250, 1978.

[61] K. R. Williams and R. S. Muller, "Etch rates for micromachining processes," *Journal of Microelectromechanical Systems*, vol. 5, no. 4, pp. 256 – 269, 1996.

[62] H. Seidel, L. Csepregi, A. Heuberger, and H. Baumgaertel, "Anisotropic etching of crystalline silicon in alkaline solutions. I. orientation dependence and behavior of passivation layers.," *Journal of the Electrochemical Society*, vol. 137, no. 11, pp. 3612 – 3626, 1990.

[63] H. Seidel, L. Csepregi, A. Heuberger, and H. Baumgaertel, "Anisotropic etching of crystalline silicon in alkaline solutions. II. influence of dopants," *Journal of the Electrochemical Society*, vol. 137, no. 11, pp. 3626 – 3632, 1990.

[64] U. Schnakenberg, W. Benecke, and P. Lange, "TMAHW etchants for silicon micromachining," in *Proceedings Transducers 1991*, (San Francisco, USA), pp. 815 – 818, 1991.

[65] G. Findler, J. Muchow, M. Koch, and H. Munzel, "Temporal evolution of silicon surface roughness during anisotropic etching processes," in *Proceedings IEEE Micro Electro Mechanical Systems 1992*, (Travemünde, Germany), pp. 62 – 66, IEEE, 1992.

[66] B. Klöck, S. D. Collins, N. F. de Rooij, and R. L. Smith, "Study of the electrochemical etchstop for high precision thickness control of silicon membranes," *IEEE Transactions on Electron Devices*, vol. 36, no. 4, pp. 663 – 669, 1989.

[67] E. Peeters, D. Lapadatu, R. Puers, and W. Sansen, "PHET, an electrodeless photovoltaic electrochemical etchstop technique," *Journal of Microelectromechanical Systems*, vol. 3, no. 3, pp. 113 – 123, 1994.

[68] H. Robbins and B. Schwartz, "Chemical etching of silicon I," *Journal of the Electrochemical Society*, vol. 106, pp. 505 – 508, 1959.

[69] H. Robbins and B. Schwartz, "Chemical etching of silicon II," *Journal of the Electrochemical Society*, vol. 107, pp. 108 – 111, 1960.

[70] B. Schwartz and H. Robbins, "Chemical etching of silicon III," *Journal of the Electrochemical Society*, vol. 108, pp. 365 – 372, 1961.

[71] B. Schwartz and H. Robbins, "Chemical etching of silicon IV," *Journal of the Electrochemical Society*, vol. 123, pp. 1903 – 1909, 1976.

[72] J. B. Lasky, "Wafer bonding for silicon on insulator technologies," *Appl. Phys. Letters*, vol. 48, no. 1, pp. 78 – 80, 1986.

[73] C. Harendt, W. Appel, H. G. Graf, B. Höfflinger, and E. Penteker, "Wafer bonding and its application to SOI fabrication," in *Proceedings Micromechanics Europe 1990*, (Berlin, Germany), pp. 81 – 86, 1990.

[74] R. Stengl, T. Tan, and U. Gösele, "A model for the silicon wafer bonding process," *Japan. Journal of Applied Physics*, vol. 28, pp. 1735 – 1741, 1989.

[75] B. Müller and A. Stoffel, "Bonding of silicon wafers at low temperature," in *Proceedings Micromechanics Europe 1990*, (Berlin, Germany), pp. 87 – 90, 1990.

[76] J. A. Folta, C. E. Hunt, and S. N. Farrens, "Low temperature wafer bonding of surfaces using a reactive sputtered oxide," *Journal of the Electrochemical Society*, vol. 141, no. 8, pp. 2157 – 2160, 1994.

[77] R. Legtenberg, S. Bouwstra, and M. Elwenspoek, "Low temperature glass bonding for sensor applications," in *Proceedings Micromechanics Europe 1990*, (Berlin, Germany), pp. 94 – 99, 1990.

[78] H. J. Quenzer, W. Benecke, and C. Dell, "Low temperature wafer bonding for micromechanical applications," in *Proceedings IEEE Micro Electro Mechanical Systems 1992*, (Travemünde, Germany), pp. 49 – 55, IEEE, 1992.

[79] W. H. Ko, J. T. Summito, and G. J. Yeh, "Bonding techniques for microsensors," in *Microsensors* (R. S. Muller, R. T. Howe, S. D. Senturia, R. L. Smith, and R. M. White, eds.), pp. 198 – 208, New York: IEEE Press, 1991.

[80] M. Nese and A. Hanneborg, "Anodic bonding of silicon to silicon wafers coated with aluminium, silicon oxide, polysilicon or silicon nitride," *Sensors and Actuators*, vol. A 37 - 38, pp. 61 – 67, 1993.

[81] P. Krause, M. Sporys, E. Obermeier, K. Lange, and S. Grigull, "Silicon to silicon anodic bonding using evaporated glass," in *Proceedings Transducers 1995*, (Stockholm, Sweden), pp. 228 – 231, 1995.

[82] C. Linder, L. Paratte, M. A. Gretillat, V. P. Jaecklin, and N. F. de Rooij, "Surface micromachining," *Journal of Micromechanics and Microengineering*, vol. 2, pp. 122 – 132, 1992.

[83] L. Svenson, J. A. Plaza, M. A. Benitz, J. Esteve, and E. Lora-Tamayo, "Surface micromachining technology applied to the fabrication of a FET pressure sensor," in *Proceedings Micromechanics Europe 1995*, (Copenhagen, Denmark), pp. 112 – 115, 1995.

[84] C. J. Kim, A. P. Pisano, R. S. Muller, and M. G. Lim, "Polysilicon microgripper," *Sensors and Actuators*, vol. A 33, pp. 221 – 227, 1992.

[85] J. M. Younse, "Mirrors on a chip," *IEEE Spectrum*, vol. 30, no. 11, pp. 27 – 31, 1993.

[86] W. Kuehnel and S. Sherman, "A surface micromachined silicon accelerometer with on-chip detection circuitry," *Sensors and Actuators*, vol. A 45, pp. 7 – 16, 1994.

[87] M. Mehregany, S. D. Senturia, J. H. Lang, and P. Nagarkar, "Micromotor fabrication," *IEEE Transactions on Electron Devices*, vol. 39, no. 9, pp. 2060 – 2069, 1992.

[88] M. Ross and K. S. J. Pister, "Micro-windmill for optical scanning and flow measurement," *Sensors and Actuators*, vol. A 46 – 47, pp. 576 – 579, 1995.

[89] J. W. Judy, T. Tamagawa, and D. L. Polla, "Surface-machined micromechanical membrane pump," in *Proceedings IEEE Micro Electro Mechanical Systems 1991*, (Nara, Japan), pp. 182 – 186, 1991.

[90] N. Takeshima, K. J. Gabriel, M. Ozaki, J. Takahashi, H. Horiguchi, and H. Fujita, "Electrostatic parallelogramm actuators," in *Proceedings Transducers 1991*, (San Francisco, USA), pp. 63 – 66, 1991.

[91] P. R. Scheeper, J. A. Voomthuyzen, W. Olthuis, and P. Bergveld, "Investigation of attractive forces between PECVD silicon nitride microstructures and an oxidized silicon substrate," *Sensors and Actuators*, vol. A 30, no. 3, pp. 231 – 239, 1992.

[92] T. Abe, W. C. Messner, and M. L. Reed, "Effects of elevated temperature treatments in microstructure release procedures," *Journal of Microelectromechanical Systems*, vol. 4, no. 2, pp. 1057 – 7157, 1995.

[93] T. J. Brosnihan, J. M. Bustillo, A. P. Pisano, and R. T. Howe, "Embedded interconnect and electrical isolation for high-aspect-ratio, SOI inertial instruments," in *Proceedings Transducers 1997*, (Chicago, USA), pp. 637 – 640, 1997.

[94] C. Marxer, C. Thio, M. A. Gretillat, O. Anthamatten, R. Baettig, B. Valk, P. Vogel, and N. F. de Rooij, "Vertical mirrors fabricated by deep reactive ion etching for fiber optic switching applications," *Journal of Microelectromechanical Systems*, vol. 6, no. 3, pp. 277 – 285, 1997.

[95] B. P. van Drieenhuizen, N. I. Maluf, I. E. Opris, and G. T. A. Kovacs, "Force-balanced accelerometer with mG resolution, fabricated using silicon fusion bonding and deep reactive ion etching," in *Proceedings Transducers 1997*, (Chicago, USA), pp. 1229 – 1230, 1997.

[96] S. Rotich, J. G. Smith, A. G. R. Evans, and A. Brunnschweiler, "Micromachined thin solar cells with a novel light trapping scheme," in *Proceedings Micromechanics Europe 1997*, (Southampton, UK), pp. 144 – 147, 1997.

[97] K. D. Budd, S. K. Dey, and D. A. Paine, "Sol-gel processing of PbTiO3, PbZrO3, PZT, and PLZT thin films," *Brit. Ceram. Proc.*, vol. 36, pp. 107 – 121, 1985.

[98] G. Yi, Z. Wu, and M. Sayer, "Preparation of PB(Zr,Ti)O3 thin-films by sol-gel processing - electrical, optical and electro-optic properties," *Journal of Applied Physics*, vol. 64, pp. 2717 – 2724, 1988.

[99] Y. L. Tu and S. J. Milne, "A study of the effects of process variables on the properties of PZT films produced by a single-layer sol-gel technique," *Journal of Materials Science*, vol. 30, no. 10, pp. 2507 – 2516, 1995.

[100] D. A. Barrow, T. E. Petroff, and M. Sayer, "Fabrication of thick, sol gel PZT films: applications to macroscopic piezoelectric devices," *Mat. Res. Soc. Symp. Proc.*, vol. 360, pp. 103 – 108, 1995.

[101] M. Lukacs, M. Sayer, D. Knapik, R. Candela, and F. S. Foster, "Novel PZT films for ultrasound biomicroscopy," in *Proceedings of the IEEE Ultrasonics Symposium*, (San Antonio, USA), pp. 901 – 904, 1996.

[102] K. Brooks, D. Damjanovic, A. Kholkin, I. Reaney, N. Setter, P. Luginbuhl, G. A. Racine, N. F. de Rooij, and A. Saaman, "PZT films for micro-pumps," *Integrated Ferroelectrics*, vol. 8, pp. 13 – 23, 1995.

[103] M. Koch, N. Harris, A. G. R. Evans, N. M. White, and A. Brunnschweiler, "A novel micromachined pump based on thick-film piezoelectric actuation," in *Proceedings Transducers 1997*, (Chicago, USA), pp. 353 – 356, 1997.

[104] M. Koch, A. G. R. Evans, and A. Brunnschweiler, "The dynamic micropump driven with a screen printed PZT actuator," in *Proceedings Micromechanics Europe 1997*, (Southampton, England), pp. 116 – 119, 1997.

[105] Z. H. Fan and D. J. Harrison, "Micromachining of capillary electrophoresis injectors and separators on glass chips and evaluation of flow at capillary intersections," *Analytical Chemistry*, vol. 66, pp. 177 – 184, 1994.

[106] J. Vollmer, H. Hein, W. Menz, and F. Walter, "Bistable fluidic elements in LIGA technique for flow control in fluidic microactuators," *Sensors and Actuators*, vol. A 43, pp. 330 – 334, 1994.

[107] R. Rapp, W. K. Schomburg, D. Maas, J. Schulz, and W. Stark, "LIGA micropump for gases and liquids," *Sensors and Actuators*, vol. A 40, no. 1, pp. 57 – 61, 1994.

[108] H. Baudry, "Screen printing piezoelectric devices," *Hybrid Circuits*, vol. 14, pp. 71 – 74, 1987.

[109] M. Prudenziati, B. Morten, and G. de Cicco, "Piezoelectric thick film materials and sensors," *Microelectronics International*, vol. 38, pp. 5 – 11, 1995.

[110] P. G. Dargie, N. R. Harris, N. M. White, J. K. Atkinson, and R. P. Sion, "Characterisation of screen printable piezoelectric thick-films," in *Sensors and their Applications VIII*, (Glasgow, UK), pp. 201 – 206, 1997.

[111] J. E. Brignell, N. M. White, and A. W. J. Cranny, "Sensor applications of thick-film technology," *IEE Proceedings*, vol. 135 I, no. 4, pp. 77 – 84, 1995.

[112] R. Maas, M. Koch, N. Harris, N. M. White, and A. G. R. Evans, "Thick-film printing of PZT onto silicon," *Materials Letters*, vol. 31, pp. 109 – 112, 1997.

[113] N. M. White, "An assessment of thick-film piezoresistors on insulated steel substrates," *Hybrid Circuits*, pp. 23 – 27, 1989.

[114] D. J. Harrison, A. Manz, and P. G. Glavina, "Electroosmotic pumping within a chemical sensor system integrated on silicon," in *Proceedings Transducers 1991*, (San Francisco, USA), pp. 792 – 795, 1991.

[115] L. M. Roylance and J. B. Angell, "A batch-fabricated silicon accelerometer," *IEEE Transactions on Electron Devices*, vol. ED-26, pp. 1911 – 1917, 1979.

[116] J. Mohr, P. Bley, C. Burbaum, and U. Wallrabe, "Fabrication of microsensor and microactuator elements by the LIGA process," in *Proceedings Transducers 1991*, (San Francisco, USA), pp. 607 – 609, 1991.

[117] W. K. Schomburg, R. Ahres, W. Bacher, S. Engemann, P. Krehl, and J. Martin, "Long-term performance analysis of thermo-pneumatic micropump actuators," in *Proceedings Transducers 1997*, (Chicago, USA), pp. 365 – 368, 1997.

[118] V. Hessel, W. Ehrfeld, H. Möbius, T. Richter, and K. Russow, "Potentials and realization of microreactors," in *Proc. Int. Symp. on Microsystems, Intelligent Materials and Robots*, (Sendai, Japan), pp. 45 – 48, 1995.

[119] W. Menz and P. Bley, *Mikrosystemtechnik für Ingenieure.* Weinheim: VCH Verlagsgesellschaft, 1993.

[120] S. C. Terry, J. H. Jerman, and J. B. Angell, "A gas chromatographic air analyzer fabricated on a silicon wafer," *IEEE Transactions on Electron Devices*, vol. 26, pp. 1880 – 1886, 1979.

[121] E. Bassous, H. H. Taub, and L. Kuhn, "Ink jet printing nozzle arrays etched in silicon," *Applied Physics Letters*, vol. 31, pp. 135 – 137, 1977.

[122] A. Emmer, M. Jansson, J. Roeraade, U. Lindberg, and B. Hök, "Fabrication and characterization of a silicon microvalve," *Journal of Microcolumn Separation*, vol. 4, no. 1, pp. 13 – 15, 1992.

[123] M. Esashi, S. Shoji, and A. Nakano, "Normally closed microvalve and micropump fabricated on a silicon wafer," *Sensors and Actuators*, vol. 20, no. 1 - 2, pp. 163 – 169, 1989.

[124] H. T. G. van Lintel, F. C. M. van de Pol, and S. Bouwstra, "Piezoelectric micropump based on micromachining of silicon," *Sensors and Actuators*, vol. 15, no. 2, pp. 153 – 167, 1988.

[125] T. Gerlach, "A simple micropump employing dynamic passive valves made in silicon," in *Proceedings Micro System Technology 1994*, (Berlin, Germany), pp. 1025 – 1033, 1994.

[126] E. Stemme and G. Stemme, "Valveless diffuser/nozzle based fluid pump," *Sensors and Actuators*, vol. A 39, no. 2, pp. 159 – 167, 1993.

[127] S. Shoji, S. Nakagawa, and M. Esashi, "Micropump and sample injector for integrated chemical analysis systems," *Sensors and Actuators*, vol. A 21, pp. 189 – 192, 1990.

[128] J. G. Smits, "Piezoelectric micropump with three valves working peristaltically," *Sensors and Actuators*, vol. A 21 - 23, pp. 203 – 206, 1990.

[129] S. T. Cho and K. D. Wise, "A high performance micro flowmeter with built-in self test," in *Proceedings Transducers 1991*, (San Francisco, USA), pp. 400 – 403, 1991.

[130] V. Gass, B. H. vander Schoot, and N. F. de Rooij, "Nanofluid handling by microflow sensor based on drag force measurement," in *Proceedings IEEE Micro Electro Mechanical Systems 1993*, (Fort Lauderdale, FL, USA), pp. 167 – 172, IEEE, 1993.

[131] M. Elwenspoek, T. S. J. Lammerink, R. Miyake, and J. H. J. Fluitman, "Towards integrated microliquid handling systems," *Journal of Micromechanics and Microengineering*, vol. 4, pp. 227 – 245, 1994.

[132] A. Desai, D. Boekenkamp, X. Yang, Y. C. Tai, E. Marxluff, and S. Mayo, "Microfluidic sub-millisecond mixers for the study of chemical reaction kinetics," in *Proceedings Transducers 1997*, (Chicago, USA), pp. 167 – 170, 1997.

[133] D. Liepmann and J. D. Evans, "Micro-fluidic mixer," *Polymeric Materials Science & Eng. Procs. ACS Divsion of Polymeric Materials Science & Engineering*, vol. 76, pp. 549 – 550, 1997.

[134] N. Schwesinger, T. Frank, and H. Wurmus, "A modular microfluid system with an integrated micromixer," in *Proceedings Micromechanics Europe 1995*, (Copenhagen, Denmark), pp. 144 – 147, 1995.

[135] M. de la Guardia, "Biochemical sensors: The state of the art," *Microchimica Acta*, vol. 120, pp. 243 – 255, 1995.

[136] M. Richter, A. Prak, J. Naundorf, M. Eberl, H. Leeuwis, P. Woias, and A. Steckenborn, "A chemical microanalysis system as a microfluid system demonstrator," in *Proceedings Transducers 1997*, (Chicago, USA), pp. 303 – 306, 1997.

[137] B. H. vander Schoot, S. Jeanneret, A. vanden Berg, and N. F. de Rooij, "Modular setup for a miniaturized chemical analysis system," *Sensors and Actuators*, vol. B 15, pp. 211 – 213, 1993.

[138] E. Bassous, "Fabrication of novel three-dimensional microstructures by the anisotropic etching of (100) and (110) silicon," *IEEE Transactions on Electron Devices*, vol. ED 25, p. 1178, 1978.

[139] L. Kuhn, E. Bassous, and R. Lane, "Silicon charge electrode array for ink jet printing," *IEEE Transactions on Electron Devices*, vol. ED 25, p. 1257, 1978.

[140] K. E. Petersen, "Fabrication of an integrated, planar silicon inkjet structure," *IEEE Transactions on Electron Devices*, vol. 26, p. 1918, 1992.

[141] J. Chen and K. D. Wise, "A high-resolution silicon monolithic nozzle array for inkjet printing," *IEEE Transactions on Electron Devices*, vol. 44, no. 9, pp. 1401 – 1409, 1997.

[142] D. Westberg and G. I. Andersson, "A novel CMOS compatible inkjet head," in *Proceedings Transducers 1997*, (Chicago, USA), pp. 813 – 816, 1997.

[143] I. Ederer, J. Grasegger, and C. Tille, "Droplet generator with extraordinary high flow rate and wide operating range," in *Proceedings Transducers 1997*, (Chicago, USA), pp. 809 – 812, 1997.

[144] R. Furlan and J. N. Zemel, "Behaviour of microfluidic amplifiers," *Sensors and Actuators*, vol. A 51, pp. 239 – 246, 1996.

[145] T. H. Kim, C. H. Cho, and D. I. D. Cho, "A three-dimensionally silicon-micromachined fluidic amplifier device," *Journal of Micromechanics and Microengineering*, vol. 8, pp. 7 – 14, 1998.

[146] U. Gebhard, H. Hein, E. Just, and P. Ruther, "Combination of a fluidic micro-oscillator and micro-actuator in LIGA-technique for medical applications," in *Proceedings Transducers 1997*, (Chicago, USA), pp. 761 – 764, 1997.

[147] A. Benitez, J. Esteve, and J. Bausells, "Bulk silicon microelectromechanical devices fabricated from commercial bonded and etched-back silicon-on-insulator substrates," *Sensors and Actuators*, vol. A 50, pp. 99 – 103, 1995.

[148] C. L. Tsai and A. K. Henning, "Surface micromachined turbines," in *Proceedings Transducers 1997*, (Chicago, USA), pp. 829 – 832, 1997.

[149] A. Ilzhofer, B. Ritter, and C. Tsakmakis, "Development of passive microvalves by the finite element method," *Journal of Micromechanics and Microengineering*, vol. 5, no. 3, pp. 226 – 230, 1995.

[150] M. C. Acero, J. A. Plaza, J. Esteve, M. Carmona, S. Marco, and J. Samitier, "Design of a modular micropump based on anodic bonding," in *Proceedings Micromechanics Europe 1995*, (Copenhagen, Denmark), pp. 144 – 147, 1995.

[151] R. T. Howe, "Surface micromachining for microsensors and microactuators," *Journal of Vacuum Science Technology*, vol. B 6, no. 6, pp. 1809 – 1813, 1986.

[152] W. K. Schomburg and B. Scherrer, "3.5 μm thin valves in titanium membranes," *Journal of Micromechanics and Microengineering*, vol. 2, pp. 184 – 186, 1992.

[153] S. Shoji, M. Esashi, B. vander Schoot, and N. de Rooij, "A study of a high-pressure micropump for integrated chemical analysing systems," *Sensors and Actuators*, vol. A 21, pp. 189 – 192, 1990.

[154] E. H. Yang, S. W. Han, and S. S. Yang, "Fabrication and testing of a pair of passive bivalvular microvalves composed of p^+ diaphragms," *Sensors and Actuators*, vol. A 57, no. 1, pp. 75 – 78, 1996.

[155] A. Olsson, G. Stemme, and E. Stemme, "Micromachined diffuser/nozzle elements for valve-less pumps," in *Proceedings IEEE Micro Electro Mechanical Systems 1996*, (San Diego, CA, USA), pp. 378 – 383, 1996.

[156] X. N. Jiang, Z. U. Zhou, J. Yao, Y. Li, and X. Y. Ye, "Micro-fluid flow in microchannel," in *Proceedings Transducers 1995*, (Stockholm, Sweden), pp. 317 – 320, 1995.

[157] T. Lisec, H. J. Quenzer, M. Kreutzer, S. Hoerschelmann, B. Wagner, and W. Benecke, "Thermally driven microvalve with buckling behaviour for pneumatic applications," *Sensors and Materials*, vol. 8, no. 5, pp. 303 – 316, 1996.

[158] H. Jerman, "Electrically-activated normally-closed diaphragm valve," *Journal of Micromechanics and Microengineering*, vol. 4, no. 4, pp. 210 – 216, 1994.

[159] H. P. Trah, H. Baumann, C. Doring, H. Goebel, T. Grauer, and M. Mettner, "Micromachined valve with hydraulically actuated membrane subsequent to a thermoelectrically controlled bimorph cantilever," *Sensors and Actuators*, vol. A 39, no. 2, pp. 169 – 176, 1994.

[160] M. A. Huff, M. S. Mettner, T. A. Lober, and M. A. Schmidt, "A pressure-balanced electrostatically-actuated microvalve," in *Technical Digest IEEE Solid-State Sensor and Actuator Workshop*, (Hilton Head Island, SC, USA), pp. 123 – 127, 1990.

[161] D. Bosch, B. Heimhofer, G. Mück, H. Seidel, U. Thumser, and W. Welser, "A silicon microvalve with combined electromagnetic/electrostatic actuation," *Sensors and Actuators*, vol. A 37 - 38, pp. 684 – 692, 1993.

[162] J. Haji-Babaei, C. Y. Kwok, and R. S. Huang, "Integrable active microvalve with surface micromachined curled-up actuator," in *Proceedings Transducers 1997*, (Chicago, USA), pp. 833 – 836, 1997.

[163] A. Meckes, J. Behrens, and W. Benecke, "Electromagnetically driven microvalve fabricated in silicon," in *Proceedings Transducers 1997*, (Chicago, USA), pp. 821 – 824, 1997.

[164] A. K. Henning, J. Fitch, D. Hopkins, L. Lilly, R. Faeth, E. Falsken, and M. Zdeblick, "A thermopneumatically actuated microvalve for liquid expansion and proportional control," in *Proceedings Transducers 1997*, (Chicago, USA), pp. 825 – 828, 1997.

[165] M. Esashi, S. Eoh, A. Masuda, F. Enquist, T. Matsuo, I. Lundström, and S. Choi, "The fabrication of a micro valve by means of micromachining," in *Proceedings of the 6th Sensor Symposium*, (Tokio, Japan), pp. 269 – 272, 1986.

[166] M. Stehr, S. Messner, J. Ulrich, H. Sandmaier, and R. Zengerle, "A microvalve with bidirectional pump effect," in *Proceedings 10th European Conference on Solid-State Sensors 1996*, (Leuven, Belgium), pp. 845 – 848, 1996.

[167] M. Stehr, H. Gruhler, H. Straatmann, S. Messner, H. Sandmaier, and R. Zengerle, "The selfpriming VAMP," in *Proceedings Transducers 1997*, (Chicago, USA), pp. 351 – 352, 1997.

[168] A. Richter, A. Plettner, K. A. Hoffmann, and H. Sandmaier, "Electrohydrodynamic pumping and flow measurement," in *Proceedings IEEE Micro Electro Mechanical Systems 1991*, (Nara, Japan), pp. 271 – 276, 1991.

[169] S. C. Jacobson, R. Hergenroeder, L. B. Koutny, and J. M. Ramsey, "Open channel electrochromatography on a microchip," *Analytical Chemistry*, vol. 66, no. 14, pp. 2369 – 2373, 1994.

[170] Y. L. Tu and S. J. Milne, "Processing and characterization of $Pb(Zr,Ti)O_3$ films up to 10 μm thick, produced from a diol sol gel route," *Journal of Materials Research*, vol. 11, no. 10, pp. 2556 – 2564, 1996.

[171] P. H. Ansari and A. Safari, "In-situ deposition of PZT thin-films by RF magnetron sputtering," *Integrated Ferroelectrics*, vol. 7, no. 1 - 4, pp. 185 – 193, 1995.

[172] G. Zhan, T. Lo, L. Liu, and P. Tsien, "A silicon membrane micropump with integrated bimetallic actuator," *Chinese Journal of Electronics*, vol. 5, no. 2, pp. 29 – 35, 1996.

[173] W. K. Schomburg, J. Vollmer, B. Büstgens, J. Fahrenberg, H. Hein, and W. Menz, "Microfluidic components in LIGA technique," *Journal of Micromechanics and Microengineering*, vol. 4, pp. 186 – 191, 1994.

[174] R. Zengerle, W. Geiger, M. Richter, J. Ulrich, S. Kluge, and A. Richter, "Transient measurements on miniaturized diaphragm pumps in microfluid systems," *Sensors and Actuators*, vol. A 47, no. 1 - 3, pp. 557 – 561, 1995.

[175] T. Gerlach and H. Wurmus, "Working principle and performance of the dynamic micropump," *Sensors and Actuators*, vol. A 50, pp. 135 – 140, 1995.

[176] M. C. Carrozza, N. Croce, B. Magnani, and P. Dario, "Piezoelectric-driven stereolithography-fabricated micropump," *Journal of Micromechanics and Microengineering*, vol. 5, no. 2, pp. 177 – 179, 1995.

[177] M. Stehr, S. Messner, and R. Zengerle, "A new bi-directional micropump for liquids and gases," in *Actuator 96*, (Bremen, Germany), pp. 61 – 64, 1996.

[178] A. Olsson, P. Enokson, G. Stemme, and E. Stemme, "Micromachined flat-walled valveless diffuser pumps," *Journal of Microelectromechanical Systems*, vol. 6, no. 2, pp. 161 – 166, 1997.

[179] T. Gerlach, "Pumping gases by a silicon micro pump with dynamic passive valves," in *Proceedings Transducers 1997*, (Chicago, USA), pp. 357 – 360, 1997.

[180] A. Richter, A. Plettner, K. A. Hofmann, and H. Sandmaier, "A micromachined electrohydrodynamic pump," *Sensors and Actuators*, vol. A 29, no. 2, pp. 159 – 168, 1991.

[181] G. Fuhr, G. Schnelle, and B. Wagner, "Travelling wave driven microfabricated pumps for liquids," *Journal of Micromechanics and Microengineering*, vol. 4, no. 4, pp. 217 – 226, 1994.

[182] J. R. Melcher, *Continuum Electromechanics*. Cambridge, USA: MIT Press, 1981.

[183] J. R. Melcher, "Travelling wave induced electroconvection," *The Physics of Fluids*, vol. 9, pp. 1548 – 1555, 1966.

[184] S. F. Bart, L. S. Tavrov, M. Mehregany, and J. H. Lang, "Microfabricated electrohydrodynamic pumps," *Sensors and Actuators*, vol. A 21 – 23, pp. 193 – 197, 1990.

[185] G. Fuhr, R. Hagedorn, T. Müller, W. Benecke, and B. Wagner, "Pumping of water solutions in microfabricated electrohydrodynamic systems," in *Proceedings IEEE Micro Electro Mechanical Systems 1992*, (Travemünde, Germany), pp. 25 – 30, 1992.

[186] A. Manz, D. J. Harrison, J. C. Verpoorte, H. Ludi, and H. M. Widmer, "Integrated electroosmotic pumps and flow manifolds for total chemical analysis systems," in *Proceedings Transducers 1991*, (San Francisco, USA), pp. 939 – 941, 1991.

[187] J. Döpper, M. Clemens, W. Ehrfeld, S. Jung, K. P. Kämper, and H. Lehr, "Micro gear pumps for dosing of viscous fluids," *Journal of Micromechanics and Microengineering*, vol. 7, no. 3, pp. 230 – 232, 1997.

[188] A. S. Dewa, K. Deng, D. C. Ritter, and C. Bonham, "Development of LIGA-fabricated self-priming in-line gear pumps," in *Proceedings Transducers 1997*, (Chicago, USA), pp. 757 – 760, 1997.

[189] R. M. Moroney, R. M. White, and R. T. Howe, "Ultrasonically induced microtransport," in *Proceedings IEEE Micro Electro Mechanical Systems 1991*, (Nara, Japan), pp. 277 – 282, 1991.

[190] K. Y. Ng, J. Shajii, and M. A. Schmidt, "A liquid shear-stress sensor fabricated using wafer bonding technology," in *Proceedings Transducers 1991*, (San Francisco, USA), pp. 931 – 934, 1991.

[191] G. Stemme, "A monolithic gas flow sensor with polyimide as thermal insulator," *IEEE Transactions on Electron Devices*, vol. 33, no. 10, pp. 1470 – 1474, 1986.

[192] M. Esashi, "Micro flow sensor and integrated magnetic oxygen sensor using IT," in *Proceedings Transducers 1991*, (San Francisco, USA), pp. 34 – 37, 1991.

[193] R. E. Hetrick, "Vibrating cantilever mass flow sensor," *Sensors and Actuators*, vol. A 21 – 23, pp. 373 – 376, 1990.

[194] C. Yang, M. Kummel, and H. Soeberg, "A transit-time flow meter for measuring millimeter per minute liquid flow," *Rev. Sci. Instrum.*, vol. 59, no. 2, pp. 314 – 317, 1988.

[195] F. Kohl, A. Jachimowicz, J. Steurer, R. Glatz, J. Kuttner, D. Biacovsky, F. Olcaytug, and G. Urban, "A micromachined flow sensor for liquid and gaseous fluids," *Sensors and Actuators*, vol. A 41 – 42, pp. 293 – 299, 1994.

[196] T. Nishimoto, S. Shoji, and M. Esashi, "Buried piezoresistive sensors by means of MeV ion implantation," *Sensors and Actuators*, vol. A 43, pp. 249 – 253, 1994.

[197] Y. Su, A. G. R. Evans, and A. Brunnschweiler, "Micromachined silicon cantilever paddles with piezoresistive readout for flow sensing," *Journal of Micromechanics and Microengineering*, vol. 6, pp. 69 – 72, 1996.

[198] R. E. Oosterbroek, T. S. J. Lammerink, J. W. Berenschot, A. van den Berg, and M. C. Elwenspoek, "Designing, realization and characterization of a novel capacitive pressure/flow sensor," in *Proceedings Transducers 1997*, (Chicago, USA), pp. 151 – 154, 1997.

[199] T. S. Leu, A. M. Lanzillotto, M. Amabile, and R. Wildes, "Analysis of fluidic and mechanical motion in MEMS by using high-speed x-ray micro-imaging techniques," in *Proceedings Transducers 1997*, (Chicago, USA), pp. 149 – 150, 1997.

[200] C. Wagner, J. Frankenberger, and P. P. Deimel, "Optical pressure sensor based on a Mach-Zehnder interferometer integrated with a lateral a-Si:H p-i-n photodiode," *IEEE Photonics Technology Letters*, vol. 5, pp. 1257 – 1259, 1993.

[201] D. de Bruyker, A. Cozma, and R. Puers, "A combined piezoresistive/capacitive pressure sensor with self-test function based on thermal actuation," in *Proceedings Transducers 1997*, (Chicago, USA), pp. 1461 – 1464, 1997.

[202] E. Obermeier and P. Kopystynski, "Polysilicon as a material for microsensor applications," *Sensors and Actuators*, vol. 30, pp. 149 – 155, 1989.

[203] H. Sandmaier and K. Kühl, "Square-diaphragm piezoresistive pressure sensor with a rectangular central boss for low-pressure ranges," *IEEE Transactions on Electron Devices*, vol. 40, no. 10, pp. 1754 – 1759, 1993.

[204] H. Guckel, "Polysilicon planar transducer." http://bob.ece.wisc.edu/research/pt.html, 1997. Webpage of Univ. of Wisconsin, MEMS group.

[205] Y. Matsuoka, Y. Yamamoto, K. Yamada, S. Shimada, M. Tanabe, and A. Yasukawa, "Characteristic analysis of a pressure sensor using the silicon piezoresistance effect for high pressure measurements," *Journal of Micromechanics and Microengineering*, vol. 5, no. 1, pp. 25 – 31, 1995.

[206] H. Schafer, V. Graeger, and R. Kobs, "Temperature independent pressure sensors using polycrystalline silicon strain gauges," *Sensors and Actuators*, vol. 17, pp. 521 – 527, 1989.

[207] H. Terabe, H. Arashima, N. Ura, K. Suzuki, K. Ohta, and M. Ishida, "A silicon pressure sensor with stainless diaphragm for high temperature and chemical application," in *Proceedings Transducers 1997*, (Chicago, USA), pp. 1481 - 1484, 1997.

[208] S. Sugiyama, K. Shimaoka, and O. Tabata, "Surface micromachined micro-diaphragm pressure sensors," in *Proceedings Transducers 1991*, (San Francisco, USA), pp. 188 - 191, 1991.

[209] Y. Suzuki, T. Kudo, and K. Ikeda, "Accurate, cost effective absolute pressure sensor," in *Proceedings Transducers 1997*, (Chicago, USA), pp. 1493 - 1496, 1997.

[210] H. Dudaicevs, M. Kandler, Y. Manoli, W. Mokwa, and E. Spiegel, "Surface micromachined pressure sensors with integrated CMOS read-out electronics," *Sensors and Actuators*, vol. 43, pp. 157 - 163, 1994.

[211] K. Ikeda, H. Kuwayama, T. Kobayashi, T. Watanabe, T. Nishskawa, T. Yoshida, and K. Harada, "Three-dimensional micromachining of silicon pressure sensor integrating resonant strain gauge on diaphragm," *Sensors and Actuators*, vol. 21 - 23, pp. 1007 - 1010, 1990.

[212] M. Mehta, A. M. Parameswaran, and S. Payandeh, "Design of a micromachined pressure sensor for endoscopic graspers," *Proceedings of SPIE - The international society for optical engineering*, vol. 2593, pp. 2 - 8, 1995.

[213] J. C. Greenwood, "High pressure resonant sensor," in *IEE Colloquium (Digest)*, (Stevenage, UK), pp. 7/1 - 7/3, 1994.

[214] S. Hein, K. Holzner, V. Schlichting, E. Obermeier, and K. Barton, "Capacitive differential pressure sensor with high overload capability using silicon/glass technology," in *Proceedings Transducers 1997*, (Chicago, USA), pp. 1477 - 1480, 1997.

[215] K. Petersen, F. Pourahmadi, J. Brown, P. Parsons, M. Skinner, and J. Tudor, "Resonant beam pressure sensor fabricated with silicon fusion bonding," in *Proceedings Transducers 1991*, (San Francisco, USA), pp. 664 - 667, 1991.

[216] H. Guckel, C. Rypstat, M. Nesnidal, J. D. Zook, D. W. Burns, and D. K. Arch, "Polysilicon resonant microbeam technology for high performance sensor applications," in *Technical Digest- IEEE Solid-State Sensor and Actuator Workshop, 1992*, (Hilton Head Island, SC, USA), pp. 153 - 156, 1992.

[217] G. Blasquez, P. Pons, P. Menini, D. Herbst, M. Schulz, and B. Hofflinger, "Capacitive pressure sensor mock-up without compensation circuits," in *Proceedings Transducers 1995*, (Stockholm, Sweden), pp. 628 – 631, 1995.

[218] J. C. Greenwood, "Etched silicon vibrating sensor," *Journal of Physics E: Sci. Instrum.*, vol. 17, pp. 650 – 652, 1984.

[219] J. C. Greenwood and T. Wray, "High accuracy pressure measurement with a silicon resonant sensor," *Sensors and Actuators*, vol. 37 – 38, pp. 82 – 85, 1993.

[220] S. Masuda, M. Washizu, and M. Iwadare, "Separation of small particles suspended in liquid by nonuniform travelling field," *IEEE Transactions on Industrial Applications*, vol. 23, pp. 474 – 480, 1987.

[221] J. P. H. Burt, T. A. K. Al-Ameen, and R. Pethig, "An optical dielectrophoresis spectrometer for low frequency measurements on colloidal suspensions," *Journal of Physics E: Sci. Instrum.*, vol. 22, pp. 952 – 957, 1989.

[222] Y. Huang and R. Pethig, "Electrode design for negative dielectrophoresis," *Measurement Science & Technology*, vol. 2, no. 12, pp. 1142 – 1146, 1991.

[223] G. Fuhr, R. Hagedorn, T. Müller, B. Wagner, and W. Benecke, "Linear motion of dielectric particles and living cells in microfabricated structures induced by traveling electric fields," in *Proceedings IEEE Micro Electro Mechanical Systems 1991*, (Nara, Japan), pp. 259 – 264, 1991.

[224] R. R. H. Coombs and D. W. Robinson, eds., *Nanotechnology in medicine and biosciences*. Amsterdam: Gordon and Breach Publishers, 1996.

[225] X. B. Wang, Y. Huang, J. P. H. Burt, G. H. Markx, and R. Pethig, "Selective dielectrophoretic confinement of bioparticles in potential energy wells," *Journal of Physics D: Applied Physics*, vol. 26, no. 8, pp. 1278 – 1285, 1993.

[226] G. Fuhr, S. Fiedler, T. Müller, H. Glasser, T. Lisec, and B. Wagner, "Particle micromanipulator consisting of two orthogonal channels with travelling-wave electrode structures," *Sensors and Actuators*, vol. A 41 - 42, no. 1 - 3, pp. 230 – 239, 1994.

[227] Y. Huang, X. B. Wang, J. A. Tame, and R. Pethig, "Electrokinetic behaviour of colloidal particles in travelling electric fields: studies using yeast cells," *Journal of Physics D: Applied Physics*, vol. 26, no. 9, pp. 1528 – 1535, 1993.

[228] M. S. Talary, J. P. H. Burt, J. A. Tame, and R. Pethig, "Electromanipulation and separation of cells using travelling electric fields," *Journal of Physics D: Applied Physics*, vol. 29, no. 8, pp. 2198 – 2203, 1996.

[229] M. Washizu, T. Nanba, and S. Masuda, "Handling of biological cells using fluid integrated circuit," *IEEE Transactions on Industrial Applications*, vol. 25, pp. 352 – 358, 1990.

[230] T. Allen, *Particle Size Measurement.* London: Chapman and Hall, 1990.

[231] U. Larsen, G. Blankenstein, and S. Ostergaard, "Microchip Coulter particle counter," in *Proceedings Transducers 1997*, (Chicago, USA), pp. 1319 – 1322, 1997.

[232] M. Koch, A. G. R. Evans, and A. Brunnschweiler, "Design and fabrication of a micromachined Coulter counter," in *Proceedings Micromechanics Europe 1998*, (Ulvik, Norway), pp. 155–158, 1998.

[233] D. Koutsouris, R. Guillet, J. C. Lelievre, M. T. Guillemin, B. P, Y. Beuzard, and M. Boynard, "Determination of erythrocyte transit times through micropores; basic operational principles," *Biorheology*, vol. 25, pp. 763 – 772, 1988.

[234] M. C. Tracey, P. H. Kaye, and J. N. Shepherd, "Microfabricated microhaemorheometer," in *Proceedings Transducers 1991*, (San Francisco, USA), pp. 82 – 84, 1991.

[235] I. Simpson and M. Harper, "Microfluidics: Applications in chemical processing and analytical science," in *Proc. Micro and nanotechnology, the thermofluids dimension*, IMechE, 1995.

[236] R. C. Weast, ed., *CRC Handbook of Chemistry and Physics.* Boca Raton, Florida: CRC Press Inc., 1988.

[237] D. Chatelain, M. Koch, A. G. R. Evans, and A. Brunnschweiler, "Two micromixers in silicon technology," in *Proceedings Micromechanics Europe 1997*, (Southampton, UK), pp. 124 – 127, 1997.

[238] M. Koch, H. Witt, A. G. R. Evans, and A. Brunnschweiler, "Improved characterisation technique for micromixers," in *Proceedings Micromechanics Europe 1998*, (Ulvik, Norway), pp. 159–162, 1998.

[239] J. Branebjerg, P. Gravesen, J. P. Krog, and C. R. Nielsen, "Fast mixing by lamination," in *Proceedings IEEE Micro Electro Mechanical Systems 1996*, (San Diego, CA, USA), pp. 441 – 446, IEEE, 1996.

[240] H. Möbius, W. Ehrfeld, V. Hessel, and T. Richter, "Sensor controlled processes in chemical microreactors," in *Proceedings Transducers 1995*, (Stockholm, Sweden), pp. 775 – 778, 1995.

[241] R. Srinivasan, S. L. Firebaugh, I. M. Hsing, J. Ryley, M. P. Harold, K. F. Jensen, and M. A. Schmidt, "Chemical performance and high temperature characterization of micromachined chemical reactors," in *Proceedings Transducers 1997*, (Chicago, USA), pp. 163 – 166, 1997.

[242] B. Xie, B. Danielsson, P. Norberg, F. Winquist, and I. Lundström, "Development of a thermal micro-biosensor fabricated on a silicon chip," *Sensors and Actuators*, vol. B 6, pp. 127 – 130, 1992.

[243] J. Drott, K. Lindstroem, L. Rosengren, and T. Laurell, "Porous silicon as the carrier matrix in microstructured enzyme reactors yielding high enzyme activities," *Journal of Micromechanics and Microengineering*, vol. 7, no. 1, pp. 14 – 23, 1997.

[244] T. Tang, G. Ocvirk, and D. J. Harrison, "Iso-thermal DNA reactions and assays in microfabricated capillary electrophoresis systems," in *Proceedings Transducers 1997*, (Chicago, USA), pp. 523 – 526, 1997.

[245] A. Manz, Y. Miyahara, J. Miura, Y. Watanabe, H. Miyagi, and K. Sato, "Design of an open-tubular column liquid chromatograph using silicon chip technology," *Sensors and Actuators*, vol. B 1, pp. 249 – 255, 1990.

[246] G. Ocvirk, E. Verpoorte, A. Manz, and H. M. Widmer, "Integration of a micro liquid chromatograph onto a silicon chip," in *Proceedings Transducers 1995*, (Stockholm, Sweden), pp. 756 – 759, 1995.

[247] S. Hannoe, I. Sugimoto, K. Yanagisawa, and H. Kuwano, "Enhanced chromatographic performance of silicon-micromachined capillary column with clean structure and interactive plasma organic film," in *Proceedings Transducers 1997*, (Chicago, USA), pp. 515 – 518, 1997.

[248] C. S. Effenhauser, A. Manz, and H. M. Widmer, "Glass chips for high-speed capillary electrophoresis separations with submicron plate heights," *Analytical Chemistry*, vol. 65, pp. 2637 – 2642, 1993.

[249] D. J. Harrison, K. Fluri, K. Seiler, Z. Fan, C. S. Effenhauser, and A. Manz, "Micromachining a miniaturized capillary electrophoresis-based chemical analysis system on a chip," *Science*, vol. 261, pp. 895 – 897, 1993.

[250] S. C. Jacobson, R. Hergenroder, L. B. Koutny, R. J. Warmack, and J. M. Ramsey, "Effects of injection schemes and column geometry on the performance of microchip electrophoresis devices," *Analytical Chemistry*, vol. 66, no. 7, pp. 1107 – 1113, 1994.

[251] A. Manz, E. Verpoorte, D. E. Raymond, C. S. Effenhauser, N. Burggraf, and H. M. Widmer, "μ-TAS: miniaturised total chemical analysis systems," in *Proceedings of the Micro Total Analysis Systems '94 Workshop*

(A. van den Berg and P. Bergveld, eds.), pp. 5 – 28, Kluwer Academic Publishers, 1994.

[252] D. J. Harrison, K. Fluri, N. Chiem, T. Tang, and Z. Fan, "Micromachining chemical and biochemical analysis and reaction systems on glass substrates," in *Proceedings Transducers 1995*, (Stockholm, Sweden), pp. 752 – 755, 1995.

[253] M. A. Northrup, C. Gonzalez, D. Hadley, R. F. Hills, P. Landre, S. Lehew, R. Saiki, J. J. Sninsky, R. Watson, and R. Watson Jr, "A MEMS-based miniature DNA analysis system," in *Proceedings Transducers 1995*, (Stockholm, Sweden), pp. 764 – 767, 1995.

[254] J. R. Webster and C. H. Mastrangelo, "Large-volume integrated capillary electrophoresis stage fabricated using micromachining of plastics on silicon substrates," in *Proceedings Transducers 1997*, (Chicago, USA), pp. 503 – 506, 1997.

[255] W. Göpel, J. Hesse, and J. N. Zemel, eds., *Sensors: A comprehensive survey*, vol. 3. New York: VCH, 1992.

[256] I. Lauks, "Polarizable electrodes, part II," *Sensors and Actuators*, vol. 1, p. 393, 1981.

[257] W. Heywang, *Sensorik*. Berlin: Springer Verlag, 1993.

[258] D. H. Kwon, B. W. Cho, C. S. Kim, and B. K. Sohn, "Effects of heat treatment on Ta2O5 sensing membrane for low drift and high sensitivity pH-ISFET," *Sensors and Actuators*, vol. B 34, pp. 441 – 445, 1996.

[259] K. Domanski, D. L. Baldwin, J. W. Grate, T. B. Hall, J. Li, M. Josowicz, and J. Janata, "Development and calibration of field-effect transistor-based sensor array for measurement of hydrogen and ammonia gas mixtures in humid air," *Analytical Chemistry*, vol. 70, pp. 473 – 481, 1998.

[260] Z. Jarzebski, *Oxide Semiconductors*. Oxford: Pergamon Press Ltd., 1973.

[261] M. Fleischer, "Fast gas sensors based on metal oxides which are stable at high temperatures," in *Proceedings 10th European Conference on Solid-State Sensors 1996*, (Leuven, Belgium), pp. 25 – 34, 1996.

[262] W. Hellmich, G. Müller, C. B. v Braunmühl, T. Doll, and I. Eisele, "Field-effect-induced gas sensitivity changes in metal oxides," in *Proceedings 10th European Conference on Solid-State Sensors 1996*, (Leuven, Belgium), pp. 505 – 508, 1996.

[263] I. Stoev and D. Kohl, "An integrated gas sensor on silicon substrate with sensitive SnO_x layer," *Sensors and Actuators*, vol. B 2, pp. 233 – 236, 1990.

[264] P. Althainz, J. Goschnick, S. Ehrmann, and H. J. Ache, "Multisensor microsystem for contaminants in air," *Sensors and Actuators*, vol. B 33, pp. 72 – 76, 1996.

[265] P. Krebs and A. Grisel, "A low power integrated catalytic gas sensor," *Sensors and Actuators*, vol. B 13, pp. 155 – 158, 1993.

[266] J. S. Lee, J. W. Park, and S. M. Shin, "Fabrication of a micro catalytic gas sensor using thin film process and Si anisotropic etching techniques," *Sensors and Actuators*, vol. B 45, pp. 265 – 269, 1997.

[267] Y. T. Hsueh, S. D. Collins, and R. L. Smith, "DNA quantification with an electrochemiluminesce microcell," in *Proceedings Transducers 1997*, (Chicago, USA), pp. 175 – 178, 1997.

[268] M. Ralfs and J. Heinze, "Disposable optochemical sensor for the determination of chlorine concentration in the ppb range," in *Proceedings 10th European Conference on Solid-State Sensors 1996*, (Leuven, Belgium), pp. 63 – 66, 1996.

[269] B. Drapp, J. Piehler, A. Brecht, G. Gauglitz, B. J. Luff, J. S. Wilkinson, and J. Ingenhoff, "Integrated optical Mach-Zehnder interferometers as simazine immunoprobes," *Sensors and Actuators*, vol. B 39, pp. 277 – 282, 1997.

[270] O. Leistiko and P. F. Jensen, "Integrated bio/chemical optical detection microsystem: the clip-on," in *Proceedings Micromechanics Europe 1997*, (Southampton, England), pp. 180 – 185, 1997.

[271] M. P. Byfield and R. A. Abuknesha, "Biochemical aspects of biosensors," *GEC Journal of Research*, vol. 9, no. 2, pp. 97 – 117, 1991.

[272] R. S. Sethi, "Transducer aspects of biosensors," *GEC Journal of Research*, vol. 9, no. 2, pp. 81 – 96, 1991.

[273] H. I. Seo, C. S. Kim, B. K. Sohn, T. Yeow, M. T. Son, and M. Haskard, "ISFET glucose sensor based on a new principle using the electrolysis of hydrogen peroxide," *Sensors and Actuators*, vol. B 40, pp. 1 – 5, 1997.

[274] R. L. C. Chen, B. W. Chang, and H. C. Chang, "Conductometric flow-injection glucose biosensing system with an immobilized glucose-oxidase mini-reactor," *Journal of the Chinese Chemical Society*, vol. 44, pp. 423 – 426, 1997.

[275] W. Wei, R. H. Wang, L. H. Nie, and S. Z. Yao, "Rapid determination of dimethoate with a surface acoustic wave impedance sensor system," *Analytical Letters*, vol. 30, pp. 2641 – 2653, 1997.

[276] B. J. Luff, J. S. Wilkinson, J. Piehler, U. Hollenbach, J. Ingenhoff, and N. Fabricius, "Integrated optical Mach-Zehnder biosensor," *Journal of lightwave technology*, vol. 16, pp. 583 – 592, 1998.

[277] T. H. Wilmshurst, *Signal Recovery from noise in electronic instrumentation*, ch. 3. Adam Hilger (Publisher), 1985.

[278] C. Di Natale, A. Macagnano, R. Paolesse, E. Tarizzo, A. D'Amico, F. Davide, and T. Boschi, "A comparison between an electronic nose and human olfaction in a selected case study," in *Proceedings Transducers 1997*, (Chicago, USA), pp. 1335 – 1338, 1997.

[279] R. Hartinger, M. Irsiegler, H. E. Endres, S. Drost, K. Rieblinger, and G. Ziegleder, "Portable and modular electronic nose for olfactometric measurements," in *Proceedings 12th European Conference on Solid-State Sensors 1998*, (Southampton, England), pp. 1111 – 1114, 1998.

[280] W. Gumbrecht and K. Abraham-Fuchs, "Multisensor analysis system for liquids based on micro-electrochemical transducers," in *Proceedings 10th European Conference on Solid-State Sensors 1996*, (Leuven, Belgium), pp. 777 – 780, 1996.

[281] W. Gumbrecht, D. Peters, W. Schelter, W. Erhardt, J. Henke, J. Steil, and U. Sykory, "Integrated pO_2, pCO_2 and pH sensor system for online blood monitoring," *Sensors and Actuators*, vol. B18 – 19, pp. 704 – 708, 1994.

[282] S. Savoy, J. J. Lavigne, S.-J. Yoo, J. Wright, M. Rodriguez, and A. Goodey, "Solution-based analysis of multiple analytes by a sensor array: toward the development of an electronic tongue," in *Proceedings of the SPIE*, vol. 3539, (Boston, USA), pp. 17 – 26, 1998.

[283] B. H. vander Schoot, E. M. J. Verpoorte, S. Jeanneret, A. Manz, and N. F. de Rooij, "Microsystems for analysis in flowing solutions," in *Proceedings Micro Total Analysis Systems*, (Twente, The Netherlands), pp. 181 – 190, 1994.

[284] J. H. Fluitman, A. van den Berg, and T. S. Lammerink, "Micromechanical components for μTAS," in *Proceedings Micro Total Analysis Systems*, (Twente, The Netherlands), pp. 73 – 83, 1994.

[285] T. S. J. Lammerink, V. L. Spiering, M. Elwenspoek, and A. van den Berg, "Modular concept for fluid handling system," in *Proceedings IEEE Micro Electro Mechanical Systems 1996*, (San Diego, CA, USA), pp. 389 – 394, 1996.

[286] C. Gonzalez, S. D. Collins, and R. L. Smith, "Fluidic interconnects for modular assembly of chemical microsystems," *Sensors and Actuators*, vol. B 49, pp. 40 – 45, 1998.

[287] C. G. J. Schabmueller, M. Koch, A. G. R. Evans, and A. Brunnschweiler, "Micromachined chemical reaction system realized on a microfluidic circuit board," in *Proceedings 12th European Conference on Solid-State Sensors 1998*, (Southampton, England), pp. 571 – 574, 1998.

[288] V. Gass, B. H. vander Schoot, S. Jeanneret, and N. F. de Rooij, "Integrated flow-regulated silicon micropump," in *Proceedings Transducers 1993*, (Yokohama, Japan), pp. 1048 – 1051, 1993.

[289] M. Pinget, N. Jeandidier, F. Ortega, D. Wix, and J. Margetaux, "Multicentre trial of a programmable implantable insulin pump in type 1 diabetes," *The International Journal of Artificial Organs*, vol. 18, no. 6, pp. 322 – 325, 1995.

[290] D. O. Keefe, C. O. Herlihy, Y. Gross, and J. G. Kelly, "Patient-controlled analgesia using a miniature electrochemically driven infusion pump," *British Journal of Anaesthesian*, vol. 6, pp. 843 – 846, 1994.

[291] P. Arquint, B. H. van der Schoot, and N. F. de Rooij, "Combined blood gas sensor for pO_2, pCO_2 and pH," in *Proceedings Micro Total Analysis Systems*, (Twente, The Netherlands), pp. 191 – 194, 1994.

[292] N. Chiem, C. Colyer, and D. J. Harrison, "Microfluidic systems for clinical diagnostics," in *Proceedings Transducers 1997*, (Chicago, USA), pp. 183 – 186, 1997.

[293] D. J. Harrison, K. Fluri, Z. Fan, and K. Seiler, "Integration of analytical systems incorporating chemical reactions and electrophoretic separation," in *Proceedings Micro Total Analysis Systems*, (Twente, The Netherlands), pp. 105 – 115, 1994.

[294] S. Drost, W. Woermann, P. Woias, B. Ross, O. Koester, W. Konz, B. Edler, W. Schuhmann, L. Meixner, and R. Ferreti, "Microanalytical system for environmental control," in *Proceedings Transducers 1997*, (Chicago, USA), pp. 931 – 934, 1997.

[295] I. Moser, G. Jobst, P. Svasek, M. Varaham, and G. Urban, "Rapid liver enzyme assay with miniaturized liquid handling system comprising thin film biosensor array," *Sensors and Actuators*, vol. B 44, pp. 377 – 380, 1997.

[296] P. Svasek, G. Jobst, G. Urban, and E. Svasek, "Dreidimensionale Mikrosystemtechnik mit Trockenresisten," in *ÖVE Schriftenreihe: Berichte der Informationstagung ME95*, (Oesterreichischer Verband fuer Elektrotechnik, Vienna), p. 327, 1995.

[297] A. Manz and H. Becker, "Parallel capillaries for high throughput in electrophoretic separations and electroosmotic drug discovery systems," in *Proceedings Transducers 1997*, (Chicago, USA), pp. 915 – 918, 1997.

[298] M. Leach, "Discovery on a credit card?," *Drug Discovery Today*, vol. 2, no. 7, pp. 253 – 254, 1997.

[299] L. Bousse, "Whole cell biosensors," *Sensors and Actuators*, vol. B 34, pp. 270 – 275, 1996.

[300] C. H. Mastrangelo, M. A. Burns, and D. T. Burke, "Microfabricated devices for genetic diagnostics," *Proceedings IEEE*, vol. 86, no. 8, pp. 1769 – 1784, 1998.

[301] R. C. Anderson, G. J. Bogdan, Z. Bamiv, T. D. Dawes, and K. Roy, "Microfluidic biochemical analysis system," in *Proceedings Transducers 1997*, (Chicago, USA), pp. 477 – 480, 1997.

[302] A. T. Woolley, D. Hadley, and P. Landre, "Functional integration of PCR amplification and capillary electrophoresis in a microfabricated DNA analysis device," *Analytical Chemistry*, vol. 68, pp. 4081 – 4086, 1996.

[303] M. A. Burns, C. H. Mastrangelo, and T. S. Sammarco, "Microfabricated structures for integrated DNA analysis," *Proceedings Natl. Acad. Sci. USA*, vol. 93, pp. 5556 – 5561, 1996.

Index